Physics in Medical Diagnosis

PHYSICS AND ITS APPLICATIONS

Series Editors

E.R. Dobbs
University of Sussex

S.B. Palmer
Warwick University

This series of short texts on advanced topics for students, scientists and engineers will appeal to readers seeking to broaden their knowledge of the physics underlying modern technology.

Each text provides a concise review of the fundamental physics and current developments in the area, with references to treatises and the primary literature to facilitate further study. Additionally, texts providing a core course in physics are included to form a ready reference collection.

The rapid pace of technological change today is based on the most recent scientific advances. This series is, therefore, particularly suitable for those engaged in research and development, who frequently require a rapid summary of another topic in physics or a new application of physical principles in their work. Many of the texts will also be suitable for final year undergraduate and postgraduate courses.

1. **Electrons in Metals and Semiconductors**
   R.G. Chambers
2. **Basic Digital Electronics**
   J.A. Strong
3. **AC and DC Network Theory**
   Anthony J. Pointon and Harry M. Howarth
4. **Nuclear and Particle Physics**
   R.J. Blin-Stoyle
5. **Thermal Physics**
   Second edition
   C.B.P. Finn
6. **Vacuum Physics and Techniques**
   T.A. Delchar
7. **Basic Electromagnetism**
   E.R. Dobbs
8. **Quantum Mechanics**
   Second edition
   Paul C.W. Davies and David S. Betts
9. **Statistical Physics**
   Second edition
   Tony Guénault
10. **Photothermal Science and Techniques**
    D.P. Almond and P.M. Patel
11. **Physics in Medical Diagnosis**
    T.A. Delchar
12. **Atomic Physics**
    D.G.C. Jones

Physics in Medical Diagnosis

T. A. Delchar

Department of Physics
University of Warwick
UK

CHAPMAN & HALL

London · Weinheim · New York · Tokyo · Melbourne · Madras

Published by Chapman & Hall, 2–6 Boundary Row, London SE1 8HN, UK

Chapman & Hall, 2–6 Boundary Row, London SE1 8HN, UK

Chapman & Hall, GmbH, Pappelallee 3, 69469 Weinheim, Germany

Chapman & Hall USA, 115 Fifth Avenue, New York, NY 10003, USA

Chapman & Hall Japan, ITP-Japan, Kyowa Building, 3F, 2-2-1
Hirakawacho, Chiyoda-ku, Tokyo 102, Japan

Chapman & Hall Australia, 102 Dodds Street, South Melbourne, Victoria
3205, Australia

Chapman & Hall India, R. Seshadri, 32 Second Main Road, CIT East,
Madras 600 035, India

Typeset in 10/12 Times by Blackpool Typesetting Services Limited, UK
Printed in Great Britain by St Edmundsbury Press, Bury St Edmunds

ISBN 0 412 61680 7

A catalogue record for this book is available from the British Library

Library of Congress Catalog Card Number: 96-71160

♾ Printed on permanent acid-free text paper, manufactured in accordance with ANSI/NISO Z39.48-1992 and ANSI/NISO Z39.48-1984 (Permanence of Paper)

To Carol, David and Elizabeth,
who encouraged me to write this book,
and put up with me while I did so.

Contents

Preface

It is often assumed that physics made its first contribution to medicine at the end of the nineteenth century, with the introduction of X–rays. In fact, physics has had a role in medicine extending over some 400 years, to the beginning of the seventeenth century. One of the reasons for this long association is that physics and medicine, in forms which we might recognize now, were both 'born' at about the same time and, perhaps surprisingly, new developments in physics have always been considered for their usefulness in clinical diagnosis.

In the early seventeenth century it was not unusual for the physician to have interests in physics, while the physicist might have interests in medicine. The physician Santorio Santorio and the physicist/mathematician, Galileo Galilei (who were friends) are the earliest, and the most important, examples of such people. Of course the titles 'physician' and 'physicist' are not really appropriate, since each profession was then a far cry from that which we would recognize today. In the eighteenth century we find another important influence on the relationship between physics and medicine – the polymath, Hermann Boerhaave. For a decade, Boerhaave simultaneously occupied the chairs of botany, chemistry and medicine in the Faculty of Medicine at the University of Leiden, and dominated the teaching and practice of medicine throughout Europe and beyond. His influence persists to this day since he encouraged the application of physics to medical diagnosis, and even the modern medical curriculum owes much to him. As time went by both physics and medicine developed rapidly and became too specialized for any one person to dominate them both; nevertheless, physics has continued to provide solutions to problems in medical diagnosis.

We should not overlook the accomplishments of the Ancient Greeks which lay behind the work started in the seventeenth century, and it is humbling to realize that, as early as about 420 BC, the Greek scholar Hippocrates (the father of modern medicine) understood the importance of a symmetrical temperature distribution between the right and left sides of the human body, although he had only his hands with which to judge temperature. He also understood that there is something special about the brain, and that the brain and its structures control behaviour in humans. It is only in the twentieth century, however, two and a half millennia later, that we have been able to use physics to show the basis for these beliefs.

This book examines the role of physics in medical diagnosis with the topics introduced, for the most part, in the order of their appearance in medical use; there are, however, some necessary deviations from the proper historical sequence. Due to this emphasis on diagnosis, applications of physics in the treatment of the human condition are omitted. Topics have been grouped so that initially the reader is given an account and explanation of those techniques of diagnosis which depend on the intrinsic properties of the body. Of these, three are considered: first, the techniques which measure the body core temperature (the oldest application of physics in medicine) and the associated infrared radiation (thermography); second, the information which can be obtained from the innate electrical activity of the human body; and third, the information arising from the magnetic fields accompanying this electrical activity. These signals appear on the diagnostic scene in the techniques of electromyography, electroencephalography and electrocardiography, and their magnetic equivalents, magnetomyography, magnetoencephalography and magnetocardiography. The remaining diagnostic techniques all use an external agency, such as X-rays, ultrasound pressure waves, γ-radiation or radiofrequency radiation, and give either information on body metabolism or images of body structures. Perhaps the most striking advance which has been made in the twentieth century is the ability to produce images of 'slices' through a chosen region of the body (tomograms). This advance is based on developments in the theory of imaging, so the mathematical basis of tomography is established before particular applications of tomography are described in subsequent chapters.

This book arose, in part, from a course which has been taught to third year undergraduates at the University of Warwick for some years. Consequently, the descriptions of the diagnostic contributions of physics to medicine are presented at a level appropriate to second or third year undergraduates and beyond, although hopefully the material will also be accessible to readers whose primary background is not physics. To assist these readers, the book has been structured so that the more theoretical or technical sections (identified by an asterisk) may be omitted. The emphasis throughout is upon the underlying physics of the techniques and the choice of techniques includes all of the important ones in use today. Where possible, an attempt has been made to give the physics its appropriate historical and technological background. Medical terminology is introduced in bold type in the first instance, and explained in a glossary.

The reader may recognize some omissions, but within the space restrictions of a book of this sort, omissions are inevitable. In the main, diagnostic techniques which have not entirely emerged from the research stage, and which do not have commercial equipment available for their application and support, have been excluded; for example, applied potential tomography. It is the writer's hope that the reader will finish this book, not only with a grasp of the physics involved in modern medicine, but also an admiration for the pioneers in this field, amongst whom number some nineteen Nobel Prize winners.

The author is indebted to many people for their help, either by critical comment on parts of the manuscript in its early stages, or the provision of medical illustrations, or both. In particular, thanks must go to Dr J.J. Shenkman and Dr M.J.A. Smith for their critical comments, to Professor E.F.J. Ring, Dr H. Singh and Dr D. Taylor, who provided illustrations drawn from their work in thermography, cardiology and nuclear medicine respectively, and finally to Mr A. Ashley, Radiology Manager at the Walsgrave Hospital, Coventry, who filled in the remaining gaps in the gallery of medical images. Acknowledgement is also due to Daniel Brown, my former editor at Chapman & Hall, who provided the initial impetus for this book.

Finally, thanks must also go to Jeanette Inness–Chaytor, who typed (and retyped) the manuscript and put up with all the changes (and without whom this book would not exist), and to Steve Carpenter, who carried out all of the photographic work.

T.A. Delchar
Coventry
August 1996

Registered trademarks

Various registered trademarks appear in this book. They are listed below, together with the company to which they are registered.

Aquadag®	Acheson Graphite
Araldite®	Ciba Geigy
Freon®	E.I. duPont de Nemours & Co Inc
Perspex®	ICI

1

From thermometry to thermography

1.1 INTRODUCTION: THERMOMETRY

Although physics has an important role in the treatment of many illnesses, it is its role in diagnosis which is the central theme of this book. We begin our consideration of the role of physics in the diagnosis of disease by looking at those diagnostic techniques which depend upon the intrinsic or endogenous activity of the human body. We can then move on to examine the techniques which operate by exposing the human body to external agencies, which may range from low energy radio waves to ionizing radiation of one sort or another.

One of the most familiar, yet also one of the oldest, examples of physics applied to medicine is the measurement of body core temperature. Clinical interest in the idea of degrees of hot and cold is of considerable antiquity and can be found, for example, in the writings of the great physician Galen in the second century AD. However, it was not until 1400 years later that anyone could actually measure the body temperature. It required the invention of the air thermometer (possibly by Galileo) and the introduction of thermometric scales, before it became possible to set the notions of degrees of hot and cold on a quantitative basis. The pioneer was Santorio Santorio, professor of anatomy at Padua and a friend of Galileo who, in 1612, introduced the air thermometer into clinical practice. Santorio's earliest instrument was probably uncalibrated and not therefore a thermometer, but a thermoscope. Over the next 250 years, the development of temperature measuring devices proceeded in parallel with the gradual establishment of a proper clinical basis for temperature measurement. In particular, the use of air as a thermometric medium was abandoned in favour of mercury.

An important part of the development of a scientific basis for temperature measurements on the human body was the fixing of temperature scales, and the work of Fahrenheit in this connection was very significant. On the clinical side, the physician and polymath, Hermann Boerhaave of Leiden, who knew Fahrenheit and used mercury thermometers made by him, was probably a seminal influence. Students of Boerhaave carried an interest in the clinical use of temperature measurement to new appointments throughout Europe, and their students similarly. In 1868, the rules for clinical thermometry were finally set out by Carl Wunderlich of Leipzig in his classic work on temperature in disease '*Das Verhalten der Eigenwärme in Krankheiten*', or literally 'the behaviour of body heat in illness'. This work summarized the results of temperature measurements in thousands of clinical cases. By this time the temperature sensitive device itself had progressed from the unwieldy air thermometer of Santorio (Figures 1.1(a), (b)) to the compact, mercury-based maximum thermometer with which we are now familiar. The first example of a compact, mercury-based thermometer was introduced in 1870 by Sir Clifford Allbutt, and closely resembled the outline of the modern instrument shown in Figure 1.1(c).

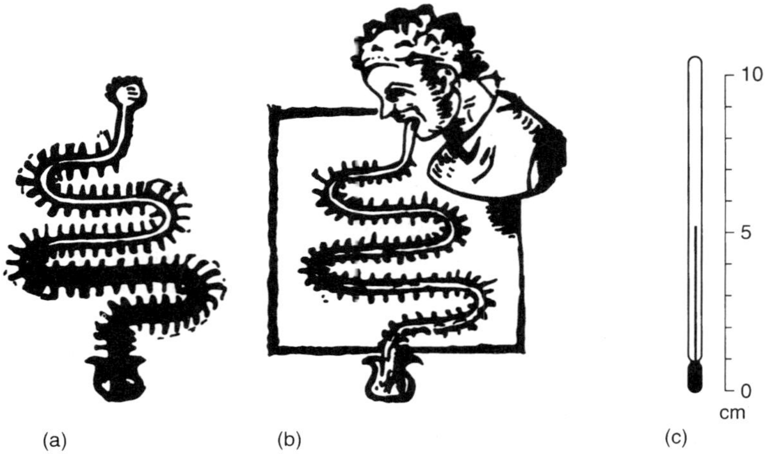

Figure 1.1 (a) Clinical air thermometer as depicted by Santorio in 1625. (b) Santorio's thermometer in use. (c) Modern clinical maximum thermometer with constricted mercury column, for comparison (scale shows usual size). It first reached this size in about 1870.

1.2 TEMPERATURE VARIATION IN HEALTH AND SICKNESS

To appreciate the basis for later work on the temperature distribution of the human body it is necessary to understand how the body core temperature is controlled and why it is usually stable. Body temperature is held within a very narrow range of values so that the body's enzyme systems can function effectively. This range is regulated at 36.8 ± 0.6 °C although, in addition, the measured value depends on both the time of day and the position on the body at which the measurement is made. Body core temperatures have a maximum value in the evening at around 6 p.m. and a minimum value at around 3 a.m.; the readings at these times differ by about 0.5 °C. The conventional clinical thermometer is used in one of three places – the **rectum**, the mouth or the armpit – and the temperature recorded for these places will normally be 37.4 °C, 36.9 °C and 36.7 °C, respectively. The highest temperature recorded (for the rectum) reflects bacteriological activity there.

The body core temperature, also called the deep tissue temperature, is usually measured by means of a standard, mercury-based, clinical thermometer (or its modern digital equivalent), inserted into one of the convenient 'apertures' or 'pockets' in the body mentioned above. This arrangement, although commonplace, is not always entirely satisfactory. An alternative approach was introduced in 1973 by A.J. Solman and J.C.P. Dalton, which permits the measurement of the deep tissue temperature from any point on the skin surface, although the time required to record the temperature is much longer than that required using a standard clinical thermometer.

It has been shown, using thermocouples mounted inside hypodermic needles, that the core temperature is reached some 10–15 mm below the skin surface, at least in normal subjects. This is because the skin acts like a layer of insulation. The body core temperature can be measured by establishing the condition that the net heat flux through the skin in a particular area is zero. This is done by introducing an auxiliary heater, together with temperature sensors, to monitor the temperature gradient. The actual experimental arrangement is shown in Figure 1.2(a), (b), and comprises a composite structure some 60 mm square, made up from four layers of rubber, interleaved with which are two thermistor temperature sensors and a thin foil heater.

Starting at the skin surface we have a very thin layer of silicone rubber which provides electrical isolation for a thermistor sensor

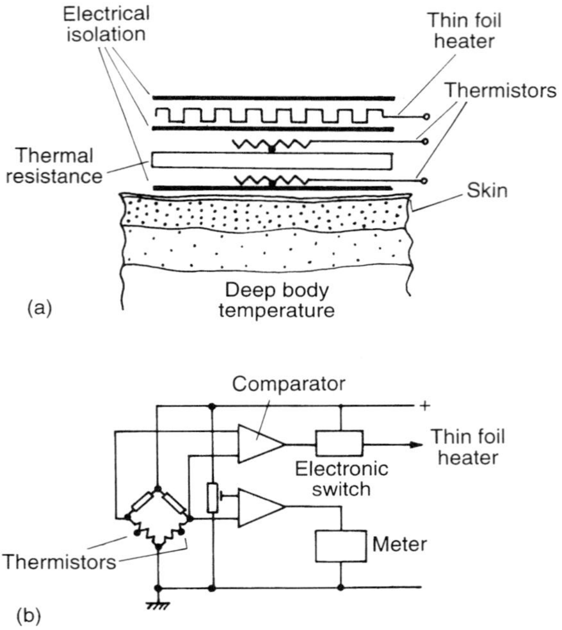

Figure 1.2 Schematic diagram showing (a) the structure of a body core temperature measuring pad, together with (b) associated control and measuring electronics.

mounted upon it. This is followed by a more substantial layer of rubber which introduces significant thermal resistance. Its function is to produce a temperature difference between its front and back faces during heat flow. Mounted upon the back face of this more substantial layer is a second thermistor isolated from a thin film heater by yet another layer of rubber. The whole structure is closed off by a final, thin layer of rubber which provides electrical isolation.

The two thermistors are connected in opposing arms of a bridge circuit feeding a high gain, dc amplifier (operational amplifier, or op-amp) functioning as a comparator. The output from the op-amp drives an electronic switch which provides power to the thin film heater. In addition, the thermistor closest to the skin surface has its resistance, and hence temperature, sensed by an additional op-amp. During operation the thin film heater provides heat to the multilayer package until the difference voltage sensed by the comparator becomes essentially zero. In this state there is no temperature gradient across the sensing package and no net heat flow, neither can there be a temperature gradient between the skin and the body core, assuming negligible edge effects and one-dimensional heat flow in the pad. Under these conditions the temperature recorded by the skin thermistor is also the body core temperature, and is displayed on the output meter.

This instrument is very simple to use since it may be placed on any part of the patient, and the deep body temperature is displayed after 10–20 min. The rather long response time can be shortened by arranging the circuitry so that the thin film heater maintains the sensing pad at approximately 37 °C, before application to the patient. The alternative approach, which involves increasing the heater power dissipation, is not acceptable owing to the possibility of burning the patient if the system malfunctions.

Body temperature is determined by the balance between the rate of heat production and heat loss. In order to maintain a constant body core temperature the rate of heat loss from the body must be carefully controlled. Heat production within the body occurs in two ways, namely by muscular exercise (including shivering) and by metabolic processes in all the tissues, but particularly in the liver. The source of heat in the metabolic processes is the conversion of carbohydrate, fat and excess protein into energy, and in a resting man occurs at a rate of approximately 80 W. Heat loss from the human body may occur by some combination of radiation, conduction, convection or evaporation. Radiation is the most important mechanism, accounting for some 55% of the total heat loss. We shall see, later in this chapter, that further diagnostic information may be obtained from measurements of the characteristics of this radiation.

The control of body core temperature is achieved by altering the volume of blood flowing through the skin tissues. Temperature sensing is performed in part of the brain (the **hypothalamus**) by specialized cells which are capable of detecting a temperature change in the arterial blood supply of as little as 0.1 °C. The blood supply comes in this instance from the **internal carotid artery** of the brain (Figure 4.28 shows an X-ray image of this artery). If the blood is cooler than it should be, the hypothalamus sends impulses to the **vasomotor centre** in the **medulla oblongata** at the base of the brain. The vasomotor centre is part of the autonomic, or involuntary, nervous system and it, in turn, sends impulses via the sympathetic nerves to adjust the diameter of the blood vessels. Increased numbers of impulses from the vasomotor centre cause **vasoconstriction** of the **arterioles** in the skin. This has the effect of reducing skin blood flow and consequent heat loss. A decrease in the stimuli causes relaxation of the arterial muscle, **vasodilation**, and this increases both blood flow and heat loss. If necessary, the temperature regulating centre can augment heat production by initiating shivering, using signals sent along the motor nerves to the muscles. The temperature control system is shown schematically in Figure 1.3. If the body core temperature rises, the process described above is simply reversed and heat loss by radiation, conduction and convection is increased. Further heat loss can be achieved, if necessary, by evaporation due to sweating; this will be initiated via the autonomic nervous system.

In illness, the body temperature regulating mechanism still operates, but the temperature is now stabilized above the normal value because the temperature sensing cells of the hypothalamus are set at a higher temperature. **Pyrogens** in the blood supply to the hypothalamus have this effect and the symptoms are those of a fever. **Antipyretics**, such as aspirin, operate by restoring the original temperature setting.

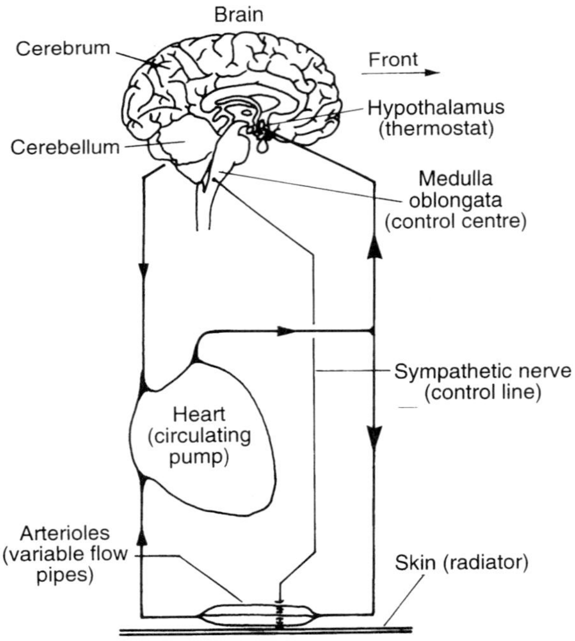

Figure 1.3 Schematic diagram of the temperature control system of the human body. The lungs are omitted from the blood supply loop.

If the air surrounding the human body is cold, the skin temperature may fall as low as 17 °C, but despite this the body core temperature will remain at the value already defined. On the other hand, the sensation of temperature depends directly on the skin temperature and therefore skin blood flow, so that dilated blood vessels in the skin give a feeling of warmth and constricted blood vessels a feeling of cold, regardless of the actual temperature. For comfort, though, the optimum skin temperature is 33 °C.

1.3 THE BASIS FOR TEMPERATURE MAPPING

The preceding discussion has considered the magnitude and stability of the body core temperature and how stability is achieved by control of the volume of blood flowing in the skin. In a healthy human, therefore, we might expect that, although skin temperature will not be uniform, it will reflect the internal structure and be bilaterally symmetric about an imaginary axis running down the centre of the body. For an erect person this axis would be defined by a plumb line suspended from the centre of the forehead, with features equidistant from this line exhibiting much the same temperature. Thus, cheek bones on the face, for example, should have the same temperature. Figure 1.4 shows that the skin temperature distribution actually found confirms this expectation. Interestingly, even this concept is not new since Hippocrates, in the fifth century BC, had instructed that 'the physician shall examine to see whether one side is hotter than the other', although he had only his hands with which to judge temperature. Persons who are unwell will not necessarily show this symmetric temperature distribution and the reasons for this stem from disturbances to the blood flow from causes other than those connected with maintaining the body core temperature. It is instructive to summarize the more important reasons for these departures from the normal.

We can distinguish two kinds of variation, those which lead to temperature increase and those which cause temperature decrease. Any local problem which leads to artificially enhanced blood supply will result in a temperature rise. Amongst this kind of problem we may note cell division at abnormal rates (as in tumours), localized centres of infection, or centres of irritation. Situations which lead to skin temperatures below the surrounding value are normally associated with failure, partial or otherwise, in the blood flow. Reasons for this include deposits in the arteries, such as may be associated with heart disease, degenerative disease, or physical damage (as in burns). Table 1.1

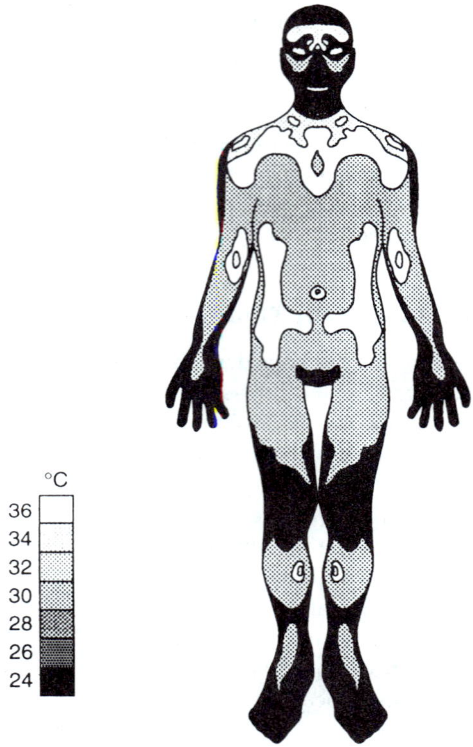

Figure 1.4 Stylized diagram of the bilaterally sensitive skin temperature distribution in a male subject (anterior view). The measurements were made after the subject had rested for 20 min in a 20 °C ambient. (Courtesy of Professor E.F.J. Ring.)

summarizes the causes of variation in the temperature distribution and indicates the magnitude and sense of the typical temperature change observed.

In principle, therefore, a new method of diagnosis can be based on measurement and mapping of the skin temperature, and this was actually attempted by E. Seguin as early as 1876, using a mercury thermometer with a large, flattened bulb, offering a contact area of approximately 650 mm^2. In practice, the use of this thermometer was unsatisfactory because application of the thermometer to the skin altered the local blood flow and, therefore, the measured temperature. An alternative method for determining the skin temperature, in fact of

Table 1.1 *Summary of temperature variations from the normal associated with various clinical conditions*

Clinical condition	*Temperature change* (°C)
Deep vein thrombosis	+2
Arthritis	+2
Tumours (malignant)	+1–4
Tumours (benign)	+1–2
1st degree burn	+1
3rd degree burn	−3
Smoking	−2
Frostbite	−3
Arterial disease of limbs	−3–+4

mapping the skin temperature, is based on measuring the infrared emission from the body as a consequence of its temperature. This technique was first deployed clinically in the UK by K. Lloyd-Williams *et al.* in 1961, following its initial demonstration by R.N. Lawson in Canada in 1956.

1.4 THE HUMAN BODY AS AN INFRARED SOURCE

In 1934 J.D. Hardy showed that the spectral distribution of radiation from the human body is essentially that of a perfect black body. Consequently, wideband measurements of the radiative output from the human body enable the body temperature to be mapped, and temperature irregularities revealed, by remote sensing. The total radiant emittance of a perfect black body is described by the Stefan–Boltzmann law which may be written as

$$R(T) = \sigma A T^4 \tag{1.1}$$

where $R(T)$ is the total power, in watts, radiated into a hemisphere, σ is the Stefan–Boltzmann constant (5.67×10^{-8} W m^{-2} K^{-4}), A is the effective radiating area and T is the absolute temperature of the radiating surface. For a perfect black body the emissivity ε is unity, although for surfaces which depart from perfect radiative behaviour, ε will be <1 and we then write equation 1.1 as

$$R(T) = \varepsilon(T)\sigma A T^4 \tag{1.2}$$

The rate of heat loss between two surfaces at temperatures T_1 and T_2 with emissivities $\varepsilon_2(T_1)$ and $\varepsilon_2(T_2)$ is given by

$$\Delta R = \sigma A(\varepsilon_1(T_1)T_1^4 - \varepsilon_2(T_2)T_2^4) \tag{1.3}$$

and equation 1.3 effectively describes the heat loss which forms the basis of thermal imaging. In fact $\varepsilon(T)$ is a slowly varying function of T; at body temperatures, however, for a given surface, it may be considered to be a constant.

The spectral distribution of radiation emitted from a surface depends on its temperature and is described by Planck's radiation law. Thus, the energy density $u(\lambda, T)$ in the wavelength range λ to $\lambda + d\lambda$ is given by

$$du(\lambda, T) = \frac{8\pi hc}{\lambda^5} \frac{d\lambda}{[\exp(hc/\lambda k_B T) - 1]} \qquad (1.4)$$

where h is Planck's constant k_B Boltzmann's constant and c the velocity of light; it is plotted in Figure 1.5. One of the most striking features of a set of spectral distribution curves for a black body at different temperatures is that each has a definite maximum, and this shifts towards shorter wavelengths as the temperature rises. The value of the maximum wavelength, in μm, is given by

$$\lambda_{max} T = 2898 \qquad (1.5)$$

This is just Wien's displacement law. Skin temperatures are in the range 25–35 °C so that using equations 1.4 and 1.5 we might expect the skin to radiate in the range from 4 to 50 μm with maximum emission at 9.5 μm, and this accords exactly with experiment. In practice, therefore,

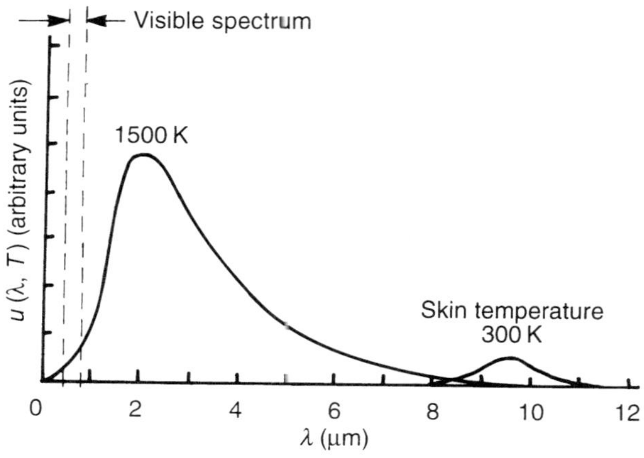

Figure 1.5 Black body energy density distribution as a function of wavelength λ, with temperature as a parameter. The energy density axis is not to scale for both curves.

the human body radiates principally in the near infrared, although it should be noted that there will also be some radiative output in the adjacent microwave region of the electromagnetic spectrum. This latter is the subject of section 1.13.

Human skin, although closely similar to a perfect black body in its radiative properties, has an emissivity of 0.98 ± 0.01. The emissivity is affected principally by surface structure so that surface contaminants can alter the emissive properties and thus the apparent surface temperature. For a surface at temperature T_0 and having an emissivity ε_0, the radiant power emitted will be proportional to $\varepsilon_0 T_0^4$. A black body, on the other hand, emitting the same radiant power, will have a temperature T_b and $\varepsilon_b = 1$. Consequently we can write

$$\varepsilon_0 T_0^4 = T_b^4 \tag{1.6}$$

and

$$\varepsilon_0 = \left(\frac{T_b}{T_0}\right)^4 \tag{1.7}$$

From equation 1.7 we note that a decrease in skin emissivity of 1% reduces the apparent temperature by approximately 0.8 K.

A further problem with the human body as a black body is that electromagnetic theory shows that maximum emission will always occur at right angles to the skin surface; that is emissivity varies with viewing angle. However, this restriction will only lead to significant problems if the viewing angle exceeds 60° from the normal; in clinical practice this situation seldom arises. It is clear that a viewing system which can measure and record the infrared output from the human body, in a stable temperature environment, permits the mapping of the body surface temperature. The resulting image, or thermogram, will reveal any change from the normal pattern of temperature distribution. We will now consider the various means by which this objective may be achieved.

1.5 INFRARED IMAGING

In order to produce a thermal image of the body surface we must be able to detect the radiant emission from that surface. Any temperature measuring device of this sort consists of a system for collecting radiation from a well defined field of view and either some means of detecting and converting this radiation into an equivalent electrical signal, or direct representation of the thermal image. There are two types of imaging system which operate in this latter way, namely the evapograph and the cholesteric liquid crystal display.

Turning to the imaging systems of the first sort, which convert the infrared image into electrical signals, the build-up of a thermal image is achieved by means of either an optical scanning system or a pyroelectric vidicon television tube. A monitor supported by an image-processing unit displays the image, together with a temperature reference scale. Detection of the infrared radiation in an optical scanning system may be carried out using thermal detectors, such as thermocouples or semiconductor bolometers, both of which depend on temperature rise in the absorbing element. They are, however, inherently less sensitive than photon detectors and have longer time constants. Photon detectors respond rapidly to radiation flux changes so that time constants of a few microseconds are typical, unfortunately they detect only a small fraction of the total radiation emitted from the object's surface. The ideas which form the basis of the various imaging systems are perhaps best described and explained within the context of the system in which they are used. We begin with the evapograph, an infrared viewing system which provides direct conversion of the infrared image into a visible image.

1.6 IMAGING WITH THE EVAPOGRAPH

The principle of evaporography is the basis of one of the oldest methods of heat detection and certainly one of the oldest types of thermal imaging device. Its origins lie in the work, in 1840, of Sir J.F.W. Herschel, son of Sir W. Herschel (who discovered infrared radiation). Sir John Herschel coined the word 'thermogram'. The idea was adapted in the 1920s by the spectroscopist M. Czerny to record infrared spectra, and improved further in the 1960s for thermal imaging.

The evapograph is a thermal imaging device which converts an infrared image into a visible image by differential evaporation or condensation of a liquid, usually a hydrocarbon, or more recently a silicone oil, on a thin membrane. The evapograph is shown schematically in Figure 1.6. Radiation from the infrared source is collected and focused by a lens which will be germanium or silicon, since glass is essentially opaque to infrared radiation at the wavelengths in question. The detector in this system is a thin nitrocellulose membrane placed at the image plane (Figure 1.7). This membrane is about 0.1μm thick, so that white light reflected from the front and rear surfaces of the

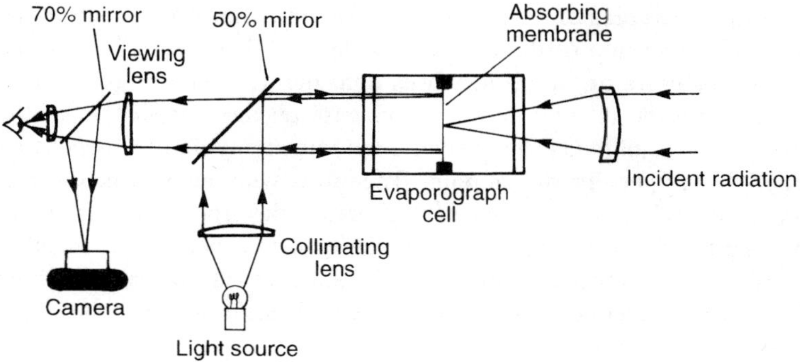

Figure 1.6 Simplified schematic diagram of an evapograph system showing the essential components.

membrane presents the first-order yellow interference colour. The front surface of the membrane, that is to say the surface facing the lens, is coated with a thin layer of gold-black which will act as a near perfect absorber of radiation. The membrane is mounted within an enclosure or cell, separating it into two halves, and the air is evacuated from both halves. The side of the cell away from the collecting lens system is lined with blotting paper, or similar absorbent material, moistened with a few drops of the chosen oil. It is so arranged that this half of the cell can be heated by an electric heater which slips over the outside surface of the cell.

The irradiance at the membrane, due to the infrared flux focused upon it, is absorbed by the gold-black to produce a heating effect which

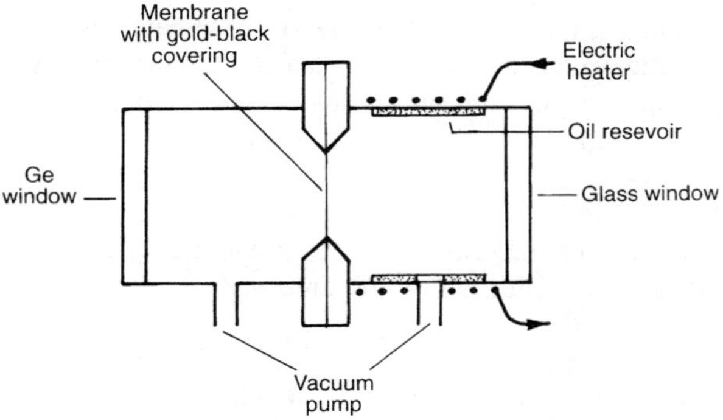

Figure 1.7 Schematic diagram showing the essential elements of an evapograph cell.

will vary according to the amount of radiation falling on that part of the film. If the section of the cell containing the oil source is heated, the vapour pressure of the oil increases until the dew point is reached, at which time the oil will begin to condense on the rear surface of the blackened membrane. The condensation rate at any point is dependent upon the temperature at that point. Therefore, with the passage of time, an oil film will be formed whose thickness varies from point to point in a manner related to the thermal pattern on the membrane. In other words the differences in infrared radiation focused on the membrane have been converted into differences in a physical quantity, the thickness of the oil film.

In order to view the resulting pattern, the membrane must be illuminated with white light, which reveals the pattern in different interference colours according to the oil film thickness. This is accomplished by a collimated beam of light from a white light source, reflected into the viewing cell via a half silvered mirror (50% transmission). An additional 70% transmission viewing mirror, also placed in the optical path, allows for direct viewing and simultaneous photographic recording. The layout of the viewing system is shown in Figure 1.6.

1.7* THEORETICAL SENSITIVITY OF THE EVAPOGRAPH

The evapograph is an energy proportional transducer, which means that the response is proportional to the total energy that has been received since the imaging process began. We now ask the question: What is the smallest temperature difference that can be detected in a reasonable length of time, say 20 s, by the resulting change in oil film thickness? It turns out this is a question which we can answer. We begin by assuming a background temperature T_B of 300 K and an oil of vapour pressure p, molecular weight M and latent heat per kilogram L. The collision frequency per unit area ν, of oil vapour molecules with the film is given by the Hertz–Knudsen equation

$$\nu = p/(2\pi m k_B T_B)^{1/2} \tag{1.8}$$

where k_B is the Boltzmann constant and m the mass per molecule. The mass transported in unit time to unit area is then

$$m\nu = p(m/2\pi k_B T_B)^{1/2} = G \tag{1.9}$$

and is, of course, a function of temperature. This result may be expressed in terms of the molecular weight M and the gas constant R as

$$G = p(M/2\pi R T_B)^{1/2} \tag{1.10}$$

Since the oil film is in equilibrium with its vapour, i.e. the steady state, this expression must also represent the amount of oil evaporating in unit time from unit area. If the temperature of the film is raised locally by ΔT, then there will be an excess quantity of material ΔG leaving unit area of the oil film in unit time given by

$$\Delta G = G(T_B + \Delta T) - G(T_B) \tag{1.11}$$

Substituting for G we can rewrite equation 1.11 as

$$\Delta G = (M/2\pi R)^{1/2} \cdot \Delta T \frac{\partial}{\partial T_B}\left(\frac{p}{T_B^{1/2}}\right) \tag{1.12}$$

We also know that the vapour pressure and temperature of the oil are related through the Clausius–Clapeyron equation,

$$\frac{dp}{dT_B} = \frac{pLM}{RT_B^2} \tag{1.13}$$

Carrying out the differentiation in equation 1.12 and substituting for dp/dT_B from equation 1.13 yields

$$\Delta G = \left(\frac{M}{2\pi R}\right)^{1/2} \Delta T \cdot p \frac{(2LM - RT_B)}{2RT_B^{5/2}} \tag{1.14}$$

The energy E required to evaporate the minimum detectable change in oil thickness, in a reasonable time, is just the mass evaporated per unit area multiplied by the latent heat of evaporation; for a thickness change Δd and an oil density ρ we can write that

$$E = \Delta d \rho L \tag{1.15}$$

The quantity represented by equation 1.15 is also that obtained if we multiply equation 1.14 by $L\,\Delta t$, where Δt is the time interval during which energy is received, so that

$$\Delta T = \frac{2R\,\Delta d \rho T_B^{5/2}}{(2\pi R)^{1/2}\,\Delta t M^{1/2} p(2LM - RT_B)} \tag{1.16}$$

If the physical constants for a given imaging oil are known, and a value for Δd determined then ΔT can be computed.

In order to obtain some sort of guidance as to the correct choice of oil for a sensitive viewing system, it is useful to introduce a figure of merit $\mathcal{M}$ for an oil. To do this we take the energy required to evaporate unit thickness of oil from unit area of membrane surface. This quantity is simply the latent heat of vaporzation multiplied by the density of the oil. In practice we need to use unit optical thickness so that we must divide the answer by the oil refractive index n. The figure of merit is then

$$\mathcal{M} = L\rho/n \qquad (1.17)$$

This figure should, of course, be as small as possible for the greatest temperature sensitivity.

The silicone polymers available as Dow-Corning silicone 200 fluids offer the best performance overall; in particular, the octamer of tetramethyl siloxane has a figure of merit of $7.3 \times 10^7\ \mathrm{J\,m^{-3}}$. The minimum resolvable temperature difference (MRTD) which may be detected with an evapograph system is around 0.2 K.

1.8 THERMAL IMAGING USING LIQUID CRYSTALS

We have now considered an example of an infrared viewing system which provides direct conversion of the infrared image into a visible image. A simpler, and cheaper, method of producing a visible image which is, at the same time, a temperature map, is by means of a liquid crystal, intermediate layer. It will be useful to outline the nature of liquid crystals in general so as to gain a better understanding of this means of image production.

Liquid crystals are examples of mesomorphs, that is materials which are neither liquids nor crystalline solids; positional order is found in one or sometimes two dimensions rather than the customary three of a conventional crystal. There are three main types of liquid crystal, named originally by Friedel, as smectic, nematic and cholesteric; it is, in fact, the last sort in which we shall be interested. A liquid crystal material is made up from arrays of organic (generally synthetic) molecules, whose main characteristic is that the overall molecular shape approximates to that of a Frankfurter sausage, that is, a very elongated or rod-shaped molecule. For minimum free energy, molecules must be packed with their long axes parallel. Figure 1.8 shows how this restriction works out in practice, so that for the smectic liquid crystal, all of the long axes are parallel. In addition, we see that the molecules align in columns between the layers and we have pseudo two-dimensional order. If this last restriction is relaxed, we simply have all of the long axes parallel and no further order; the result is a one-dimensional array, termed by Friedel a nematic. For both the smectic and nematic materials we can specify the direction of the molecular axes by a unit vector $\hat{\boldsymbol{a}}$, which is fixed spatially.

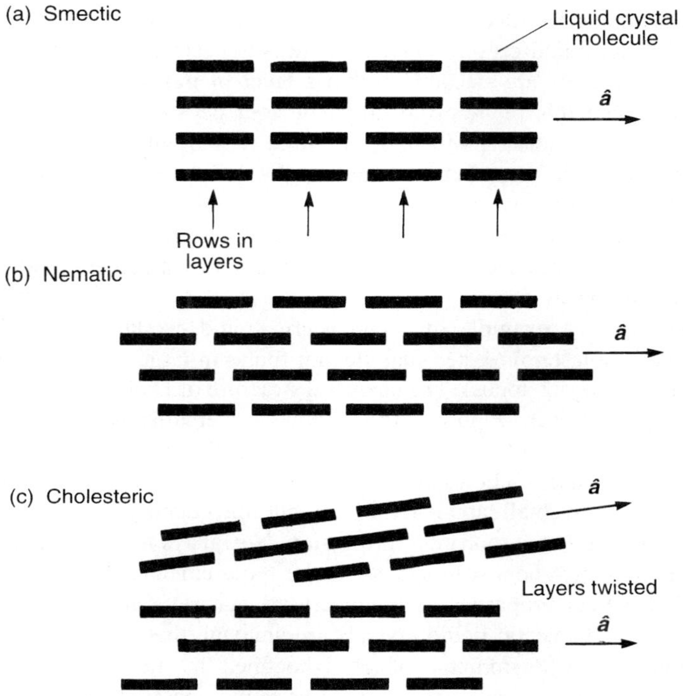

Figure 1.8 Basic ordering of liquid crystal molecules, showing the molecular arrangement for (a) smectic (b) nematic and (c) cholesteric liquid crystals. The cholesteric liquid crystal is a form of nematic. The unit vector $\hat{\boldsymbol{a}}$ identifies the direction of the molecular axes in each layer of molecules and has only one direction for smectic and nematic materials.

The third member of the liquid crystal family is the cholesteric liquid crystal, so named because one of the earliest examples was based on the naturally occurring material, cholesterol. The molecular architecture of a cholesteric liquid crystal is subtly different from that of the nematic and smectic materials. The difference is that the Frankfurter-like molecule now carries a small molecular protrusion at one end. The effect of this protrusion is to alter the way in which the molecules stack. Since the stacking arrangement for any liquid crystal represents the minimum free energy configuration, this is reached for the cholesteric liquid crystal if each layer of molecules has the vector defining its molecular axis rotated, or twisted slightly, from that of the preceding layer. The twist is required to accommodate the molecular protrusion, and has the effect that the unit vector defining the direction of the long axis of the liquid crystal molecule is now not fixed, but rotates as the molecule layers are stacked. Within a layer of molecules all axes are parallel, but this relationship does not hold between adjacent layers. The structure is now like a nematic liquid crystal within any layer, but with a twist of axes between layers. We have just one-dimensional order. It turns out that this simple steric view of the reason for cholesteric behaviour is not always true; cholesteric behaviour can be introduced by replacement of a hydrogen atom by a deuterium atom in some instances.

The earliest example of a cholesteric liquid crystal is the natural product cholesterol (a steroidal alcohol found in large quantities in the brain and in gall stones). The chemical structure of cholesterol is shown in Figure 1.9(a). Figure 1.9(b) shows schematically, the twisted arrangement of molecules which leads to a preferred helical conformation. In the case of cholesterol itself, the twist of axes between adjacent layers is quite small, about 15′ of arc, but this quantity depends on the exact details of the molecular protrusion. Nowadays it can be ‘tailored’ by the organic chemist to suit particular requirements.

The optical properties of liquid crystal materials are unusual. The nematic and smectic liquid crystals are uniaxial, that is they possess a unique axis of symmetry which is defined by the unit vector $\hat{\boldsymbol{a}}$. Refractive indices measured with polarizations parallel to, or perpendicular to, $\hat{\boldsymbol{a}}$ will be different. Further, their optical absorbance α, which is given by

$$\alpha = \log I_0/I \tag{1.18}$$

where I_0 is the intensity of the incident radiation and I that of the transmitted radiation, is also a function of the direction of the electric vector. For a cholesteric liquid crystal, if we specify that the axis of the helical structure is coincident with the z axis, then we can characterize the helix in the following manner, which defines the components of $\hat{\boldsymbol{a}}$ resolved along the three orthogonal axes

$$a_x = \cos(q_0 z + \theta) \tag{1.19}$$

$$a_y = \sin(q_0 z + \theta) \tag{1.20}$$

$$a_z = 0 \tag{1.21}$$

In equations 1.19 and 1.20, θ is an arbitrary angle and q_0, which has the units of reciprocal length, determines the overall characteristics of the

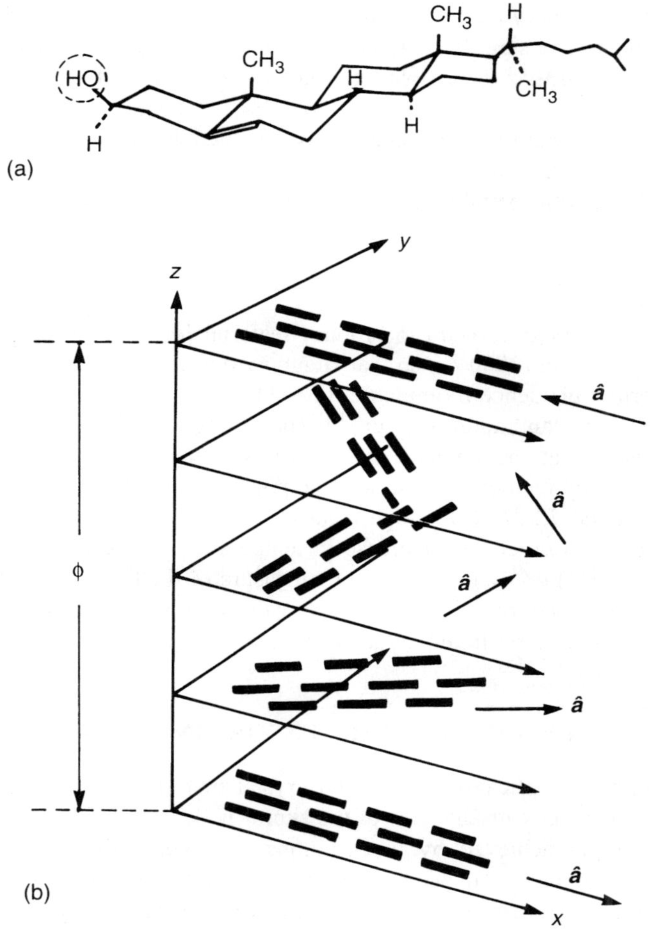

Figure 1.9 (a) Conformation of the cholesterol molecule upon which many cholesteric liquid crystal materials are based. The six-membered rings have a flattened chair conformation; the five-membered ring is planar. The molecule can be lengthened further by reacting the OH group with organic acids to form esters, e.g. cholesterol nonanoate. (b) Helical structure of a cholesteric liquid crystal showing one half rotation of the molecular axes.

helix. Thus we note from Figure 1.10b that the helix has a spatial period ϕ defined by

$$\phi = \pi/|q_0| \tag{1.22}$$

The pitch is $2\pi/|q_0|$, or twice the spatial period since $-\hat{\boldsymbol{a}}$ and $+\hat{\boldsymbol{a}}$ are equivalent. The sign given to q_0 determines the sense of the helix, so that q_0 is positive for a right-hand helix and negative for a left-hand helix.

If a light beam of wavelength λ is sent parallel to the z axis then, because the material is optically anisotropic, the optical properties are modulated with spatial period ϕ. This can give rise to Bragg reflections provided that

$$2\phi = k\lambda \qquad (k \text{ an integer})$$

It is quite straightforward to produce material for which ϕ is approximately 300 nm (3000 Å), so that Bragg scattering occurs in the visible for normal incidence illumination ($k = 1$).

The important additional quality possessed by cholesteric liquid crystals is that the pitch of the helical structure is very sensitively dependent on temperature. In fact, the pitch is a decreasing function of T, that is $\mathrm{d}\phi(T)/\mathrm{d}T < 0$, which means that the molecular layers move towards each other with increase in temperature. If we determine the quantity $(1/\phi)(\mathrm{d}\phi/\mathrm{d}T)$ we gain some measure of the temperature sensitivity possessed by these materials; values of around $100\ \mathrm{K}^{-1}$ are typical. Cholesteric liquid crystals are termed thermochromic.

1.9 LIQUID CRYSTAL SENSORS IN PRACTICE

For medical applications the liquid crystal material must be in good thermal contact with that part of the skin it is desired to image. This is most simply achieved by first painting the skin with a coating of colloidal graphite (Aquadag), to provide a black viewing background, followed by a sprayed-on coating of the cholesteric liquid crystal material in a volatile organic solvent. When the solvent has evaporated a uniform coating of liquid crystal material remains which, when viewed by white light incident normally, exhibits interference colours which vary according to the underlying skin temperature. Unfortunately, this approach is not altogether suitable for clinical use and has been largely abandoned in favour of microencapsulated liquid crystals which exhibit greater longer term stability.

The layer of liquid crystal material on the skin adopts the conformation shown in Figure 1.10(a). Since the helix pitch decreases with increasing temperature, the interference colours are the reverse of those which one would instinctively assign, thus blue corresponds to a skin

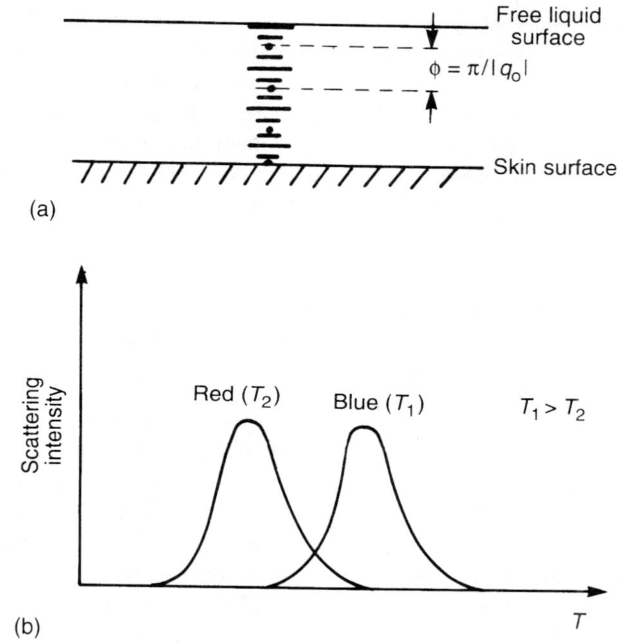

Figure 1.10 (a) Conformation of a liquid crystal array on the human skin showing how the helical structure builds with its axis perpendicular to the skin surface. (b) Intensity of light scattered from a cholesteric liquid crystal layer on the skin, as a function of temperature, with scattering wavelength as the parameter.

temperature which is higher than that for red. The intensity distribution as a function of temperature is shown in Figure 1.10(b). Temperature resolution for any thermographic system may be expressed in two ways. For liquid crystal displays it is the minimum resolvable temperature difference (MRTD) which is the appropriate choice; a value of 0.1 K is achievable.

A common choice of cholesteric liquid crystal for medical applications is cholesterol nonanoate, (that is the ester of cholesterol with nonanoic acid) although other molecules of this type, or indeed mixtures of molecules, may be used. In each case, the absolute temperature range of the film is approximately 3 K, and the response time is usually around 20–30 s.

The method of application described above has two advantages, it is cheap and does not alter the skin temperature; it does however have a particular disadvantage in that it is difficult to restore the skin to its original clean condition. This latter problem may be overcome by using a liquid crystal encapsulated in a thin plastic film with a black background (these are now widely available). Although the liquid crystal material is protected, and there is no contamination of the skin, it can alter the temperature distribution which it is being used to measure, since it is not easy to apply the encapsulated material without exerting uneven skin pressure.

1.10 INFRARED IMAGING WITH AN OPTICAL SCANNING SYSTEM

In an optical scanning system, the scene is viewed by an optical system capable of transmitting and focusing infrared radiation in the range, say 8–14 μm. For such a system, lenses with a high refractive index are advantageous, since they confer high spatial resolution; the lenses should have high transmittances as well. Although glass is essentially opaque in the wavelength range required, a germanium or silicon lens is transparent, but its transmittance is low (only 47% at 2 μm). Antireflection coatings increase the value to 95%. As a consequence, both germanium and silicon find use as infrared lenses and/or windows, but always with an antireflection coating.

The mechanical scanning system employs a combination of lenses, rotating prisms, rotating multisided mirror-drums or rocking mirrors. The exact arrangement varies from one manufacturer to another. One of the earliest versions of an optically scanned imaging system was the Barnes Thermograph introduced in the 1950s. Modern scanning thermographs differ only in the details of the optical path and the types of radiation detector used.

Figure 1.11 shows schematically how the image from the patient is scanned mechanically. Radiation from the patient is collected by a germanium lens and directed onto a vertical scan mirror. This then reflects the radiation onto further focusing mirrors before the horizontal scan is carried out using a reflecting, 10-sided polygon capable of rotation on its axis. Also in the viewing path is a reference surface at a known temperature, the radiation from which is incorporated in the overall view as a colour scale of temperature. Finally, the radiation is focused onto the detector element which, in early thermographs, would be a bolometer based on thermocouples or perhaps a thermistor. The bulk of medical thermography has been carried out with single-element detectors which have the advantage of both electronic and mechanical

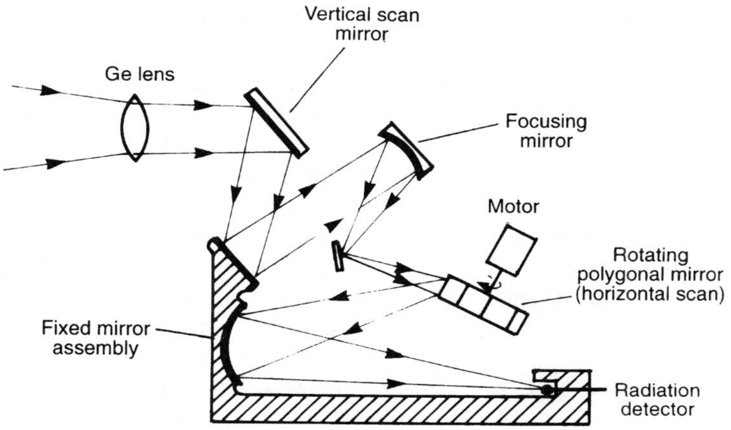

Figure 1.11 Mechanically scanned thermographic system shown schematically.

simplicity. However, a multielement array of n similar detectors gives a signal which increases in proportion to n whilst the associated noise goes up only as $n^{1/2}$. An additional advantage possessed by array detectors is that scan speeds are higher than those from single detectors. In a clinical context this means that high resolution, real-time, imaging is possible, which permits precise focusing on the skin surface and continuous observation of thermal changes

1.11* INFRARED DETECTOR CHARACTERISTICS

In scanning systems the detector characteristics dominate the imaging performance. This performance is usually specified in terms of the minimum resolvable temperature difference (MRTD), as already noted, or alternatively, as the noise equivalent temperature difference (NETD), the latter being the temperature difference for which the signal-to-noise ratio at the input to the display is unity. Although the temperature resolution is important, so also is the angular resolution and field of view. Angular resolution may lie in the range from 3 mrad down to as little as 0.5 mrad; 1–3 mrad is typical. The field of view can range from 5° × 5° to as much as 40° × 60°. The MRTD is currently 0.08 °C.

An important group of detectors are those in which photon absorption results in the release of charge carriers (usually electrons) in proportion to the intensity of the incident radiation. Many different materials have been used as photon detectors in medical imaging, including indium antimonide (InSb), cadmium mercury telluride (CdHgTe) and lead telluride (PbTe). These photon detectors may be used in either the photoconductive or photovoltaic mode. In the former, the change in conductance appears as a voltage change across a series load resistor fed from a bias voltage; this voltage is in turn amplified. In the photovoltaic mode (open circuit) an emf is generated across the device without the need for external bias or a load resistor. Usually, these detectors have to be operated with cooling to 77 K in order to reduce thermal detector noise; this is achieved either directly with liquid nitrogen, by means of Joule–Thompson cooling, or by a Stirling cycle cooling engine.

A detector's ability to detect small input signals is determined by its responsivity and noise. The responsivity is the ratio of the output voltage to radiant input power, expressed in volts per watt. The output voltage due to incident radiation is a very small fraction of the DC bias voltage across the detector and is measured by exposing the detector to chopped radiation from a calibrated source. The alternating voltage component at the chopping frequency is then recorded. The spectral responsivity is defined as the rms signal voltage per unit rms radiant power and is specified in terms of peak wavelength and chopping frequency of the measuring system.

The figure of merit usually used is the noise equivalent power or NEP. This is defined as the rms value of the sinusoidally modulated radiant power falling on a detector which will give rise to an rms signal voltage V_s equal to the rms noise voltage V from the detector. The temperature of the black body radiation source must be specified and is usually 500 K. The reference bandwidth is usually 1 or 5 Hz, and the centre frequency is normally 90, 400, 800 or 900 Hz. We express the noise equivalent power (NEP) for a 500 K source, 900 Hz chopping frequency and 1 Hz bandwidth as NEP (500 K, 900, 1). The units of NEP are W $\mathrm{Hz}^{-1/2}$. Related to NEP is the detectivity D which is the reciprocal of NEP, and for many detectors the NEP is proportional to the square root of the detector area, A. It is customary to use a normalized detectivity D^*, or specific detectivity, where measurements have been made normalized to a unit bandwidth. We define D^* as

$$D^* = (A\,\Delta f)^{1/2}/\mathrm{NEP} \tag{1.23}$$

because D^* varies with the wavelength of the incident radiation and the frequency at which the noise is measured, these are always stated in parenthesis whenever D^* is specified. As an example, D^* (5 μm, 800 Hz,

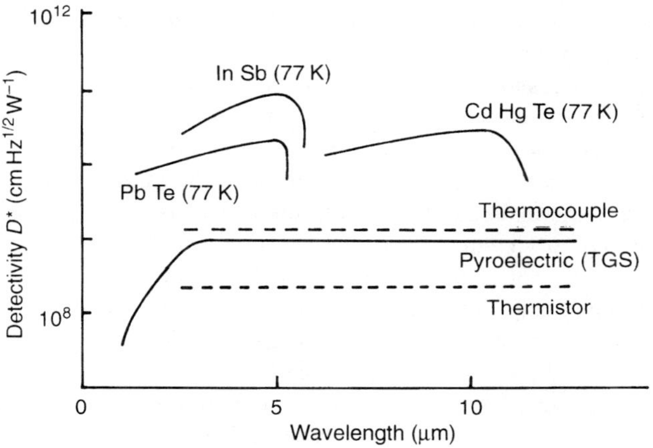

Figure 1.12 Normalized detectivity D^*, as a function of wavelength, for a range of infrared detector materials and systems.

1) is interpreted as meaning that the given value was measured at 5 μm, with noise measured at 800 Hz, and the measurement was finally normalized to unit bandwidth. D^* is a figure of merit which enables one to calculate the theoretical maximum detectivity when performance is noise limited; it has units $Hz^{1/2}\ W^{-1}$. Figure 1.12 shows how normalized detectivity varies with wavelength for a range of photon detectors. A conventional thermocouple and a thermistor are shown for contrast. Also included for comparison is the pyroelectric detector which forms the basis of the last imaging system we shall consider.

1.12 THE PYROELECTRIC VIDICON

The pyroelectric vidicon depends on the pyroelectric effect exhibited by certain ferroelectric crystals, for example barium titanate or tri-glycine sulphate (TGS). These materials undergo a change in spontaneous polarization when subjected to a change of temperature. If the polarization axis is perpendicular to the faces of the crystal slab, then this change is equivalent to an increment in surface charge equal to $P\,\delta T_m$, for unit area; P is the pyroelectric coefficient of the material and δT_m the change in temperature. If a slice of such a material is used as the target in a vidicon tube, then the video signal current $i(t)$ from unit area is given by:

$$i(t) = \frac{P}{\delta t}[T_m(t) - T_m(t - \tau_f)] \qquad (1.24)$$

where δt is the element scan time, T_m the element temperature and τ_f the integration time.

This equation shows that the vidicon is only responsive to changes in the radiation level which alter T_m. This has two important consequences: first, the background radiation corresponding to the ambient temperature has no effect, and second, once the target of the vidicon has come to thermal equilibrium with the viewed object, T_m will remain constant and no viewing signal will be produced. This tells us that the viewing signal must be modulated in order to view an object (a patient). Generally, this modulation is achieved by interposing a chopper disc between the vidicon face and the viewed object.

A vidicon tube is shown schematically in Figure 1.13. It comprises an evacuated envelope containing an electron gun, surrounded by electron focus and scan coils. The tube itself is closed at one end by a germanium window, antireflection coated to give good transmission in the 8–14 μm band. The seal between the germanium window and the vidicon tube is achieved by means of an indium ring which also acts as both mechanical support for, and electrical connection to, the pyroelectric target plate. The target plate is a thin slice, 30 μm thick, of triglycine sulphate cut with the pyroelectric axis normal to the viewing face. On the side of the target exposed to the incident radiation there is an electrically conducting layer of material chosen to be a good absorber of thermal radiation. The electron gun consists of an indirectly heated cathode, a control grid G_1 and accelerator grid G_2. G_3 is a wall anode whilst G_4 is a fine mesh grid placed behind the target plate to ensure that the electric field to the rest of the plate is uniform.

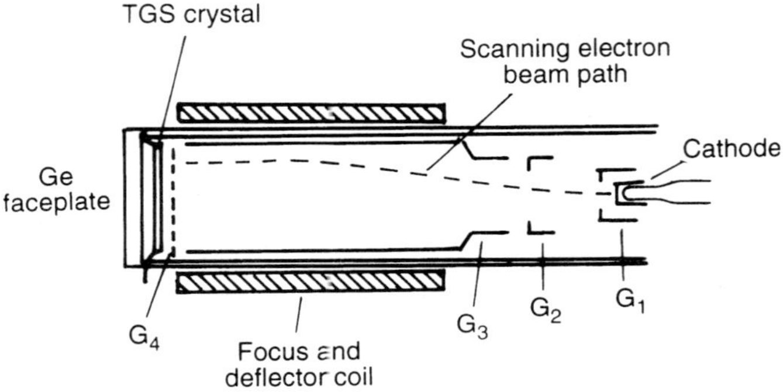

Figure 1.13 Schematic view of section through a pyroelectric vidicon tube based on triglycine sulphate (TGS).

In operation, the target plate is maintained at a uniform potential by the conducting layer, so that the incident thermal radiation leads to local potential differences on the rear side of the target which serve to modulate the scanning electron beam current. Unfortunately, these potential differences can be of either polarity, depending on whether the temperature of the target is high or low. As negative potential differences cannot be sensed by the electron beam, it is necessary to provide a positive bias current or pedestal current, upon which these negative potential differences may be superimposed. This is done by admitting a small amount of gas to the vidicon tube during manufacture, so that the scanning electron beam generates a small, background, positive ion current. During thermal scanning, the mesh G_4 is maintained at a slightly higher potential than G_3 so that positive ions generated at the rear of the tube are retained there. Electrodes G_2 and G_3 will typically be run at 200 V with respect to the cathode at 0 V, whilst G_4 is set at 210–230 V. The output signal currents are small, amounting to only a few hundred picoamps per degree centigrade change in object temperature.

Figure 1.14 shows the experimental layout of the pyroelectric vidicon. A germanium, high transmission, lens is used to focus the thermal image onto the 20 mm diameter, pyroelectric image plate and the incident radiation is chopped by a sectored disc rotating to give a chopping frequency of 25 Hz. The ferroelectric potentials developed by

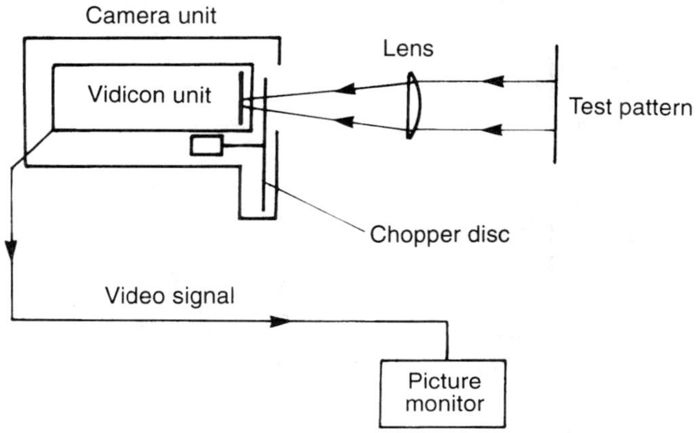

Figure 1.14 Experimental layout of pyroelectric vidicon viewing a thermal test pattern in the form of a mask, etched to give a bar pattern, in front of a large-area black body source. It measures MDTD.

the chopped incident infrared radiation serve to modulate the electron beam current received by the target plate. The resulting current flowing in the external circuit is amplified before display on a conventional TV monitor. Image quality is inferior to that obtainable using photon detector systems although the MRTD for a pyroelectric vidicon can be 0.1 °C. On the other hand these units do not require cryogenic cooling, and importantly they have full compatibility with standard video cameras so that overlay of thermal and video images is possible.

1.13 THERMOGRAPHY AT MICROWAVE FREQUENCIES

In the preceding sections we have surveyed the mapping of the body surface temperature by means of the infrared radiation emitted by the body, that is measurements concentrated at and around a wavelength of 9.5 μm. In fact, the black body radiation spectrum of the human body also extends into the radiofrequency part of the electromagnetic spectrum, although this is not evident from Figure 1.5 If, however, we plot the specific intensity of the black body radiation against frequency on a logarithmic scale (Figure 1.15), we note that there is some radiation output in the microwave region of the radiofrequency spectrum, that is the frequency range from 1 GHz (30 cm) up to 300 GHz (1 mm). This is the part of the black body radiation spectrum immediately adjacent to the infrared region. At 1 GHz though, the radiation intensity is reduced by a factor of approximately 10^8 from its maximum value at around

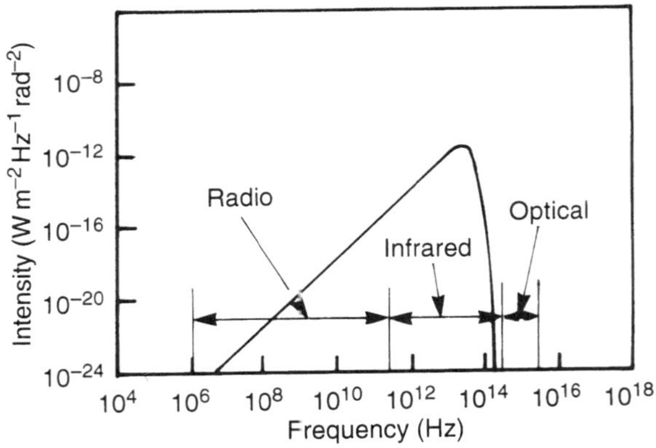

Figure 1.15 Black-body radiation intensity plotted against radiation frequency showing the optical, infrared and radiofrequency regions. The microwave region is that part of the radiofrequency spectrum immediately adjacent to the infrared.

10 μm, and this would appear to preclude the possibility of using such radiation as a means of sensing temperature. However, microwave radiometers, developed primarily for radioastronomy, can easily detect radiation at this intensity level.

There is a further advantage in moving to the microwave region of the black body spectrum and that is the significant penetration depth (transparency) which human tissue possesses at these and lower frequencies. Figure 1.16 shows how the penetration depth varies with frequency and is plotted as the depth at which the intensity has fallen to 1/e (0.37) of its original value, against frequency in MHz. We shall see that this 'transparency' is doubly important in that it not only permits the measurement of thermal radiation emitted by deep lying organs, but it also provides another 'window' by means of which we can 'look' into the body. Magnetic resonance imaging (MRI) techniques which utilize this window and embrace the frequency range 1–100 MHz form the subject of Chapter 8.

We can measure the radiative output from the human body by comparing its output with that of a thermal radiation source running at

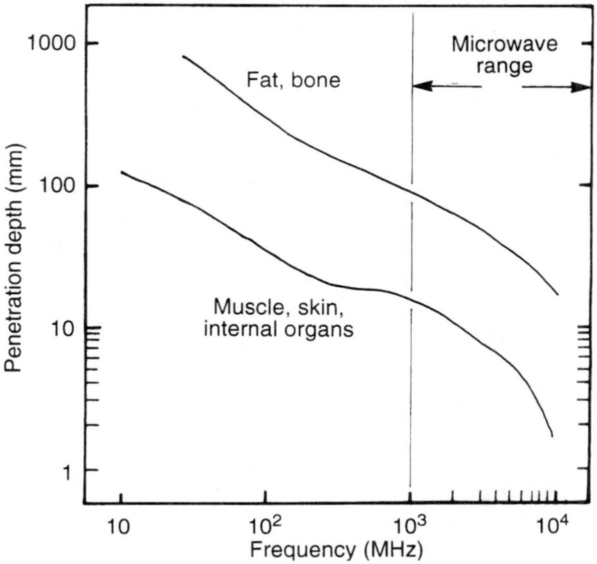

Figure 1.16 Penetration depth, that is depth at which radiation intensity has dropped by 1/e, plotted against frequency, for the radiofrequency range of the electromagnetic spectrum. The microwave region is identified.

a carefully maintained, constant temperature near room temperature. The radiation from these two sources is measured alternately, perhaps at a switching frequency of around 10 Hz. The resulting square-wave signal is synchronously detected and fed to a LED display, giving a digital readout of temperature. The same basic principle is employed in the Barnes thermograph, where the switching between a reference temperature and the input temperature is accomplished using a sectored disc running at a known temperature.

Although the Barnes thermograph gives a temperature map directly, microwave thermography only generates a figure for the body organ temperature immediately beneath the receiving antenna. This is because the microwave radiation from the body must be collected by a rectangular waveguide, with dimensions of the order of 10 mm × 20 mm, which is placed flush against the skin in order to minimize reflective loss at the skin–air interface. Mapping therefore has to be carried out point by point. Against this disadvantage, compared with conventional infrared thermography, microwave thermography can pick up temperature disturbances at much greater tissue depths than is possible with the former technique. The temperature sensitivity nevertheless is approximately 0.1 K for a radiation collection time of approximately 5 s; this is strictly comparable to the other thermographic systems.

1.14 THERMOGRAPHY IN PRACTICE

We have now outlined the basis of thermography and the means by which thermograms are obtained. A wide range of physical phenomena have been utilized in obtaining thermographic images, ranging from differential evaporation through thermochromic liquid crystals arrays to pyroelectric crystals detectors and microwave antennas. The displays from most of these systems are in the form of a false colour image, either because they are derived by interference methods or because they are displayed electronically on a colour VDU. Although some of the techniques described were first used in the 1950s or 1960s, their modern equivalents are capable of very impressive performance. It is possible to image with a spatial resolution such that tiny droplets of perspiration emerging from sweat pores can be seen, while a temperature resolution of approximately 0.08 K is obtainable. Figure 1.17 shows a high resolution thermogram of fingers.

Thermographic investigations are usually made in one of two ways. The patient is viewed against a constant background temperature, generally in the range 19–26 °C, depending upon the investigation. The

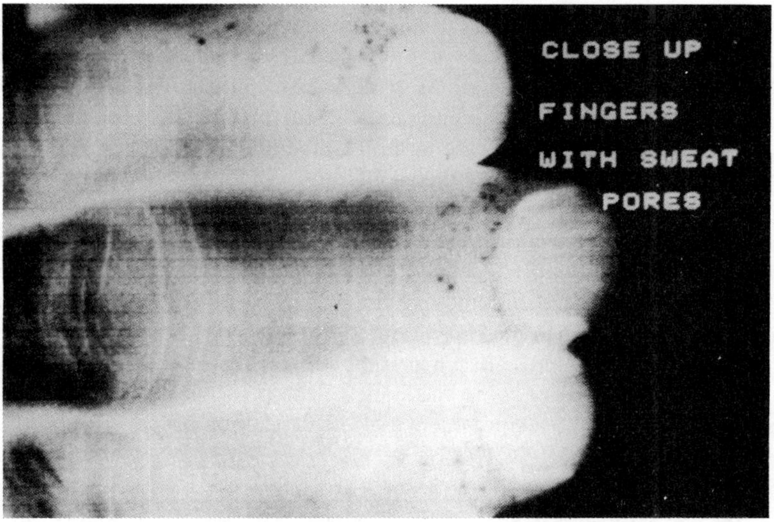

Figure 1.17 High resolution thermogram of fingers showing individual sweat pores and cooling by evaporation around the nailbed. (Courtesy of Professor E.F.J. Ring.)

patient must, of course, reach equilibrium with the surroundings before viewing begins. Alternatively, viewing of a patient is conducted after imposition of a thermal stress. The thermal stress may be applied to a localized area, such as a limb, by getting the patient to place a hand or a foot in a bath of hot (or cold) water for a short time, followed by viewing of the thermal image. Clinical information is derived from the way in which the limb heats up after cooling or cools down after heating. For more general application of a thermal stress it may be sufficient to expose the body to an air stream from a fan. Generally speaking, it is vascular disturbances, or suspected vascular disturbances, which are imaged by these means.

We can summarize the range of clinical situations which may be investigated by thermography as:

1. vascular disorder
2. inflammatory conditions such as rheumatoid arthritis and stress fractures
3. oncological studies
4. assessment of pain and trauma
5. dermatological conditions.

Vascular disorder may take many forms and reveal itself as hypothermal patterns (temperature below normal) or hyperthermal patterns (temperature above normal). A hyperthermic effect which is readily revealed by thermographic imaging is **deep vein thrombosis**. This condition leads to an increase in limb temperature so that, for example, a leg suspected of having this condition will image hotter than the remaining leg, which provides a contrast and calibration (Figure 1.18). The higher than normal temperature exhibited by the leg on the right of the image is thought to result from the local release of vasoactive chemicals associated with the formation of the **venous thrombosis**, this causes an increase in the resting blood flow. Peripheral arterial disease is also common and can cause **ischaemia**. Skin-temperature studies show hypothermal patterns and indicate the presence of arterial **stenosis**, which may be a consequence of the deposition of lipid (fat) in the

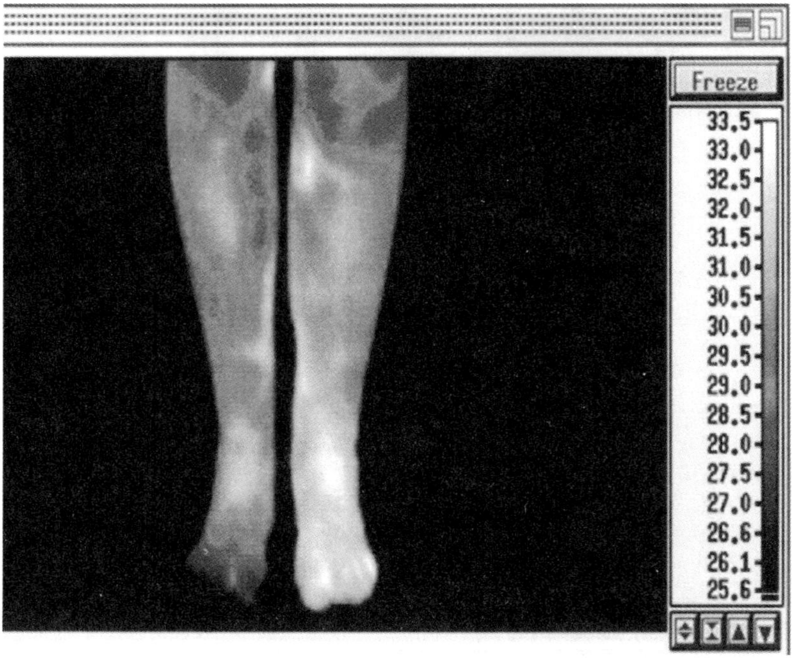

Figure 1.18 Thermogram of both legs of a subject with suspected deep vein thrombosis. Comparison with Figure 1.4 shows that the left leg in the image has a normal temperature difference whilst the right leg is hotter than normal. (Courtesy of AGEMA Ltd.)

arteries brought on, for example, by a disease such as **diabetes mellitus** which causes **hyperlipemia**.

Inflammatory conditions such as arthritis are accompanied by **over-perfusion** of the tissue (increased blood supply) and a consequent increase in skin temperature. Thermography is able to distinguish between deep-seated inflammation and a more superficial involvement. It can also be used to evaluate the effects of drug therapy. Figure 1.19 shows an infrared thermogram of both knees of a subject. The one on the right of the image is inflamed by arthritis, the other is normal.

One of the earliest demonstrations of the clinical use of thermographic imaging was the imaging of tumours both benign and malignant. Malignant tumours tend to be warmer than benign tumours owing to a higher metabolic rate and vascular changes around the tumour. In general, the larger the temperature differential between the tumour and its surroundings, the poorer the prognosis. The treatment of malignant tumours by chemotherapy or radiotherapy can be monitored

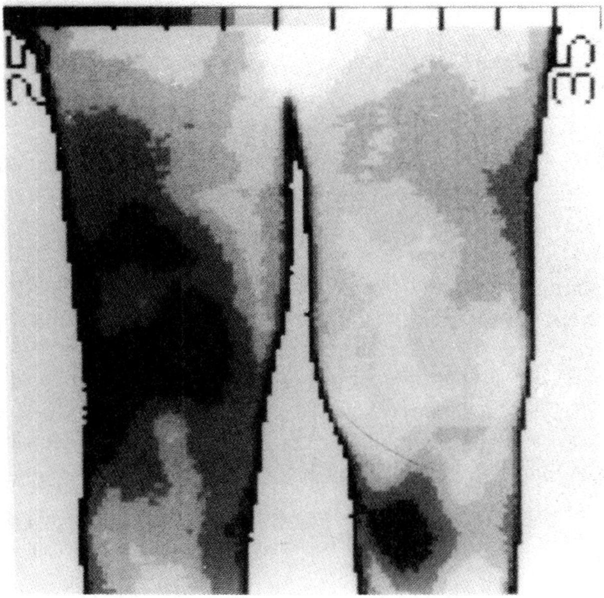

Figure 1.19 Thermogram of both knees of a subject. The one on the right of the image is inflamed with arthritis, the other is normal. (Courtesy of Professor E.F.J. Ring.)

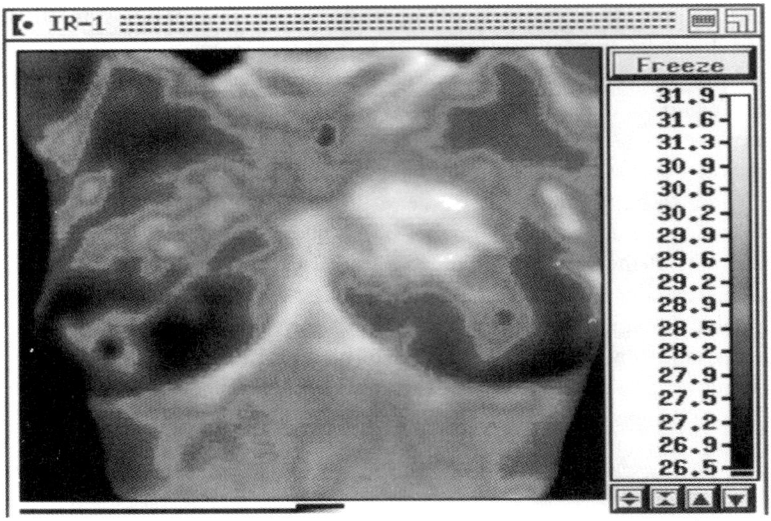

Figure 1.20 Thermogram of a breast tumour (right-hand side of image). (Courtesy of AGEMA Ltd.)

thermographically, so that temperature drops of 1 °C or more are consistent with tumour regression. The image from a malignant breast tumour is shown in Figure 1.20.

Thermography is a sensitive means of diagnosing stress fractures of long bones and may detect these before the fracture becomes apparent in radiographs. Figure 1.21 shows a thermographic image of a stress fracture.

In summary, thermal measurements on the human body, either of core temperature or surface temperature distribution, represent entirely risk-free diagnostic techniques, yet at the same time embrace both one of the earliest applications of physics in medicine and one of the oldest ideas in medical diagnosis. In the next chapter we shall examine the diagnostic information which arises from the intrinsic electrical activity of the human body.

PROBLEMS

1. The peak radiation intensity from the human body is found to occur at 9.5 μm. Assuming that the human body radiates like a perfect black body, calculate the average skin temperature to which this radiative output corresponds.

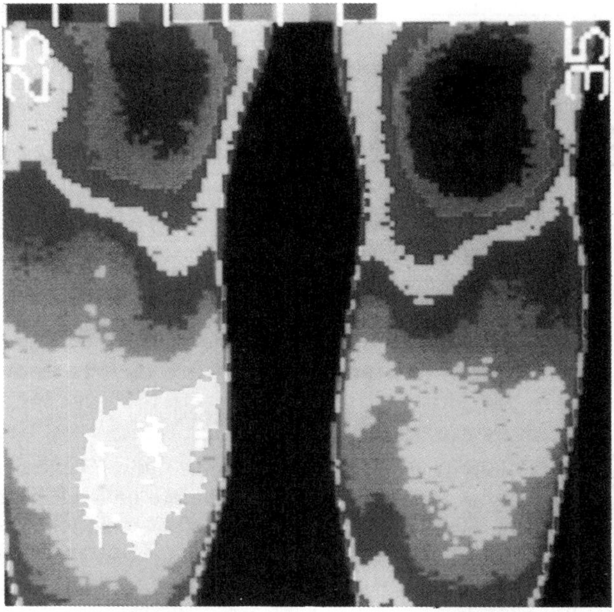

Figure 1.21 Themogram of a stress fracture of the upper right **tibia** (left-hand side of image) showing the elevated tissue temperature local to the fracture. (Courtesy of Professor E.F.J. Ring.)

2. A section of the body, imaged thermographically, has a recorded temperature of 33.9 °C, but it is known to be contaminated by talc which reduced the emissivity. Independent measurements reveal that the actual skin temperature is 30 °C. What is the skin emissivity?
3. Hexadecane oil may be used as an imaging liquid in evapographic imaging systems. Given that this oil has a refractive index of 1.433 and a figure of merit of 193×10^6 J m^{-3}, calculate the minimum detectable energy input, in J m^{-2}, if the minimum detectable change in oil thickness 10 nm.
4. Calculate the heat loss power density due to radiation from the human skin when the skin temperature is 32 °C and the ambient is 20 °C. Assume that the emissivity of both the skin and its surroundings is unity.

2

Electrical signals from the human body

2.1 INTRODUCTION

Most electrical activity in the body takes place in the central nervous system, especially in the brain. Nowadays we are quite clear about the importance of the brain and, interestingly, so also was Hippocrates as far back as the fifth century BC. In the intervening years the significance of the brain has not always been appreciated, and there was even a time (the Middle Ages) when it was believed to be composed of mucus, which drained out through the nose when one had a cold!

More recently, with the advent of the electronic computer, it has been tempting to think of the brain as some sort of ultra-compact computer, albeit operating on an electrochemical/chemical basis rather than the strictly electronic basis of the man-made machines. However, although the brain may in one sense be considered to function in a digital rather than an analogue mode, the analogy cannot be extended any further since the brain is not a 'clocked' structure, nor does it have a stored 'program' or a static memory location. Furthermore, it tolerates redundancy (cell death) without noticeable loss of operating efficiency. The electronic computer, on the other hand, generally has no capacity to restore function when it is lost.

The brain must be organized in a very different fashion from the modern computer, and that includes computers operating with so-called neural networks whose similarity to the human brain is limited to **synapse**-like interconnections. Unfortunately, the brain is the least well understood structure in the human body, despite its unique importance. The question of how the brain stores, retrieves and processes information remains essentially unanswered. In large measure this is a consequence of its extreme complexity, since there are estimated to be around 10^{11} nerve cells (**neurons**) in the brain, with 10^{10} in the **cerebral cortex** (the outer layer of the brain, or grey matter). The cerebral cortex is a sheet of grey tissue 2–4 mm thick comprising interwoven neural networks involving some 10^{15} interconnections or **synapses**. Although most regions of the cerebral cortex have been mapped functionally, the details of cortical organization and structure still elude us, and there are even suggestions that it may have a quantum basis. These difficulties should not surprise us since they are associated with what is believed to be the most complex structure in the whole universe.

In marked contrast to an electronic computer, the brain has a vital role in the synthesis of a range of specialist chemicals. Some of these are the **neurotransmitters**, which determine and sustain cerebral function: there are at least 50 different neurotransmitters known. Chemical synthesis is very energy dependent, with the consequence that cerebral energy consumption occurs at a rate which makes the brain, in metabolic terms, the most active organ in the body. The average rate of energy expenditure by the brain is some 20 W; this figure should be compared with the total body thermal dissipation of 80 W (Chapter 1).

It is the cells of the brain, acting as the nervous system, which control the functions of the body. Although it is now clear that the brain is a cellular structure, this was not always obvious since brain cells have a complex, asymmetric, three-dimensional structure which makes it hard to distinguish one cell from another. The cells of the brain may be divided into two basic groups: nerve cells or neurons, and a much more numerous group of cells which surround the neurons and fill the spaces between them, called **neuroglia** (nerve glue) or **glial cells**. Glial cells can themselves be further divided into several subclasses based largely on their appearance.

The nervous system is itself divided into central and peripheral components, where the **central nervous system** is that part located within the spinal column and skull. In the central nervous system the main types of glial cells are **astrocytes** and **oligodendrocytes**. Astrocytes, as the name implies, are star-like in appearance with numerous long arms radiating out from a central body. Oligodendrocytes also have a central body, but the radial arms tend to be shorter and more branched than those of astrocytes. It is now known that glial cells, in particular oligodendrocytes and **Schwann cells** (another class of glial cell), play an essential role in the electrical activity of neurons, but despite the fact that glial cells comprise 90% of the cells of the brain, their other functions are not yet well defined. It is probable that they have evolved multiple functional roles, but it is only their role in neuron signalling which is understood at all well.

Within the brain the nerve cells, or neurons, are the individual signalling, or information transfer, elements. The signalling may be intracellular, involving the movement of information from one part of the cell to another, or intercellular, where information transfer is from one cell to another. This ability to transfer information is essential so that the body can sense the details of its environment, transfer these details to the brain for processing and, finally, generate a response. The mode of operation of the neuron is now well established as electrical in nature and the signals transmitted in the nervous system are also known to be electrical; this electrical basis for neuronal activity was first identified around 1790.

In the eighteenth century the experimental work of first Galvani, and then Volta, showed that the interaction of body tissue with electricity brought about changes. Galvani's classic experiments with dissected frog muscles and static electricity machines demonstrated convulsive contractions of the frog's limbs. Galvani explained this as a discharge of 'nerveoelectrical fluid' previously accumulated in the frog muscle. Volta, on the other hand, supposed that the electricity came not from animal power, but from contact with the dissimilar metals of his voltaic pile. He showed that the frog muscle need not be included in the circuit: electricity excited the nerve, and the nerve the muscle. The conflict between the ideas of Galvani (1791) and Volta (1792) was not resolved until 1843, when the work of Emil du Bois-Reymond showed that the nervous impulse is accompanied by local differences in electrical potential and that these differences propagate in the form of a wave. This work was carried further by Hermann von Helmholtz, a friend and contemporary of du Bois-Reymond. Helmholtz's thesis of 1842 demonstrated the connection between nerve cell and nerve fibre. Although Helmholtz is chiefly remembered as a physicist, his first job was as an army surgeon and his early research work was on the nervous system; in 1850 he measured the speed of nerve conduction.

Work on the characterization of nerve signals within the body coincided with the early stages of development of the galvanometer as a sensitive measuring instrument; indeed, it required the galvanometer for its success. The work of du Bois-Reymond was in fact conducted with the astatic, moving magnet galvanometer (Figure 2.1(a)), while the string galvanometer (Figure 2.1(c)), introduced by Einthoven in 1903, played an important role in establishing the pattern of electrical signals associated with the nervous system, particularly those involved with the heart. Einthoven's work led to the introduction of the electrocardiograph, for which he was awarded the Nobel Prize (Physiology or

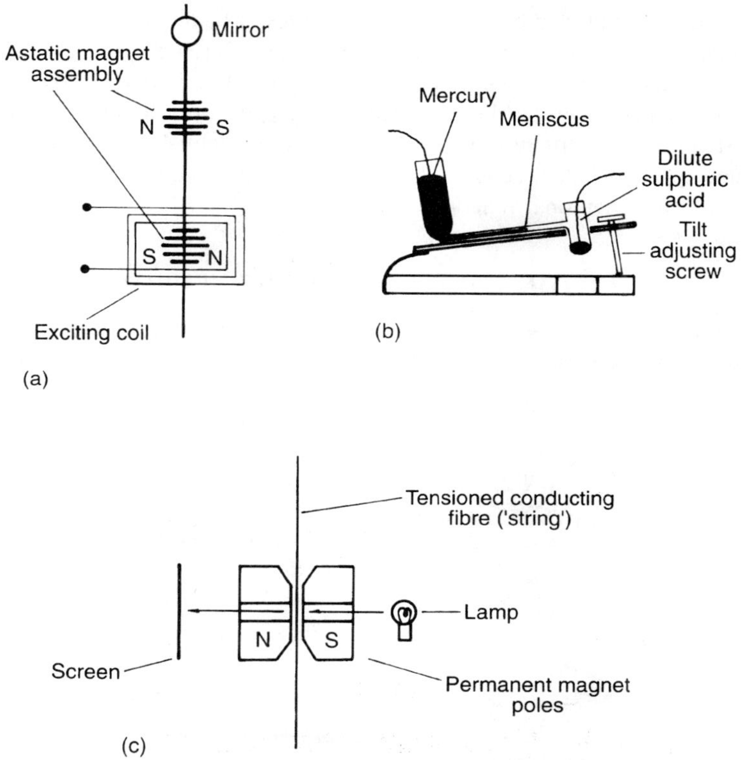

Figure 2.1 (a) Outline diagram of the astatic, moving magnet galvanometer of W. Thompson (Lord Kelvin), used to make the earliest measurements of the electrical behaviour of the nervous system. (b) The capillary electrometer of Lippmann, capable of sensing voltage changes of as little as 1 mV, by observation of the mercury meniscus in a tilted glass capillary. Its use was largely discontinued when the string galvanometer was introduced by Einthoven. (c) Outline of the basic structure of a string galvanometer. The vibration or movement of the conducting fibre between the magnet poles is illuminated by the lamp, through holes in the magnet poles, and recorded on the screen.

Medicine) in 1924. An alternative to the galvanometer, the capillary electrometer (Figure 2.1(b)), was also in use until around 1907. It was very sensitive despite being rather crude. Work on the actual mechanism by which nerve signals are generated required the use of modern electronic techniques. In the hands of Hodgkin and Huxley these techniques led to the award of a Nobel Prize (Physiology or Medicine) in 1963 for the elucidation of the electrical basis of nerve conduction.

Neurons and glial cells have some features in common with all other cells in the human body. However, there are differences which lead to the particular and special properties of neurons. It is important to begin our consideration of the electrical activity of the human body by examining the structure of the neuron before establishing how particular types of glial cells (Schwann cells and oligodendrocytes) are involved in neuronal function.

2.2 THE STRUCTURE OF THE NEURON

Figure 2.2(a) shows a section through a typical neuron revealing its overall asymmetric shape and characteristic cell structures or **processes**.

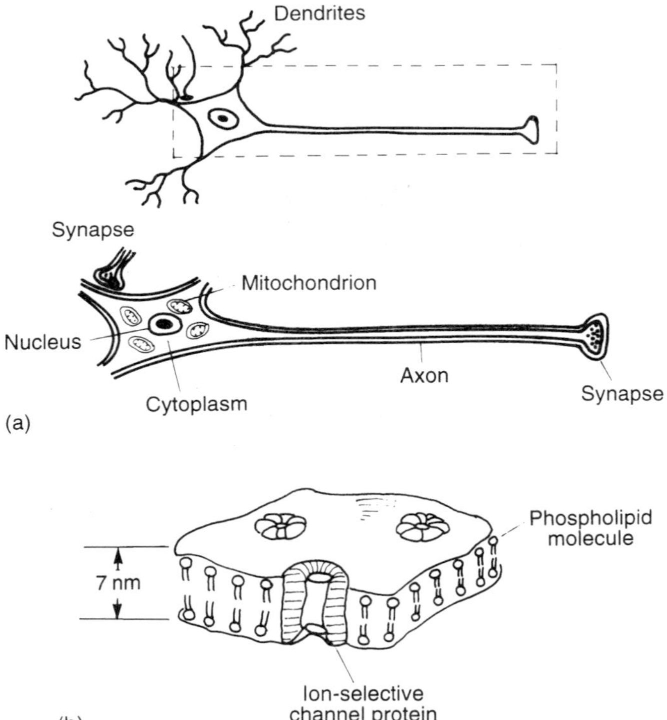

Figure 2.2 (a) Section through a typical neuron, the part outlined in the dashed box is shown in more detail underneath. (b) Cut-away view of a segment of neuron membrane showing the arrangement of phospholipid molecules and ion-selective channel proteins.

The neuron comprises a cell body, or **soma,** with various, different, but connected, entities. We shall contrast the structure of neurons with that of ordinary cells in order to emphasize the special features of the former. If we begin with the cell body we find that it contains, in common with all cells, a nucleus storing the genetic material DNA. The neuron, again in common with all cells, is enclosed by a membrane consisting of a double layer of phospholipid molecules. The membrane acts as a barrier, preventing the contents of the cell, known as the cytoplasm, from mixing freely with the extracellular space or plasma. We shall see that the membrane is actually an active structure and not just a retaining sac.

Figure 2.2(b) shows a schematic diagram of a segment of membrane in which the layers of phospholipid molecules are shown as tadpole-like structures. Charged, hydrophilic headgroups (circles) are in contact with the aqueous solution inside and outside the cell, and form the membrane surfaces. Hydrophobic tails on the phospholipid molecules form the middle of the membrane and confer high electrical resistance. Spanning the membrane at occasional intervals are ion-selective channel proteins which have time and voltage dependent conductances. The ion selectivity is conferred by a voltage (field) sensitive hydrophilic pore or channel which extends over the length of the protein, and spans the membrane.

A membrane is typically about 7 nm (70 Å) thick and, under usual operating conditions, supports an intense electric field of the order of $10^7\ \mathrm{V\,m^{-1}}$. This magnitude of electric field is close to the value which results in dielectric breakdown. The membrane channel proteins have to operate in the presence of this high electric field; indeed, it is their sensitivity to field changes which lies at the heart of their conduction and 'switching' capabilities. We may think of these ion channels as analogous to the transistors of a modern electronic system.

In common with all other cells, the neuron contains **mitochondria** to supply the cell's energy requirements. The energy requirements of a neuron are, however, particularly large, in order to maintain the trans-membrane ionic gradients which are essential for the neuron signalling function. In consequence, the neuron is generously endowed with mitochondria. There are other components enclosed by the membrane which are also found to some extent in all cells, but we will not, for the moment, concern ourselves with these. All of these components are described by the generic term **organelle.** The features described above are those which the neuron possesses in common with all other cells; we now consider those features which make the neuron unique, that is, those features which permit signalling, both intracellular and intercellular.

The specialized features of the neuron which permit information transfer, either out of or into the neuron, are the **axon**, the **dendrite** and the **synapse**. The synapse is the point of information transfer between neurons. The axon (Figure 2.2(a)) is a thin, constant-diameter, tube-like structure which emerges from a conical shaped thickening of the body of the neuron and may extend for distances ranging from micrometres to metres: it always terminates in a synapse, or synapses in the case of branched axons. Specialized proteins in the axonal membrane allow it to transmit electrical signals rapidly along its length to the synapse; the role of the axon must be considered that of signal transmission.

Apart from the axon the other significant structures attached to the neuron are the dendrites. Dendrites are neuronal structures which tend to be thicker and much shorter than axons. They are usually tapered and highly branched, giving rise to a 'tree-like' structure (indeed the word is derived from the Greek *dendron*, meaning tree). Like the axonal membrane, the membrane of the dendrite contains a set of special proteins which enable the dendrite to carry out a particular function. In this case, however, the proteins are configured for the reception and integration of information, so that the role of the dendrite is that of signal reception, although this is not always exclusively so; we may think of the dendrite as the 'aerial' or 'receiver' for the neuron. There may well be several 'dendritic trees' attached to a neuronal body; on the most complex and beautiful type of neuron, the **Purkinje cell**, which is characteristic of the **cerebral cortex**, the number of 'twigs' or spines is estimated to be about 120 000.

The spines or 'twigs' of the dendritic tree are the synaptic input sites at which the neuron receives information from another cell or cells. The neuron has evolved an unique and highly specialized structure, the synapse, to carry out this task. The name is derived from the Greek word *synapsis* meaning connect. The synapse, at the point of contact with other cell structures, allows unidirectional information transfer, although it does not actually touch the dendrite, but is separated by a gap of 20–30 nm (200–300 Å) . The basis for our ideas on the synapse was laid at the beginning of the twentieth century by the English physiologist Charles Sherrington, who coined the name.

2.3 NEURONAL PROPERTIES

We have seen that the structure which permits neurons to transmit electrical signals to other neurons is the axon. For much of our knowledge of the properties of axons we are indebted to the squid

Loligo, whose giant axon (which may be up to 1 mm in diameter) can be removed from the animal; the **axoplasm** may be extruded and replaced by a suitable saline solution whilst still preserving the electrical properties of the axon membrane. The majority of axons are surrounded by a so-called **myelin sheath** which is formed by the tight wrapping of glial cells around the axon during its development. Oligodendrocytes form the wrapping for axons in the central nervous system and Schwann cells form the wrapping for axons of the peripheral nervous system. Each Schwann cell forms myelin around a single axon; oligodendrocytes can myelinate as many as 50 axons. Neither the oligodendrocytes nor the Schwann cells form a continuous covering for the axon. Instead, at intervals there are gaps in the covering known as the **nodes of Ranvier** after the neurologist M.L. Ranvier (Figure 2.3(a)). We may think of the axonal structure as rather like a tube of semipermeable or porous material, around which, at regular intervals, are wound many turns of an insulating tape. The gaps between the tape-wound sections correspond to the nodes of Ranvier already mentioned, while the many turns of insulating tape correspond for example to individual Schwann cells (Figure 2.3(b)). Our analogy gives no feeling for the scale of these structures. In fact, the length of myelin sheath may be only about 1 mm, the width of a Schwann cell, and the gap several micrometres.

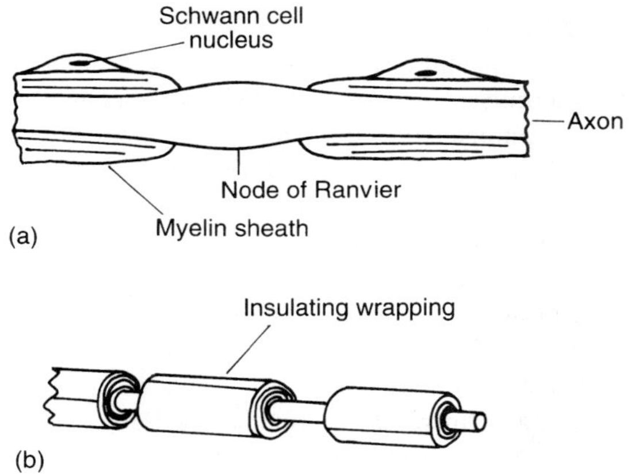

Figure 2.3 (a) Cross-section through a myelinated axon of the peripheral nervous system showing the wrappings of Schwann cells and the gaps between them which constitute the nodes of Ranvier. (b) Simulation of a myelinated axon from wrappings of an insulating tape spaced out along a porous tube or pipe.

The multiple membrane layer which forms the myelin sheath acts as an excellent insulator for the axon, so that an electric current can flow across the membrane of the axon only at the gaps in the myelin sheath, the nodes of Ranvier. This structure has profound implications for the speed of transmission of signals along the axon, as we shall see; these transmission properties will be considered further in our discussion of the action potential (section 2.5). Axons may be as much as a metre long in humans and even longer in larger animals so that the question arises: What is the mechanism of axonal information transfer which permits the transmission of information over long distances without degradation?

Neurons show a potential difference across the neuronal membrane. This membrane potential arises, in all cells, by virtue of the fact that ions are transported across the membrane via channel proteins and that the membrane has different permeability for different ions. The principal ions involved are sodium and potassium. In the steady state, that is when the axon is not electrically active, the value of the membrane potential lies in the range -70 to -90 mV, depending on species, the minus sign signifying that the cytoplasm, or cell content, is negative with respect to the plasma. This is the **resting potential**, V_r. If we take the value of the resting potential as a benchmark we can say that if the potential is more negative than the resting potential, then the cell is **hyperpolarized**; if it is less negative, then the cell is **depolarized**. Figure 2.4(a) shows these relations graphically.

The value of the potential difference across the membrane can be measured by inserting into the cell a test electrode connected to a high input resistance voltmeter (electrometer). The electrode may take the form of either a very fine, chlorided silver wire, or a fine-tipped glass pipette filled with conducting salt solution. While the measuring electrode is outside the cell membrane, that is in the plasma, the potential difference between it and the reference electrode is zero. However, when the measuring electrode penetrates the membrane, a negative voltage of between -70 and -90 mV is recorded, with respect to the reference electrode (Figure 2.4(b)).

By means of straightforward electronics based on operational amplifiers (section 2.10) we can cause current to be injected into, or extracted from, the cell using the electrode system described above. In the former case the cell is depolarized (the membrane potential becomes less negative than V_r); in the latter case the cell is hyperpolarized (the membrane potential becomes more negative than V_r) (Figure 2.4(c)). Current injection involves the transport of positive charge into the cell,

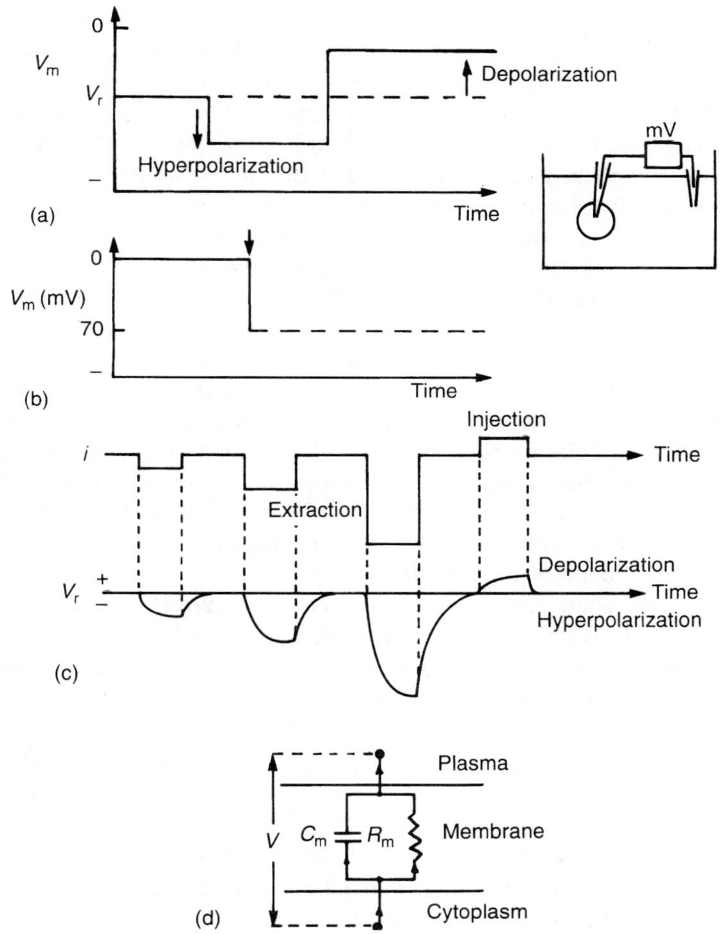

Figure 2.4 (a) Relative values of the membrane potential for the hyperpolarized and depolarized conditions as measured by a microelectrode inserted into an axon immersed in a bath of electrolyte, see inset. (b) Change in voltage accompanying the insertion of the microelectrode into an axon; the recorded change is the resting potential. (c) Changes in membrane potential V_m accompanying current injection/extraction relative to the resting potential V_r at $t = 0$. (d) Equivalent circuit for cell membrane.

whilst current extraction transports positive charge out of the cell. By convention the inward current is considered negative and the outward positive. In practice current injections or extractions occur naturally as a consequence of physiological events such as synaptic or sensory inputs. When current is extracted, leading to a hyperpolarizing change, the size of the voltage change depends only on the magnitude of the current, a passive response. Similarly, when current is injected leading to depolarization, provided that the magnitude of the current change is small, the voltage change will depend on the current, a passive response again; this behaviour is *not* found for large depolarizing currents.

It is apparent from Figure 2.4(c) that instantaneous changes in current flow to the neuron are not accompanied by instantaneous voltage changes. The reason for this lies in the structure of the membrane, which impedes the flow of ions between the intracellular and extracellular compartments; it constitutes an electrical resistance of magnitude R_m ($\Omega\,m^2$). In addition, the lipid bilayer of the membrane provides an extremely thin insulating layer between two conducting solutions, the plasma and cytoplasm. This permits the membrane to separate and store charge. The latter function means that it must be thought of as an electrical capacitor; we assign it a magnitude C_m ($F\,m^{-2}$). These two functions operate in parallel to give the passive equivalent circuit of Figure 2.4(d).

The membrane response to an instantaneous current change cannot be accompanied by an instantaneous voltage change because the membrane capacitance must charge or discharge with a characteristic time constant. Inspection of Figure 2.4(d) shows that we may write J_m, the current density, in terms of the circuit elements so that

$$J_m = C_m \frac{dV}{dt} + \frac{V}{R_m} \tag{2.1}$$

If we make an instantaneous or step function change in J_m from zero to some constant value, then the potential V across the membrane at any time t after the current change will be given by

$$V = J_m R_m \{1 - \exp(-t/\tau_m)\} \tag{2.2}$$

where $\tau_m = R_m C_m$ and $J_m R_m$ is the steady state voltage reached when t is large. Similarly the discharge characteristic will show an exponential form with the same time constant.

The values of τ_m found for biological membranes vary over a wide range. This is a consequence of variations in R_m, since C_m is found to be essentially constant for all membranes, at about $10^4\,\mu F\,m^{-2}$. The membrane time constant τ_m is a very important factor in determining the endogenous or innate activity of a neuron and how it responds to stimuli.

2.4 THE RESTING POTENTIAL

It is appropriate now to consider the origin of the resting potential upon which all other electrical activity is superimposed. We consider a membrane separating two aqueous solutions containing a single, positively charged, ionic species which can permeate the membrane; the balancing negative ion cannot readily permeate the membrane. Let us also assume that there is a substantial concentration difference between the two sides of the membrane and that, initially, the potential difference V_m, across the membrane is zero (Figure 2.5). We shall also assume that there is no net charge within the aqueous solution on either side of the membrane because the positive and negative ions balance each other. Under these conditions there is a concentration gradient for the positive (K^+) ions, which have their maximum concentration inside the cell; this concentration gradient causes positive ions to diffuse out of the cell down the concentration gradient. Once the diffusion process begins, however, charge is redistributed across the membrane and the

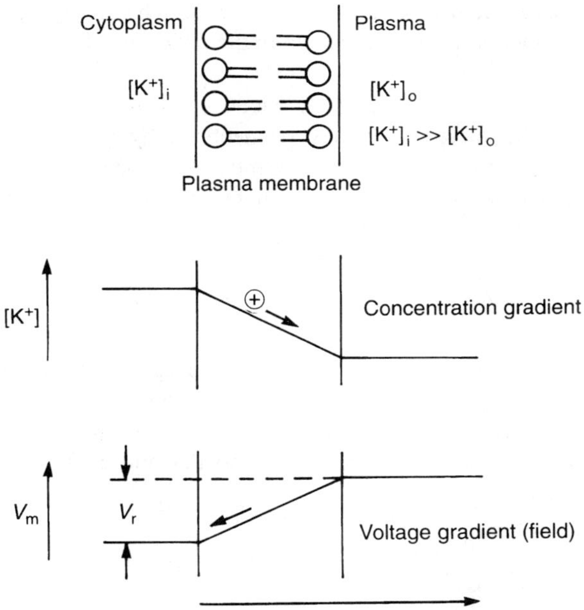

Figure 2.5 Origin of resting potential across cell membrane, a balance between the diffusion driven transport of ions down the concentration gradient and the opposing drift of ions in the voltage gradient resulting from ion transport.

inside of the cell becomes negative (loses positive potassium ions) while the outside becomes positive (gains positive ions). A potential difference of about 70 mV is then established across the membrane with an accompanying field of about 10^7 V m^{-1}.

At time zero for this process the membrane potential V_m is zero, but as soon as charge transfer commences, a potential gradient (field) builds up to oppose charge transfer, and finally the potential gradient balances the concentration gradient. At this time there is no net flow of potassium ions and the potential difference becomes constant. The potential difference is established by the transfer of a number of positive ions which is actually very small compared with the total number of ions in the intracellular and extracellular compartments of the cell, so that the potential difference results without any significant change in the ion concentration gradient. The physical situation is similar in some respects to that which exists in a semiconductor *p–n* junction in the absence of an applied potential, where a potential builds up across the junction to oppose the flow of electrons and holes down their respective concentration gradients, and a region essentially free of mobile charge, the depletion region, develops. The rectifying character of the *p–n* junction is also replicated in the *I/V* curves for K^+ transport, as one would expect.

The equilibrium potential difference E_{K^+} required to meet these conditions with electrolytes may be deduced by balancing the drift (field dependent) and diffusion (concentration dependent) components of the membrane current. Thus for potassium ions, the total current density J_{K^+} (A m^{-2}) is given by

$$J_{K^+} = -eD_{K^+}\frac{d}{dx}[K^+] + e[K^+]\mu_{K^+}\mathscr{E} \tag{2.3}$$

where e is the electron charge, D_{K^+} is the diffusion coefficient for potassium ions, $[K^+]$ is the potassium ion concentration, μ_{K^+} is the potassium ion mobility, and $\mathscr{E}$ is the electric field within the membrane. At equilibrium, the net current is zero so that the drift and diffusion components are equal and

$$D_{K^+}\frac{d}{dx}[K^+] = [K^+]\mu_{K^+}\mathscr{E} = -[K^+]\mu_{K^+}\frac{dV}{dx} \tag{2.4}$$

Substituting for D_{K^+} using the Einstein equation $D_{K^+} = (k_B T/e)\mu_{K^+}$, where k_B is the Boltzmann constant, we may integrate across the membrane between boundary conditions V_0 at $[K^+]_o$ and V_i at $[K^+]_i$. If we set $V_i - V_o = E_{K^+}$ (by convention the reference level is the external potential), then

$$E_{K^+} = \frac{k_B T}{e} \log_e \frac{[K^+]_o}{[K^+]_i} \tag{2.5}$$

This equation was originally derived by Nernst (from different considerations) in the form

$$E_{K^+} = \frac{RT}{zF} \log_e \frac{[K^+]_o}{[K^+]_i} \tag{2.6}$$

and is consequently known as the Nernst equation. Here E_{K^+} is the equilibrium potential for potassium ions, R is the gas constant per mole, z the ion valence, F the Faraday (number of coulombs per mole of ion), $[K^+]_o$ the potassium ion concentration outside the cell, and $[K^+]_i$ the potassium ion concentration inside the cell, both in mol l^{-1}. Since Boltzmann statistics apply at equilibrium, when there is no net current density, we can also obtain equation 2.5 directly, by rearranging the Boltzmann equation

$$[K^+]_o = [K^+]_i \exp\left(-\frac{e(V_o - V_i)}{k_B T}\right)$$

For any monovalent ion I^+, at room temperature, the Nernst equation reduces to

$$E_{I^+} = 58 \log_{10} \frac{[I^+]_o}{[I^+]_i} \tag{2.7}$$

(E_{I^+} is expressed in mV). Equations 2.5 and 2.6 are, of course, equivalent.

If we take the giant axon of the squid (*Loligo*), the concentration relationships can be summarized as follows follows:

1. $[K^+]_i = 20[K^+]_o$
2. $[Na^+]_o = 10[Na^+]_i$
3. $[Cl^-]_o = 14[Cl^-]_i$.

Furthermore we can give some indication of the permeability of the plasma membrane to ions when the axon is exhibiting the resting potential. These are a high permeability (p_K) to potassium, a low permeability (p_{Na}) to sodium, and a permeability (p_{Cl}) to the chlorine ion which is essentially zero. When only potassium ions are present, the value calculated for the resting potential, using the Nernst equation,

Table 2.1 *Neuron characteristics*

Cell type	*Ion*	*Equilibrium potential* (mV)	*Extracellular to intracellular ratio*	*Resting potential* (mV)
Human (skeletal muscle)	Na^+	+53	6 : 1	−80
	K^+	−95	1 : 38	
	Cl^-	−85	26 : 1	
Squid axon	Na^+	+55	10 : 1	−70
	K^+	−76	1 : 20	
	Cl^-	−104	14 : 1	

is −75 mV and is equal to the equilibrium potential E_{K^+}. The measured value is slightly less negative than this at −70 mV. Table 2.1 lists some characteristic potentials of neurons for humans, with those for *Loligo* as a comparison.

It is possible to describe the measured membrane potential more precisely than the Nernst equation allows. The Nernst equation describes the potential difference arising from the transport of a single ionic species. D.E. Goldman derived an equation to describe the membrane potential in terms of the concentration gradient and permeation properties of several different ions:

$$V_m = \frac{RT}{zF} \log_e \left\{ \frac{[K^+]_o + [p_{Na}/p_K][Na^+]_o + [p_{Cl}/p_K][Cl^-]_i}{[K^+]_i + [p_{Na}/p_K][Na^+]_i + [p_{Cl}/p_K][Cl^-]_o} \right\} \quad (2.8)$$

This equation was also derived independently by A. Hodgkin and B. Katz. When p_{Cl} and p_{Na} are zero, the membrane is, of course, only permeable to potassium ions and the Goldman–Hodgkin–Katz (GHK) equation reduces to the Nernst equation. The value of −70 mV measured for the resting potential reflects the fact that the resting permeability for sodium ions, although small, is not zero. This means that sodium ion transport also contributes to the membrane resting potential.

To summarize, V_m represents the membrane potential difference at any instant of time; V_r is the resting potential established in the steady state, while E_I^+ is the equilibrium potential for single monovalent ion flow. The resting potential exactly balances the ion flows across the membrane; at the resting potential the net current density is zero. However, in real cells the resting potential is not exactly equal to the equilibrium potential for any of the diffusing ions and no individual ion is at equilibrium. Each ion will flow down its appropriate concentration gradient and contribute a current; the membrane actually operates in the *steady state* rather than at equilibrium. Nevertheless the total ion current density is zero. This means that currents resulting from positive ion flow out of the neuron are balanced by currents from positive ion flow into the neuron.

Since only two ions are involved, the currents carried by these ions must be equal and opposite, but this seems surprising, or even unlikely, in view of the fact that the membrane permeability for potassium ions is much greater than that for sodium ions. However, the driving potential for potassium ion transport is small, just $V_m - E_{K^+}$, or about +5 mV, whilst for sodium ions $V_m - E_{Na^+}$ is quite large (approximately −125 mV). We know that $J_{K^+} = -J_{Na^+}$ so we can write, using Ohm's law,

$$(V_m - E_{K^+})/R_{K^+} = -(V_m - E_{Na^+})/R_{Na^+} \tag{2.9}$$

or

$$(V_m - E_{K^+})G_{K^+} = -(V_m - E_{Na^+})G_{Na^+} \tag{2.10}$$

where G_{K^+} and G_{Na^+} are the membrane conductances of potassium and sodium ions respectively, with units of siemens m^{-2} ($S\,m^{-2}$). The actual ion current for a given ion depends on the membrane conductance or resistance, as well as on the driving force or potential difference for the ion; this is a very important result.

If we now examine the ionic conductance we see that a sudden change in the conductance will alter the ion current and equally suddenly the steady state is destroyed. A new membrane potential is required to balance the altered ion flow and restore the steady state. It is convenient, although not precise, to think of membrane permeability and conductivity as equivalent (they are not identical, since conductance depends on ion concentration whereas permeabililty does not). We can then use the Goldman–Hodgkin–Katz equation to describe the membrane potential as the ion conductivities are changed. Figure 2.6 shows how V_m varies as the ratio G_{Na^+}/G_{K^+} is varied. The GHK equation gives us numbers for the limiting values of V_m which are actually just E_{K^+} and E_{Na^+}, the equilibrium values when either potassium or sodium ions alone are involved in transport. A selective change in the membrane conductance for a given ion will result in a change in V_m and this is, in fact, the means by which neurons produce meaningful electrical signals.

The assumptions used to derive the GHK equation, relating the resting potential to the ion permeabilities and concentrations, may also be used to derive an expression for the steady-state cationic current

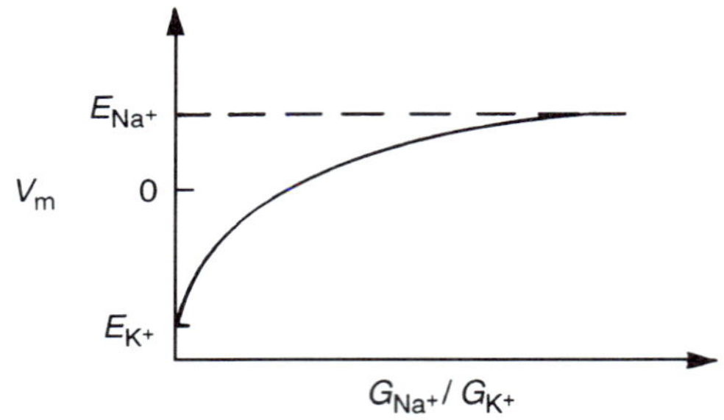

Figure 2.6 Variation in membrane potential as a function of the ratio of the conductances for sodium ions, G_{Na^+}, and potassium ions, G_{K^+}. When the sodium conductance is very low, the membrane potential approaches the equilibrium value for potassium ions. When the sodium ion conductance is much higher than the potassium conductance, the membrane potential approaches the sodium equilibrium potential, E_{Na^+}.

density, J. The GHK current density equation, expressed for sodium and potassium ions alone, is

$$J = \frac{\psi e}{[1 - \exp(-\psi)]} [p_K\{[K^+]_i - [K^+]_o \exp(-\psi)\} + p_{Na}\{[Na^+]_i - [Na^+]_o \exp(-\psi)\}] \tag{2.11}$$

Here we have permeabilities p_K and p_{Na} to K^+ and Na^+ ions respectively, and a potential difference expressed as $\psi = eV_m/k_B T$. The total current density J is, in this case, equal to $J(K) + J(Na)$. In the steady state the potential V_m becomes the resting potential V_r and corresponds to the potential required for $J = 0$ or $J(K) = -J(Na)$. This condition would eventually destroy the ionic concentration gradient if it were not for the intervention of a restoring mechanism, the sodium–potassium pump, which is the subject of section 2.14.

2.5 THE ACTION POTENTIAL

A resting neuron exhibits a steady potential difference across its plasma membrane unless current is injected into, or extracted from, the neuron. Current extraction results in hyperpolarization and the cell contents become even more negatively charged as a consequence. The situation for current injection is more complex since initially the neuron becomes depolarized and the cell contents become less negative with respect to the plasma. However, if the depolarizing current is made large enough, a critical threshold is reached at which the neuronal response is not passive. This threshold varies from neuron to neuron but is usually found in the range 10–20 mV depolarized (more positive than) V_r. The nature of the response is sketched out in Figure 2.7. It may be characterized as a large change in membrane potential, which lasts several milliseconds, superimposed on the passive response. The initial depolarization occurs rapidly; the return to a passive response, the resting potential, occurs slightly less rapidly. The active response of the membrane when the depolarization exceeds threshold is the *action potential* or nerve impulse. At its peak V_m becomes about 50 mV positive, that is the inside of the cell becomes positive relative to the

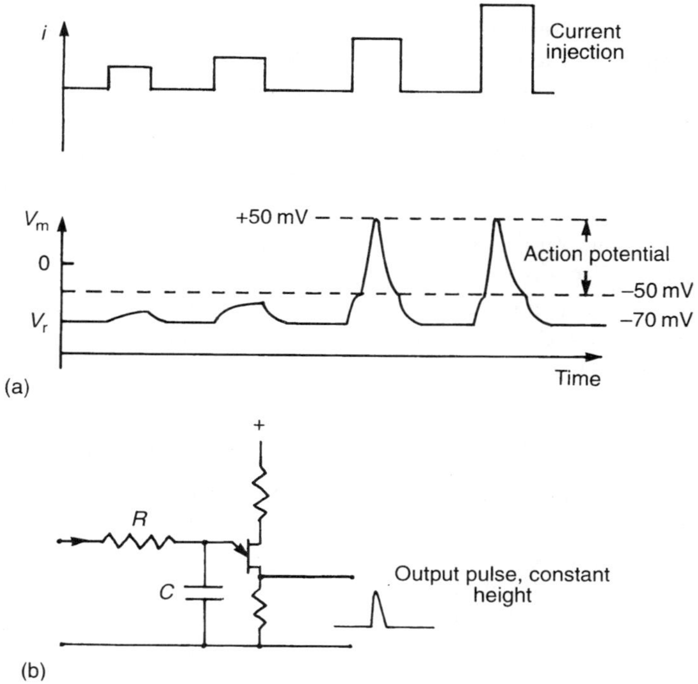

Figure 2.7 (a) Variation in membrane potential V_m with increasing current injection showing the form of the action potential. (b) Unijunction transistor circuit which exhibits behaviour analogous to that of the cell membrane during current injection, in the sense that if sufficient positive charge is delivered to the capacitor C to raise its potential to the peak point voltage of the device, then the unijunction 'fires' and short circuits capacitor C. An output pulse of constant height results, analagous to the action potential. A substantial current input will produce a stream of output pulses.

external plasma, and the membrane field changes polarity. The threshold ensures that small random polarizations do not generate action potentials.

It is important to note another property of action potentials, that is they are *all-or-none* events and the **all-or-none law** is an essential feature of signal transmission by axons. In fact depolarization produces an increase in sodium ion conductance which, in turn, produces further depolarization. The sodium ion conductance G_{Na} increases further and an intrinsically unstable, positive feedback loop is established which gives rise to the *all-or-none action potential*. Any stimulus which is large enough to produce an action potential produces the same size action potential, regardless of stimulus strength. Expressed another way, once the stimulus is above the threshold value, the amplitude of the response no longer reflects the amplitude of the stimulus. We can find an electronic analogue for this behaviour in the unijunction transistor (Figure 2.7(b)), which exhibits similar properties, although to a much shorter time scale. Indeed, the electronic analogy extends further to digital computers and systems where information is transmitted as a bit (binary digit) stream; the action potential is analogous to the binary digit (logical high). For axons, information about the strength of the stimulus clearly has to be encoded in some way, since the magnitude of the action potential is constant. How this is achieved can be understood by examining the time dependence of the action potential with respect to the original stimulus.

2.6 CURRENT TO FREQUENCY CONVERSION

The curves of Figure 2.7 show that once the depolarization exceeds a certain threshold, an action potential results. This threshold ensures that only important stimuli produce an action potential. For several milliseconds after the firing of an action potential it is impossible to evoke another action potential, no matter how large the depolarizing stimulus. The axon is said to be *refractory* to stimuli during this time, the *absolute refractory* period. Following the absolute refractory period is a *relative refractory period* during which the stimulus must be larger than normal to evoke an action potential (Figure 2.8).

We can interpret this behaviour in terms of the threshold, so that during the absolute refractory period the threshold can be considered essentially infinite, while during the relative refractory period the threshold is larger than normal. After sufficient time the threshold returns to normal. The curve of Figure 2.8 may be presented in another

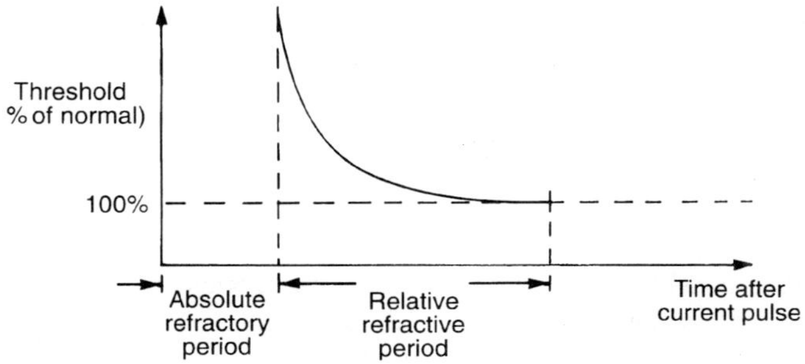

Figure 2.8 Variation of action potential threshold with time after initiation by a current pulse. For a short period of time thereafter (several ms), no action potential can be excited; the threshold is greater than normal.

way, that is, it describes the stimulus strength required to generate a second action potential, as a function of time after the first action potential. Reference to Figure 2.7(a) shows that the stronger the stimulus the shorter the time which elapses between the stimulus and the action potential; this is the *strength–latency* relationship which, combined with the refractory period, allows encoding of stimulus strength in terms of the frequency of action potentials in the axon.

Figure 2.9 shows two different depolarizing currents, that of Figure 2.9(a) being less than that of Figure 2.9(b); superimposed are the resulting action potentials. The point to note is that the frequency with

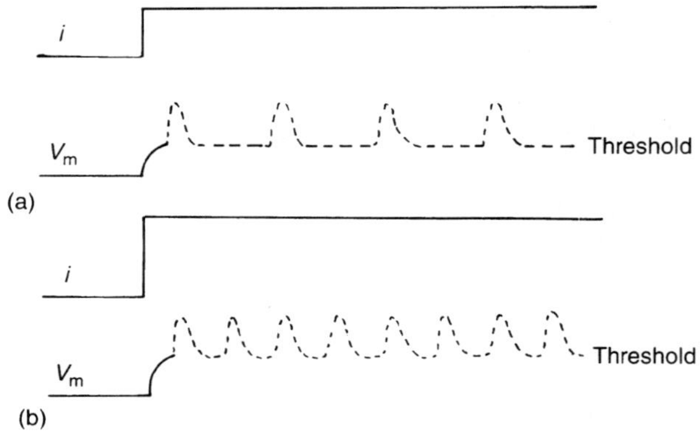

Figure 2.9 The effect of different levels of current 'injection' on the firing frequency of action potential spikes. The firing frequency depends on the extent of depolarization; thus depolarization in (b) exceeds that in (a) so that the firing frequency is higher.

which the action potential spikes appear depends on the extent of depolarization, the frequency increasing with increased depolarization. The unijunction transistor of Figure 2.7(b) has the same ability to convert current to pulse frequency.

2.7 NEURONS IN ACTION

We have considered a local change in the membrane potential at a single point on the axon. However, the particular role of an axon is the transmission of information from one neuron to another; we must now consider this in more detail. Suppose a hyperpolarizing current flows out of an unmyelinated axon at one end. The membrane potential can, in principle, be measured at intervals along the length of the axon by means of microelectrodes inserted through the membrane. Figure 2.10(a) shows the result of this sort of measurement. The membrane potential is seen to fall off with distance from the point of excitation. In fact, the potential is found to fall off exponentially with distance according to the relationship.

$$V_x = V_0 \exp(-x/\lambda) \tag{2.12}$$

where V_x is the potential a distance x from the point of excitation, V_0 is the potential at the point of excitation and λ is a membrane constant known as the space constant. It specifies the distance at which the value of V_0 has fallen to 1/e of its initial value. (For a 10 μm radius axon λ will be about 2 mm.) This phenomenon is known as passive spread and is exhibited not only by live axons, but also by dead ones; it is also seen in the theory of linear, one-dimensional, cables.

If, on the other hand, we choose to cause depolarization, then for currents which take the axon past the critical threshold, a large action potential is generated. Now equation 2.12 shows us that, in the unmyelinated axon, passive spread associated with this action potential will ensure that neighbouring regions of the axon will be raised above threshold and the electrochemical activity accompanying depolarization will spread along the axonal membrane with the change in potential; the action potential propagates without attenuation. It is able to do this because passive spread is sufficient to depolarize adjacent elements of axonal membrane above threshold. Figure 2.10(b) illustrates this effect. A beautiful and much used analogy is that of the firecracker fuse, where ignition of one point on the fuse brings the neighbouring segment above its ignition temperature, and this process continues until the whole fuse is consumed. Although passive spread of the potential is

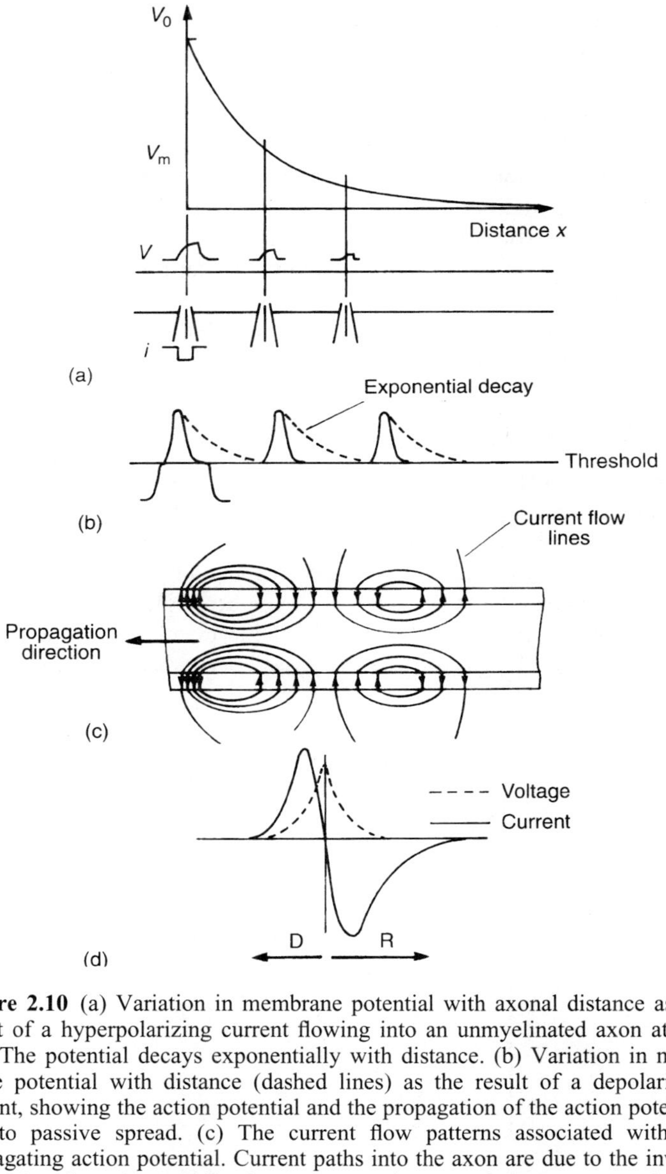

Figure 2.10 (a) Variation in membrane potential with axonal distance as the result of a hyperpolarizing current flowing into an unmyelinated axon at one end. The potential decays exponentially with distance. (b) Variation in membrane potential with distance (dashed lines) as the result of a depolarizing current, showing the action potential and the propagation of the action potential due to passive spread. (c) The current flow patterns associated with the propagating action potential. Current paths into the axon are due to the inward flow of sodium ions while the current paths out of the axon are due to potassium ions. (d) Current and voltage distribution across the propagating action potential, showing the resulting current quadrupole, broken down into its two component dipoles, D due to depolarization and R due to repolarization.

rapid in an unmyelinated axon, actual movement of the action potential along the axon is relatively slow, since it involves the transport of ions across the membrane, and a progressive change of these transport properties along the membrane.

The actual current flow patterns accompanying the propagating action potential, together with the transmembrane potential and axial, intracellular action current, are shown in Figure 2.10(c), (d). Current paths into the axon are due to the inward flow of sodium ions while current paths out of the axon are due to potassium ions. The resulting current distribution constitutes a current quadrupole made up from two current dipoles, that due to depolarization labelled D and that due to repolarization labelled R in Figure 2.10(d).

2.8* THE LINEAR CABLE MODEL OF AN AXON

In the linear cable model of an axon we visualize an axon as a cylindrical membrane surrounding an intracellular fluid phase of constant cross-sectional area (Figure 2.11(a)). We assume that the intracellular fluid impedance may be represented by an ohmic resistance and that all non-axial voltage gradients may be neglected. The voltage changes are assumed small enough that the membrane properties are linear, while the length of the cable/axon is considered infinite.

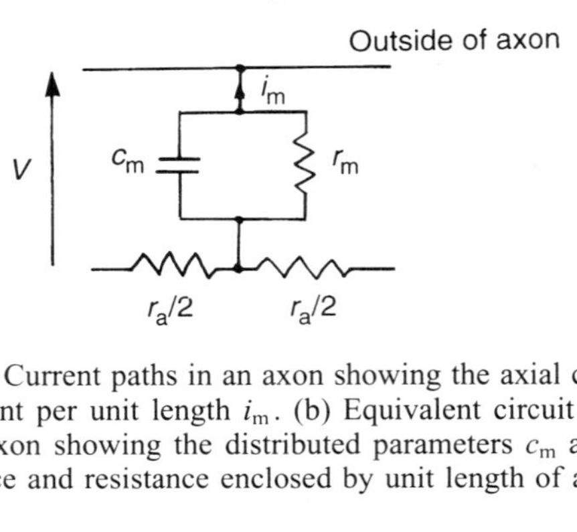

Figure 2.11 (a) Current paths in an axon showing the axial current i_a, and the membrane current per unit length i_m. (b) Equivalent circuit representation of unit length of axon showing the distributed parameters c_m and r_m, the membrane capacitance and resistance enclosed by unit length of axon.

If, for the moment, we ignore leakage current through the cell membrane, the relationship between intracellular axial current i_a and the intracellular voltage V is given by Ohm's Law,

$$\Delta V/\Delta x = -r_a i_a \tag{2.13}$$

where x is distance along the cable and r_a is the intracellular resistance per unit length of cable ($\Omega\ m^{-1}$). In fact, the membrane leakage current is not negligible, so that, even in the steady state, current leaks out across the membrane and i_a will not be constant over any significant length. Equation 2.13 must therefore be modified to the differential form describing the voltage drop ∂V across an infinitesimal distance ∂x at point x, yielding

$$\partial V/\partial x = -r_a i_a \tag{2.14}$$

We now assume that our model axon is immersed in a large volume of extracelluar fluid so that the extracellular resistance may be ignored, and V is just the deviation from the resting potential V_r. In addition, the rate of change in i_a with distance along the model axon must be equal and opposite to the membrane current per unit length of cable/axon i_m, thus

$$\partial i_a/\partial x = -i_m \tag{2.15}$$

Differentiating equation 2.14 yields

$$\partial^2 V/\partial x^2 = -r_a\ \partial i_a/\partial x \tag{2.16}$$

Substitution of equation 2.15 in equation 2.16 yields

$$\partial^2 V/\partial x^2 = r_a i_m \tag{2.17}$$

which gives a relationship for the membrane current. From equation 2.1 we can write that

$$i_m = c_m\ \partial V/\partial t + V/r_m \tag{2.18}$$

where c_m and r_m are now distributed parameters, the membrane capacitance and resistance respectively, enclosed by unit length of axon, that is $c_m = 2\pi a C_m$ and $r_m = R_m/2\pi a$, where a is the axon internal radius (Figure 2.11(b)).

Equation 2.18 can be rewritten, using equation 2.17, as

$$\frac{1}{r_a}\frac{\partial^2 V}{\partial x^2} = c_m \frac{\partial V}{\partial t} + \frac{V}{r_m} \tag{2.19}$$

(We can think of equation 2.19 as the 'cable equation'; it is in fact the equation derived by Lord Kelvin to describe the characteristics of the first Atlantic cable.)

Rearranging equation 2.19 we obtain

$$\lambda^2 \frac{\partial^2 V}{\partial x^2} = \tau_m \frac{\partial V}{\partial t} + V \tag{2.20}$$

where $\lambda = (r_m/r_a)^{1/2}$ and $\tau_m = r_m c_m$; these are just the space constant and time constant already introduced in our discussion of the axon. The units of r_m will be Ω m, and for r_a $\Omega\,m^{-1}$, thus λ has the dimensions of distance as required. It can also be seen to correspond to the distance at which r_a and r_m become equal, so that for a length x of axon, the axial resistance will be $r_a x$ whilst the transverse membrane resistance will be r_m/x. When $r_a x = r_m/x$ we see that $x^2 = r_m/r_a = \lambda^2$. The space constant λ determines the tendency for current to spread along the cable

At this juncture we can consider typical values for the axonal properties. The plasma membrane may well be of the order of 5–10 nm (50–100 Å) thick; its resistance r_m to steady current flow is of the order of 1–20 kΩ and its specific resistivity of the order of 10^8 Ω m (note: the intracellular fluid has a resistivity of approximately 1 Ω m). The very high specific resistivity of the membrane is a consequence of the fact that ions are confined to cross it only at particular pores or channels and that there are only about 50 of these per μm^2 of membrane surface. For a myelinated axon these channels are concentrated at the gaps in the myelin sheath. The specific properties of these ion channels will be discussed in section 2.11. The resistance r_a may well be around 3000 M$\Omega\,m^{-1}$ so that $r_m < r_a$ and current tends to flow out of the axon through r_m.

The derivation of equation 2.19 has assumed that r_m, r_a and c_m are independent of x, V, i and t. Outside a small range of potentials though, r_m is actually strongly dependent on V and t, so equation 2.20 must then take the form

$$\frac{1}{r_a} \frac{\partial^2 V}{\partial x^2} = c_m \frac{\partial V}{\partial t} + i_m(V, t) \tag{2.21}$$

where i_m is the membrane ionic current per unit length of axon, that is $i_m = 2\pi a J_m$ and has units of A m^{-1}. Solutions for this equation may be obtained by application of Laplace transforms; we can redefine the independent variables in terms of the dimensionless quantities $X = x/\lambda$ and $T = t/\tau_m$ so that equation 2.20 becomes

$$\frac{\partial^2 V}{\partial X^2} - \frac{\partial V}{\partial T} - V = 0 \tag{2.22}$$

The steady-state solution may be obtained by putting $\partial V/\partial T = 0$, and then

$$V_x = V_0 \exp(-x/\lambda)$$

and V_x falls exponentially towards zero; this is equation 2.12 once more. A plot of $\log_e V_x$ against x yields a straight line of slope $1/\lambda$. By this means λ and hence r_m/r_a may be determined. Equation 2.19 may be developed further by rewriting it in the form

$$\frac{\partial^2 V}{\partial x^2} = r_a c_m \frac{\partial V}{\partial t} + r_a g_m V \tag{2.23}$$

where g_m ($S\ m^{-1}$) is the membrane conductance per unit length or $1/r_m$. Now since the action potential V actually propagates without change, the wave equation applies and the term $\partial^2 V/\partial x^2$ in equation 2.23 may be replaced by $(1/\theta^2)(\partial^2 V/\partial t^2)$; the partial time derivatives then become total, leading to

$$\frac{1}{\theta^2}\frac{d^2 V}{dt^2} = r_a c_m \frac{dV}{dt} + r_a g_m V \tag{2.24}$$

where θ is the propagation speed of the action potential along an axonal membrane. The value of θ may be determined experimentally, and this type of measurement forms part of electromyography (EMG) (section 2.18). It should be noted that θ will increase with increasing temperature and is proportional to $1/\tau_m$.

2.9 THE MYELINATED AXON

For those situations where depolarization exceeds the threshold value, an action potential spike of considerable magnitude is generated. As pointed out in section 2.5, the depolarization process 'ripples' along the axon, always in the direction of membrane where depolarization has yet to exceed the threshold value, that is towards the synapse. For the myelinated axon this process is not possible owing to the presence of the insulating wrapping of oligodendrocytes or Schwann cells. However, passive spread still occurs and the potential decays exponentially as before. Its magnitude will still be large enough at the next node of Ranvier (gap in the myelin sheath) to excite a further action potential. This process repeats along the length of the axon with each node being depolarized beyond the threshold and firing an action potential (Figure 2.9(b)). In this fashion an action potential may again be propagated over the length of an axon without degradation, only now much more rapidly then before. In effect the action potential appears to jump from node to node; this is known as *saltatory conduction* (from the Latin *saltare* meaning to leap or dance). The overall result is that conduction occurs many time faster than in nonmyelinated axons of the same dimensions, since depolarization processes are confined to the nodes of Ranvier. For example, the speed of action potentials in unmyelinated axons can be less than $2\ \mathrm{m\ s^{-1}}$; for myelinated axons it can reach $100\ \mathrm{m\ s^{-1}}$. The myelinated axon actually forms a better analogy to the submarine cable than does the unmyelinated axon. The extent to which an oligodendrocyte or Schwann cell invests an axon varies with the diameter of the axon. In this way θ, which is proportional to $1/\tau_{\mathrm{m}}$, can be optimized. Thus, for a given axon diameter, the extent of myelination does not seem to be arbitrary, but is probably adjusted to give maximum transmission velocity, ensuring the most efficient use of the available volume. How this condition is achieved is not known.

We may now summarize the electrical properties of axons which permit reliable information transfer. There are four important points to make:

1. the threshold required for the generation of an action potential guarantees that small random variations in the membrane potential do not generate a signal which can propagate along the axon;
2. the all-or-none law ensures that all action potentials are the same size, again protecting the quality of transmitted information;
3. the refractory period and the strength–latency relationship convert direct current changes into a corresponding frequency code;
4. passive spread of an action potential ensures that the depolarizing feedback loop propagates along the axon without attenuation. Alternatively, for a myelinated axon, passive spread is sufficient to ensure that adjacent nodes of Ranvier are brought above threshold and fire an action potential, which now propagates from node to node by saltatory conduction, with higher speed and metabolic economy than occurs in a unmyelinated axon.

Myelination of an axon leads to a significant increase in the speed of nerve pulse conduction so that, for a given overall axon diameter, signals will be transmitted about 10 times faster in a myelinated axon than in an unmyelinated one. Thus, for a given signal transmission rate, the volume occupied by a myelinated axon will be approximately 1/100 of the volume occupied by an unmyelinated axon, yielding more efficient use of the available space. Even more efficient use of the available space is achieved in the central nervous system, where myelination of the axons is carried out by oligodendrocytes instead of Schwann cells. Each oligodendrocyte can myelinate around 50 adjacent axons, compared with the one-to-one ratio found for Schwann cells in the peripheral nervous system. The initiation of an action potential depends on depolarizing current flows charging the axonal membrane and raising the local potential to the threshold value. Ultimately, axonal membranes are voltage (actually field) sensitive.

Point 3 above actually deserves further comment since neurons differ in the pattern of action potential firing they exhibit. This difference is, of course, necessary since neurons have a wide variety of physiological functions to control, quite apart from the range of different behaviours they initiate and mediate. It is implied in the preceding discussion that a neuron will fire a single action potential of constant magnitude or perhaps a steady train of action pulses. In fact they exhibit much more complex temporal behaviour. Figure 2.12 shows some examples of this behaviour. We see that some neurons are normally silent (Figure 2.12(a)); their membrane potential remains at the resting potential unless some external stimulus triggers an action potential. Other neurons exhibit more complex, innate behaviour in which action potentials are fired *without* external stimulus so that we see regular firing 'beating' (Figure 2.12(b)) or 'bursting', (Figure 2.12(c)). This latter type of behaviour is used to generate rhythmic behaviours such as breathing, walking and chewing. The magnitude and duration of action

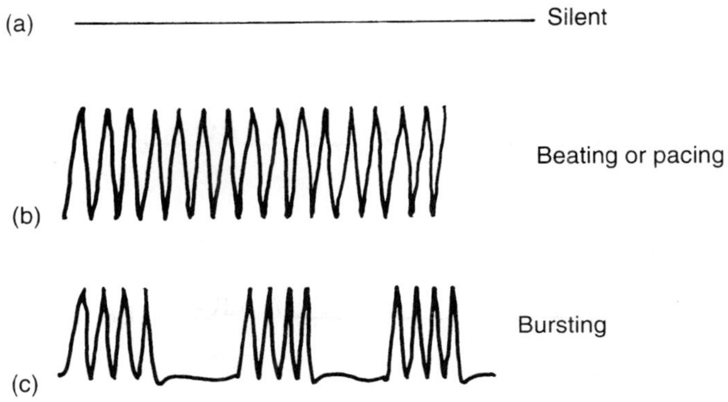

Figure 2.12 Temporal behaviour of a neuron showing the variety of behaviour which can be exhibited. (a) Silent neuron; (b) regular firing or 'beating'; (c) regular firing but for relatively short time intervals, known as 'bursting'.

potentials are capable of modulation, following synaptic or hormonal stimulation, but this topic is outside the scope of this book.

The discussion in the preceding paragraphs has focused on the behaviour of axons. If, however, we examine the neuronal cell body, we find that some cell bodies do not initiate action potentials at all. They are electrically inexcitable, yet they still carry out electrical signalling, only now in more subtle ways. Even in those cell bodies which do initiate action potentials, the more subtle mechanism may still be present and the consequent electrical activity can exhibit a more complex pattern than is found in axons.

The neuronal response to a sustained depolarizing stimulus has been described and illustrated in Figure 2.9. Unfortunately, this periodic and regular response is not always found. Instead a neuron may initiate a single action potential at the onset of stimulus and remain silent thereafter. The behaviour is known as *accommodation* to the stimulus. Alternatively, a neuron may not fire an action potential until some time after the onset of stimulation, that is, it exhibits a *delayed response*. A consequence of this delay is that a short period of stimulation will not initiate any action potentials. These responses are depicted in Figure 2.13.

2.10 THE VOLTAGE CLAMP AND WHOLE CELL PATCH

Before proceeding to a discussion of the properties of the membrane ion channels which give the neuron its unique characteristics, we need

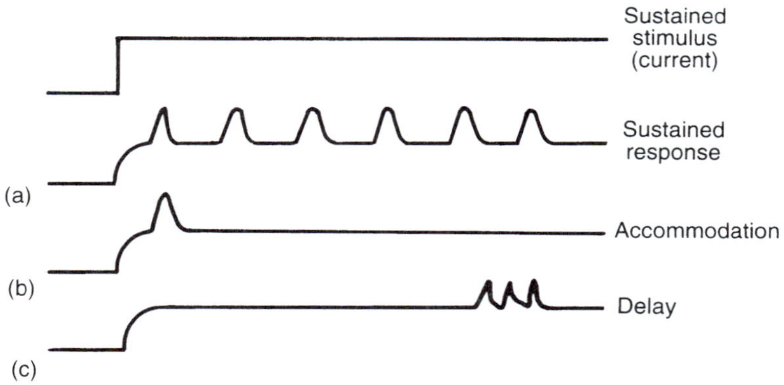

Figure 2.13 Temporal neuronal response to a sustained stimulus showing (a) sustained response, (b) accommodation, where only one action potential is fired, and (c) delay, when a group of action potentials are fired after a delay.

to explore the experimental means by which neuronal electrical characteristics have been established and, in particular, how the transmembrane voltage has been controlled and the properties of individual ion channels determined. The former point is very important since, as we shall see, the sodium and potassium conductances are themselves a function of voltage. The membrane is not in a steady state during the firing of an action potential since depolarization produces an increase in sodium ion conductance which, in turn, produces further depolarization. The sodium ion conductance G_{Na} increases further and, as already pointed out, an intrinsically unstable, positive feedback loop is established which gives rise to the regenerative all-or-none action potential.

In order to study changes in membrane conductance it is necessary to measure conductance at fixed voltages. This is achieved by means of the voltage clamp technique introduced by K.C. Cole and his co-workers in 1949. Figure 2.14(a) shows the essential features of the voltage clamp. It comprises two microelectrodes which are inserted into the cell; one is a voltage sensing electrode, the other a current passing electrode. The signal V_m from the voltage sensing electrode is applied to the inverting input of the op-amp, while a control voltage V_c is applied to the non-inverting input. Since the gain of an op-amp is of the order of 10^5, V_m cannot differ significantly from V_c without a substantial current being forced to flow into the current carrying electrode. By this means V_m is forced to be essentially equal to V_c, whilst the electrode current is measured independently. This current is equal to the ion current flowing across the cell membrane at the potential V_c.

The clamp concept described above measures whole-cell properties. In fact, it is possible to monitor the properties of a single ion channel by means of 'patch clamp' recording, in which the current flowing in a single ion channel is measured directly. This is done by means of a glass pipette with an internal diameter of the order of a micrometre or so at its tip, which is placed against the cell membrane. Suction is applied to the inside of the pipette which leads to an excellent electrical and mechanical seal (gigaseal) between the glass and the membrane. This seal is good enough to prevent the escape of ions. By this means one or more ion channels may be isolated under the pipette. The actual ion current flowing is determined by means of a sensitive, current-to-voltage converter, again based upon an operational amplifier (Figure 2.14(b)). The output voltage of the amplifier is, in this case, equal to the product $-i'R$, where i' is the channel current; currents as small as

10^{-13} A may be measured quite easily. This technique, which allows us to isolate the properties of a single ion channel, was developed by E. Neher, B. Sakmann and their co-workers in the 1970s.

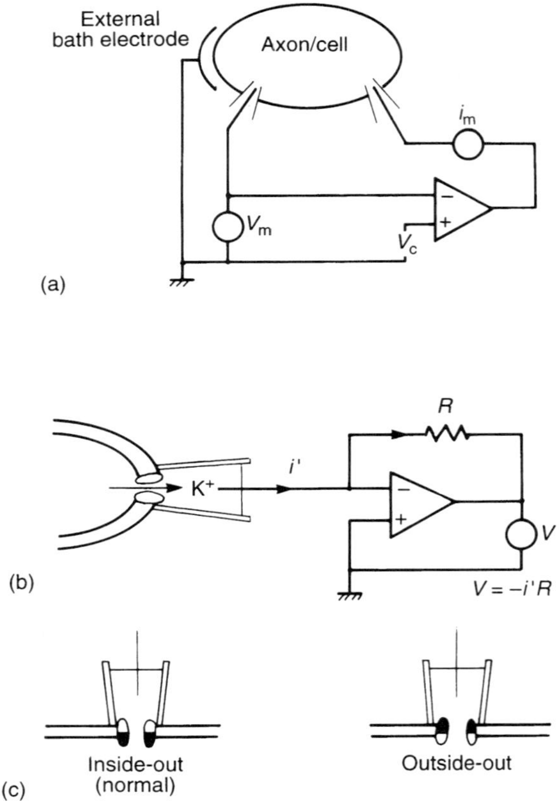

Figure 2.14 (a) The voltage clamp circuit of Cole *et al.* which uses an operational amplifier to establish and maintain V_m equal to V_c (the clamp or command voltage) whilst the current across the membrane i_m is measured independently. (b) The current to voltage converter circuit which permits the measurement of ion channel currents as small as 10^{-13} A. The ion channel current is just $-V/R$. (c) Single-ion channel, or patch clamp, technique in which an ion channel is detached on the end of the micropipette. This may be done in two distinct ways, inside-out and outside-out. In both cases the transport of selected electrolytes can be studied in detail. The seal between the micropipette and the ion channel can have an electrical resistance as high as 10^9 Ω and has come to be known as a gigaseal.

If the pipette suction is increased beyond a certain level it is possible to detach the electrode plus an ion channel. This may be done in two distinct ways, inside-out (the scenario described above) and outside-out, the reverse of above, where what was the inside of the cell membrane is in contact with the pipette orifice (Figure 2.14(c)). This technique, introduced by O.P. Hamill and co-workers in 1981, allows us to explore the properties of single ion channels in even greater detail, since now either end of an ion channel in the membrane can be exposed to a selected electrolyte. A modification of this approach is known as the whole-cell patch, wherein a pipette microelectrode is applied to a cell in the manner described for the patch clamp recording system. This is followed by destruction of the membrane under the pipette tip, either by a suction pulse or by a sudden large voltage change. The solution in the microelectrode can then be freely exchanged with the contents of the cell; the cell can be voltage clamped in the way outlined previously and its properties studied with different electrolytes.

2.11 ION CHANNELS AND MEMBRANE ION CURRENTS

The membrane can change its permeability to ions selectively, as has been outlined in our discussion of the action potential. This is a particularly striking property for any membrane and we need to have some idea of how it is achieved before we can understand the electrical properties of neurons. The permeability of the membrane is a consequence of the presence of ion channels, or hydrophilic pores, traversing the membrane. These ion channels are formed from protein structures whose conformations can be altered rapidly and whose ion transport properties can be determined by the techniques outlined in section 2.10. It is, perhaps, surprising that we can measure the ion current and activity associated with a single channel, that is a single protein molecule, but there are two important reasons why this is possible. The first reason lies in the advances made in modern electronics, particularly the development of operational amplifiers with input leakage currents which may be as small as 1 fA. The second reason lies in the fact that ion transport through a channel is actually quite rapid, so that the current flowing is in the 1–20 pA range. This current range corresponds to the transport of between 6×10^6 and 12×10^7 ions per second through the channel.

Ion transport rates of this magnitude through a single ion channel, must mean that the transport process involves diffusion, since all other processes are many orders of magnitude too slow. The model which has been developed to describe an ion channel is that of a membrane-spanning hydrophilic pore which is set in a hydrophobic plasma membrane. However, these pores are not just simple holes in the membrane; they are shaped in such a fashion that an ion, together with its strongly bound shell of water molecules, must make a tight fit with the narrowest part of the pore (a selectivity filter). Channel selectivity is achieved by narrowing the pore to suit the diameter or shape of one or other hydrated ions. Furthermore, we endow the channel proteins with the ability to modify their structures rapidly so that ions can be blocked which had hitherto been passed. We may think of this as 'gating'.

On the face of it this does not seem like an adequately selective arrangement since the smaller sodium ion must surely be able to pass through a potassium channel. However, selectivity is achieved by very subtle means. If we consider the axonal sodium channel, it has cross-section dimensions of 0.3 nm $\times$ 0.5 nm at its narrowest point and this accepts the sodium ion plus its water of hydration: it is too small, however, for a potassium ion unless it sheds its water of hydration (which requires energy). The axonal potassium channel has dimensions of 0.3 nm $\times$ 0.3 nm at its narrowest point (actually smaller in cross-section than a sodium channel), yet potassium is permeated and sodium is not. The reason for this is that a potassium ion will just pass through this channel at its narrowest point if it loses its water of hydration; it simultaneously gains energy from 'solvation' by weak polar groups on the channel protein. Although a sodium ion can fit through this channel if it loses all of its water of hydration (it is 0.08 nm smaller in diameter than a potassium ion), the energy loss involved is not balanced by an energy gain from 'solvation', since the sodium ion is too small to interact with the relatively weak polar groups lining the channel, instead it experiences electrostatic repulsion forces.

Any discussion of the properties of the pores or ion channels must include four different aspects:

1. single channel conductance, a measure of the rate at which ions pass through the channel;
2. ion selectivity, the nature (identity) of the ions whose passage is permitted;
3. gating, the opening and closing of the channel under the influence of factors such as transmembrane voltage;
4. pharmacology, the susceptibility of the channel to blocking by special compounds.

Using the patch clamp technique we can measure the properties of a single ion channel. In particular, we can measure its conductance g_i' and the reversal potential, that is the potential difference at which current flow ceases. In general this is not zero volts, as it would be with a metallic conductor. Typical responses are set out in Figure 2.15. The single channel conductance is given by $g_i' = \Delta i'/\Delta V$ and has the units of siemens (S). A typical value would be in the range 5–400 pS. For membrane conductance we have used the symbol G, where $G = \Delta J_m/\Delta V$. Here J_m is the total membrane current density, which is the sum of the individual ion channel contributions (units S m^{-2}).

The current–voltage characteristic of a single ion channel provides important information about the identity of the ions which normally flow through the open channel. In particular, the reversal potential, V_{rev},

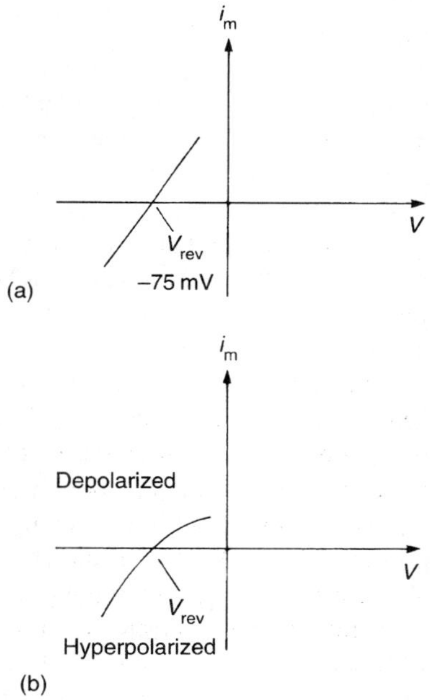

Figure 2.15 (a) Current–voltage characteristics of a single ion channel demonstrating simple ohmic behaviour. The voltage V_{rev} at which the current goes to zero is the equilibrium potential for the ion carried by the channel and affords a means by which the channel ion may be identified. (b) Non-linear current–voltage characteristic exhibited by an inward potassium channel showing anomalous rectification. This sort of current may well play an important role in regulating the resting level of neuronal activity.

or potential at which no current flow occurs, irrespective of whether the channel is 'open' or 'closed', can be compared with the equilibrium potential E deduced from the Nernst equation for a particular ion. In Figure 2.15(a) the reversal potential is shown as -75 mV and we conclude that it is potassium ions which are being carried by this channel since this value is the equilibrium potential E_{K^+} deduced from the Nernst equation. Because the patch clamp technique permits us to alter the extracellular concentration of potassium ions at will, we can also alter the value of E_{K^+} and test further for the identity of the ions flowing in the channel. The reversal potential should alter with change of ion concentration in the micropipette. Of course, if the major ion responsible for conduction is not potassium, then there will be no change in V_{rev}. Equally, if the reversal potential deviates significantly from E_{K^+} we may assume that the channel in question is not completely selective to potassium ions.

The total membrane current in a neuron must be viewed as the sum of the contributions of a class of ion channels which are permeable to potassium ions, together with a class of ion channels which are permeable to sodium. Under resting conditions the potassium channels are sometimes 'open' and the sodium channels are mostly 'closed'. The 'open' and 'closed' states are not static, but dynamic, and we should really be thinking in terms of the probability of a channel being open, together with the frequency with which it changes state. The frequency with which a channel opens or closes is, for some channels, a function of membrane potential. We can imagine that this voltage sensitivity arises because the protein forming the ion channel has some net charge and can be reshaped by an applied electric field. The electric field changes its polarity, but not to any great extent its magnitude when an action potential occurs. In the case of the sodium channel in a squid axon, the voltage sensor responsible for opening and closing the channel was shown in 1993 by R.D. Keynes and H. Meves to carry the equivalent of 0.8 of an electron charge. Although a channel is 'open' to a given ion, it does not follow that a current will flow, since, there is a difference between 'open' channel current and 'open' probability. Figure 2.15(b) shows that, for some potassium channels, a current–voltage plot does not yield a straight line and the effect of an incremental change of voltage in the hyperpolarized region has much more impact on current flow than the same change in the depolarized region. Such a channel is said to exhibit anomalous or inward rectification and this is a property of the channel when it is 'open'. An ion channel will usually spend some time open and some closed and may make repeated transitions between these states. A voltage dependent channel's kinetic behaviour is characterized by its rate of activation or deactivation. These rates may be very different for different ion channels as is shown in Figure 2.16. This difference plays an essential role in the shaping of action potentials.

To summarize our understanding of the electrical signalling in nerve cells; the flow of ion currents across the membrane lies at the heart of this activity and these ion currents arise, in turn, from the activity of individual ion channels through the membrane. These ion channels are actually specialized membrane proteins which provide a hydrophilic pore through an otherwise hydrophobic membrane. The ion channels provide a pathway down which specific ionic species may diffuse as a result of the concentration gradient. Since diffusion is rapid, the ion current for an individual pore can be measured; measurements reveal that the ionic pathways may be either 'open' or 'closed'. Interchange between these two states is mediated by factors such as the membrane potential, neurotransmitters or intracellular messengers. Each ion channel generates a current which contributes to either the hyper-polarization or depolarization of the cell. The rate at which these processes occur depends on the channel's sensitivity to voltage, or other influences, as well as its rate of activation or deactivation. The

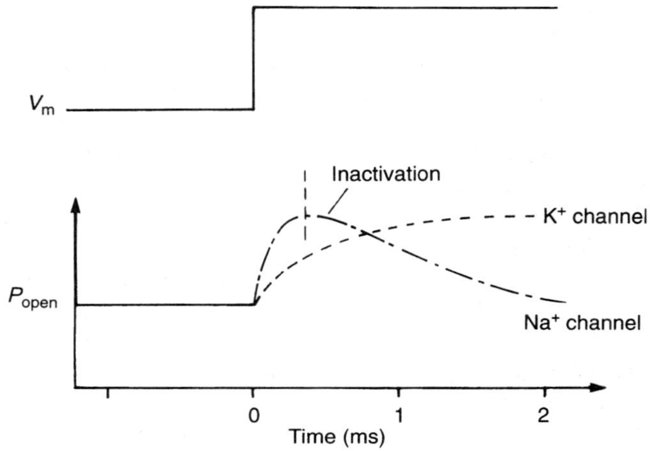

Figure 2.16 Relative response rates sodium and potassium channels showing the markedly different rates of activation, and the relatively rapid rate of deactivation, of the sodium channels. The differing temporal responses are important in shaping the action potential.

normal firing pattern of a neuron and its response to stimulation is actually an interplay between the currents carried by the different types of ion channel.

2.12 CALCIUM CHANNELS

The model we have developed to explain neuronal properties is elegant and relatively simple in that it functions with just two ions, potassium (K^+) and sodium (Na^+). Unfortunately, despite this elegance and simplicity, the picture is incomplete and must be broadened to include the presence of calcium ions and calcium channels. Calcium channels are particularly interesting because calcium is more than a simple charge carrier across the membrane. Calcium ions not only contribute to the action potential, but they also have a role as intracellular messengers. In this latter role they interact with calcium binding proteins to regulate a variety of intracellular enzymes. An additional activity is the gating of several types of ion channel while, finally, intracellular calcium controls the release of chemical neurotransmitters at synapses. We can summarize the role of calcium as that of a transducer, taking an electrical signal (depolarization) and converting it into chemical signals within the cell. This behaviour distinguishes the calcium ion from sodium and potassium ions in neuronal function.

A particularly interesting aspect of calcium ion behaviour is the calcium-dependent potassium current. Figure 2.17(a) shows the outward potassium current for a neuron when the inward sodium current is blocked during depolarization. The characteristic 'N' shape is actually due to the presence of a calcium-dependent potassium ion contribution (Figure 2.17(b)) which arises from the entry of calcium ions, through voltage-dependent calcium channels, followed by the activation of additional potassium channels. The i/V curve, due to potassium ions alone, during depolarization (Figure 2.17(c)) exhibits 'rectification'; that is, the characteristic curve increases in slope with increasing V in much the same fashion as we would see for a p–n junction. The electronic parallel may be extended further in that for certain types of diode (tunnel diodes) an additional tunnelling current is superimposed on the forward characteristic, yielding a curve (Figure 2.17(d)) which is qualitatively similar to the curve in Figure 2.17(a). Both of these curves exhibit a region of negative resistance which, in the tunnel diode, may be used for switching purposes. The negative resistance region is marked in Figure 2.17 for both the neuron and the tunnel diode. A consequence of this negative resistance region is that

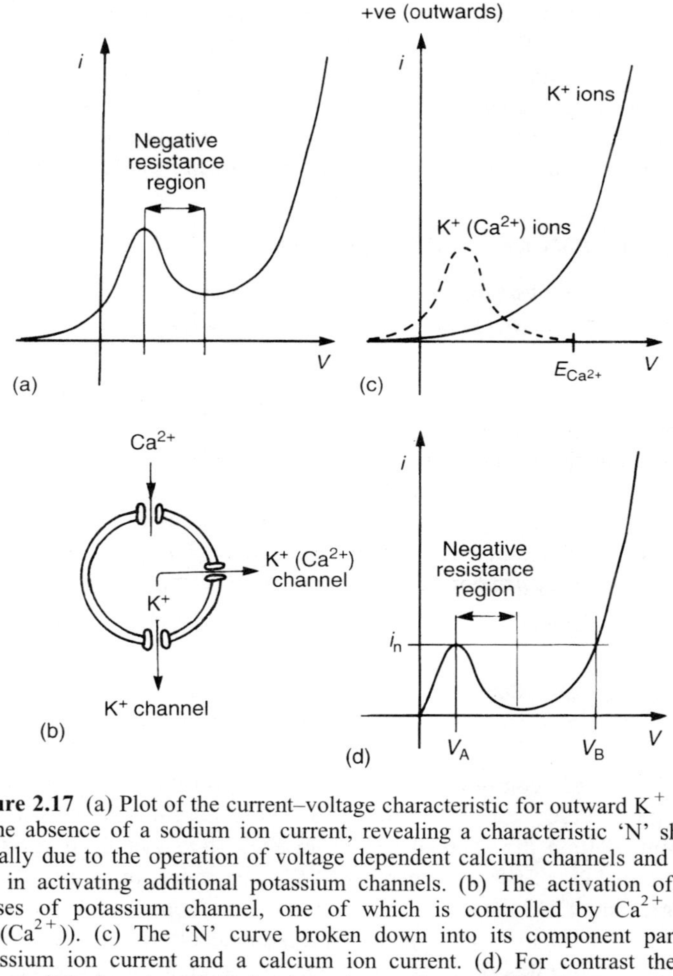

Figure 2.17 (a) Plot of the current–voltage characteristic for outward K^+ ions, in the absence of a sodium ion current, revealing a characteristic 'N' shape, actually due to the operation of voltage dependent calcium channels and their role in activating additional potassium channels. (b) The activation of two classes of potassium channel, one of which is controlled by Ca^{2+} ions ($K^+(Ca^{2+})$). (c) The 'N' curve broken down into its component parts, a potassium ion current and a calcium ion current. (d) For contrast the i/V characteristic for a tunnel diode where the usual *p–n* rectification curve has an additional 'hump' due to a contribution from tunnelling through the potential barrier. Both curves (a) and (c) exhibit negative resistance regions.

for a given current, say i_n as marked in Figure 2.17(d), any incremental increase in applied current will, under appropriate conditions, result in a transition from point A, low impedance, to point B, high impedance, without apparently traversing the intervening curve. The voltage 'switches' from V_A to V_B. One would therefore expect some similar sort of 'switching' behaviour to be associated with the K^+, Ca^{2+} ion combination, even in the absence of sodium ions.

It is not possible, within the scope of this chapter, to do any more than sketch in the behaviour of calcium channels. For more detail on calcium channel behaviour the reader is referred to one of the standard texts on the neuron, for example Levitan and Kaczmarek (1991).

2.13 ION CHANNEL SELECTIVITY AND GATING

The membrane of a single neuron may possess many different types of single channel activity, and these can be classified according to four different criteria:

1. single channel conductance (that is a measure of the rate at which ions pass through the channel);
2. ion selectivity (identifying the type of ion which is permitted passage through the open channel;
3. gating (the opening and closing of the channel under the influence of factors such as the transmembrane potential);
4. pharmacology (the influence of chemical compounds on channel properties).

We have seen in section 2.11 that single channel conductance and the reversal potential provide information on the nature of the ions flowing in a particular channel. It is convenient to think of the opening and closing of a channel as a gating process in which the channel protein changes conformation either to allow ion flow or to obstruct it. However, these two states are in dynamic equilibrium so that the amount of time the channel spends in each state will depend on the relative values of the free energies of the two states. In turn these free energies may be measured through the rate constants for channel opening and closing. The physical measurements on a single channel embrace these rate constants and reflect the amount of time a channel spends open or closed. There will always be a balance between these two states, although this balance is a dynamic one in the sense that the time spent open, compared with the time spent closed, may alter under the influence of external factors.

Perhaps the most important of these external factors is the membrane potential. Figure 2.18 shows how the amount of time spent open or closed changes for a potassium channel under the impact of voltage change. In order to explain this effect we assume that the channel protein possesses some net charge which, under the influence of the

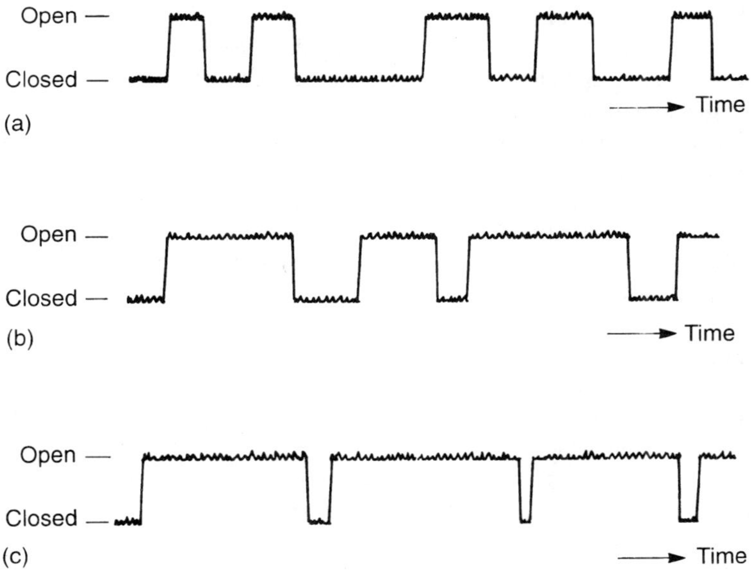

Figure 2.18 Effect of voltage on the amount of time a potassium channel spends open or closed. As the voltage becomes progressively more positive the potassium channel spends more time open.

applied electric field, allows it to change conformation and either block or unblock ion passage. As we have seen in section 2.11, this is known to be true for sodium channels. Of course, even if an ion channel is fully open, it may still not conduct if the potential equals the reversal potential. The fact that the channel is fully open may, however, be demonstrated by altering the ion concentration on one or both sides of the membrane to shift the equilibrium potential.

Figure 2.15(b) shows that the current–voltage relationship is not always linear for a single ion channel, and that such a channel is said to display rectification, but we must be clear that although this is an example of voltage dependence, it is *not* voltage dependent gating. The measurement of current–voltage dependence probably reflects blocking of current flow by non-permeant ions or may indeed be a property of particular channel proteins.

A factor which further complicates our analysis of ion channel behaviour is the rate of activation of a voltage dependent channel (Figure 2.16). Thus a sodium channel may activate fully within a millisecond of a membrane potential change, whilst a potassium channel activates more slowly. Equally important, of course, is the rate of deactivation of a channel and we see that deactivation may be not only voltage dependent, but also time dependent. This effect is shown for the sodium channel in Figure 2.16. The interplay of these effects is ultimately responsible for the shaping of action potentials. The activity of ion channels must, in many cases, be tightly controlled by the action of neurotransmitters, that is chemicals released by one neuron which influence the activity of other neurons. The gating of other channels is influenced by intracellular messenger molecules or ions, for example calcium ions.

In summary we note that the flow of ion currents across the membrane is the basis for electrical signalling in nerve cells. These ion currents arise from the activity of individual ion channels in the membrane, where specialized membrane proteins provide hydrophilic pores through which specific ions may diffuse: the driving force for diffusion is the concentration gradient. Ion channels permit extremely rapid ion transport so that individual ion currents can be measured. In turn it can be demonstrated that ion channels exist in either an open state permitting ion flow, or a closed state in which ions may not pass. The transition between these states is influenced by membrane potential, neurotransmitters or intracellular messengers. When an ion channel is active the resulting current acts either to hyperpolarize or depolarize the neuron, at a rate which is a function of the neuronal sensitivity to voltage, or other influences such as its rate of activation and deactivation. Hyperpolarization or depolarization influences the activity of other types of voltage dependent ion channels. In general, the firing of a neuron will result in a complex interplay between the various ion channels.

2.14 SODIUM AND POTASSIUM ION PUMPS

The normal operation of a neuron requires the exit of potassium ions and the entry of sodium ions. Clearly, some mechanism must exist to reverse this situation, since otherwise the neuron will eventually cease operation and the membrane potential will become zero. This problem is overcome by means of energy-driven pumps which effect the transport of sodium ions out of the cell and potassium ions into the cell. This is achieved by means of an active ion transporter known as the sodium–potassium adenosinetriphosphatase pump, in effect an ion pump which utilizes an enzyme to hydrolyse ATP (adenosine triphosphate) and uses the energy thus obtained to move, or transport, each ion against its

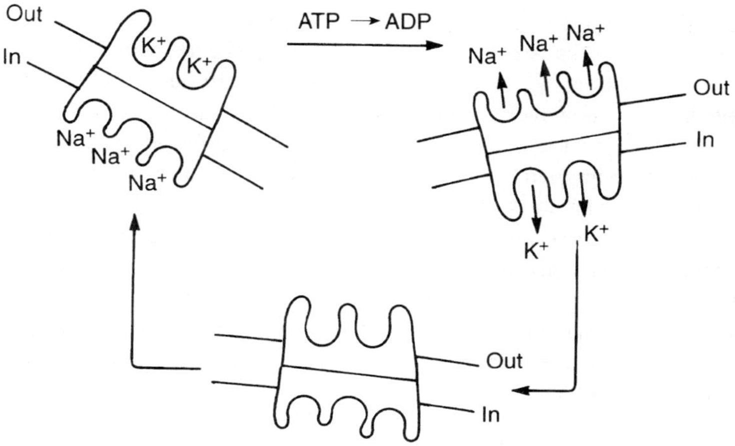

Figure 2.19 The sodium–potassium 'pump' responsible for the restoration of the sodium and potassium ion concentrations in a neuron. The idealized 'pump' molecule is here depicted as rotating about an axis in the neuron membrane and transporting three sodium ions out of the neuron for every two potassium ions transported inwards. In operation the 'pump' causes the neuron to hyperpolarize.

concentration gradient; the pump is actually a protein molecule. This transport activity may involve some sort of movement or rotation of the pump molecule in the membrane. Rotation is a particularly attractive option and is depicted in Figure 2.19. The stoichiometry of sodium–potassium–ATPase is such that three sodium ions are moved out of the neuron for every two potassium ions which are moved in, a transport ratio of 3 : 2. The net effect is that there is a outward current of one sodium ion per ATP molecule hydrolysed, and the pump is said to be 'electrogenic' because, in operation, it causes the neuron to hyperpolarize; it is also known as the sodium pump. Chloride ion moves out of the neuron passively with the sodium. The mitochondria play a role in restoring the phosphate groups lost during hydrolysis. Operating at its maximum rate, each pump can transport some 200 sodium ions and 133 potassium ions, per second, across the membrane. A typical small neuron has around 10^6 sodium pumps.

2.15 INTERCELLULAR SIGNALLING – THE SYNAPSE

The origin of the electrical signals propagating within axons and the general electrical characteristics of neurons have been surveyed in the preceding sections. It remains now to examine how the intercellular signalling process is achieved. In the normal cell, the resting potential is maintained at a stable, relatively unchanging level. The resting potential will however change in response to a wide variety of stimuli, resulting in either hyperpolarization or depolarization, as we have seen. It has been implied in our previous discussion that the change in membrane potential is uniform over the whole of the cell. In practice, this may not be so and we can find a transient shift of the membrane potential in a localized area of the cell. This local potential may result from a change in applied voltage (*electrotonic potential*) or membrane permeability (*synaptic or generator potential*). The synaptic potential occurs with change in membrane permeability induced through neurotransmitters by three possible routes:

1. an increase in potassium conductance (hyperpolarization);
2. an increase in sodium conductance (depolarization); and
3. an increase in sodium and potassium conductance which leads to depolarization, but to a lesser extent.

The generator potential occurs by mechanism 3 (Figure 2.20(a)). Figure 20(b) shows the development of the synaptic potential in response to the release of a neurotransmitter (acetylcholine) at the synaptic cleft.

When an area of membrane is depolarized during an action potential, ionic currents flow; there is also a longitudinal current flow both inside and outside the membrane. This flow of current (positive charges) towards non-depolarized regions internally, and depolarized regions externally, tends to depolarize the membrane in the areas that surround the region of the action potential. Usually, this depolarization is enough to shift the membrane potential to threshold and generate an action potential in the membrane immediately adjacent. The action potential will thus spread away from its site of initiation along an axon. Because of the refractory period, the potential cannot reverse and spread back into an area just depolarized.

The rate of conduction of the action potential along the membrane depends on the amount of longitudinal current flow and on the amount of current needed to produce depolarization in the adjacent membrane; Figure 2.21 shows the situation for a myelinated axon. The longitudinal current flow can be increased by increasing the diameter of the axon since this reduces the internal resistance of the axon. Of course, if the axon is myelinated then increased conduction speed is obtained by the ‘skipping’ action of saltatory conduction.

The local potential mentioned earlier is a graded potential, that is its amplitude is proportional to the size of the stimulus: this local potential will always be measured with respect to the resting potential. If more than one stimulus occurs simultaneously, or they at least overlap, then the resulting local potential is larger. In other words, local potentials can be summed algebraically so that hyperpolarizing and depolarizing

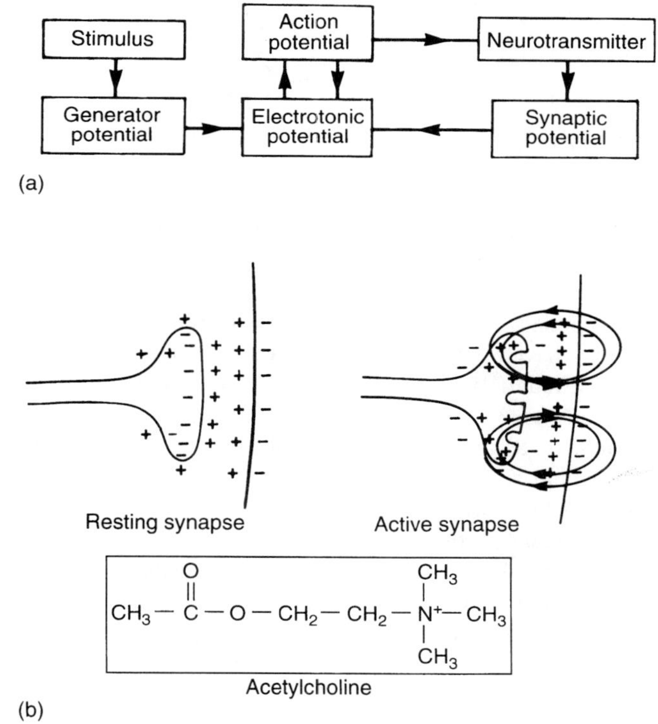

Figure 2.20 (a) Organization of the potentials which are significant in the response of a neuron to a stimulus. (b) The synaptic transmission process in a chemical synapse. In a resting synapse both the axon terminal and muscle fibre are normally polarized. In an active synapse the action potential proceeds from left to right in the diagram and depolarizes the synapse, resulting in the release of neurotransmitter, commonly acetylcholine, from the nerve terminal. The neurotransmitter diffuses across the synaptic cleft and produces a local current flow and a synaptic potential. The postsynaptic current flow paths are indicated by arrows and constitute a current dipole. Acetylcholine is shown in the box. (c) Idealized section through an electrical synapse or gap-junction showing its bidirectional nature.

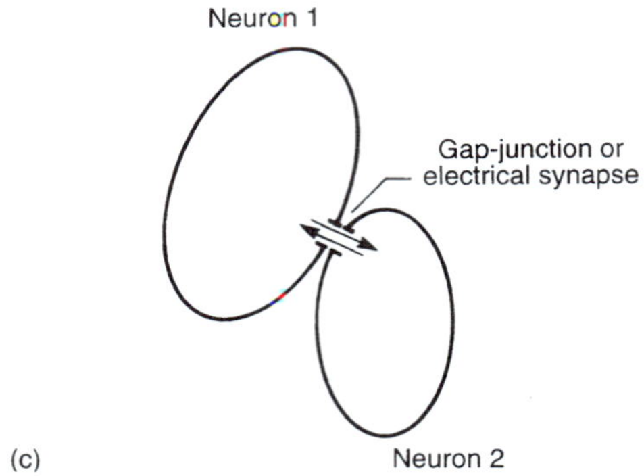

Figure 2.20 *Continued*

potentials will tend to cancel, whereas two hyperpolarizing or two depolarizing potentials will add; the latter may reach threshold and initiate an action potential, even though the individual potentials are themselves below threshold.

Local potentials have, by definition, a spatial dimension but, in addition, they have a temporal dimension since once the action potential has occurred it does not disappear immediately. Consequently, local potentials have a temporal course which outlasts the stimulus and we can think of temporal summation where two spatially coincident potentials exhibit some time overlap. Equally, we can think of spatial summation for two local potentials which are closely adjacent but not spatially coincident. In this case the potential change is not confined to the area of stimulus, but falls off over a finite distance of the membrane; a few micrometres would be typical. These qualities of temporal and spatial summation are important since they allow the neuron to

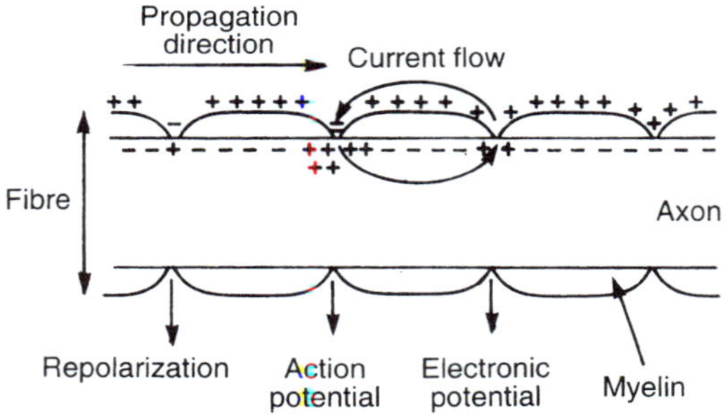

Figure 2.21 Conduction process for a myelinated axon.

integrate signals arriving at different times, on the one hand, or from different sources on the other: when either of these types of local potential reach threshold they will initiate an action potential. Figures 2.22(a) and (b) shows how temporal and spatial summation might occur and highlights the role of the synapse (axon terminal).

The synaptic terminations of most neurons are specialized to synthesize chemical agents, **neurotransmitters**, which when released act on other neurons; these structures are **chemical synapses**. The neurotransmitters are synthesized by enzymes which are manufactured in the cell body. The neurotransmitters are then transferred down the axon by an active, oxygen dependent, axoplasmic transport system and are stored in the nerve terminals in small membranous structures (**vesicles**). When the axon terminal is depolarized, neurotransmitter is released in an amount directly proportional to the change in membrane potential, that is to the magnitude of the action potential at the nerve terminal. This depolarization is termed an *end plate potential* (EPP). The

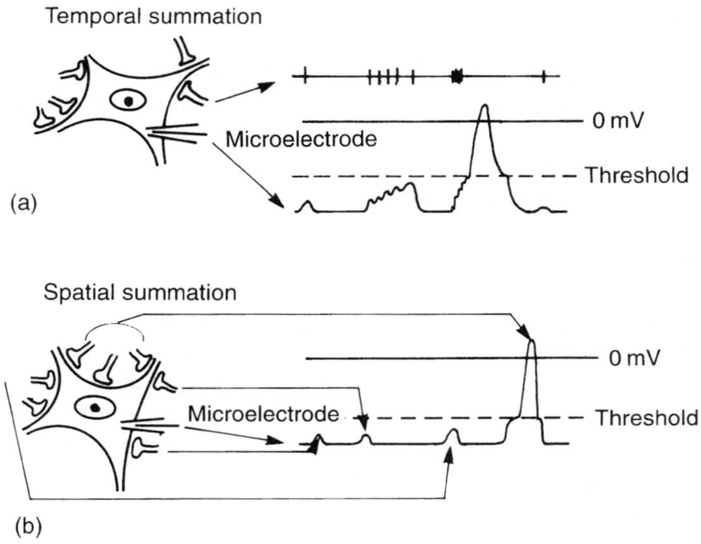

Figure 2.22 (a) The signal obtained from a microelectrode as the result of a series of stimuli which are spaced out in time. If the stimuli are close together in time then temporal summation occurs with the result that an action potential is initiated. (b) The signal obtained from a microelectrode as a result of synaptic stimuli occurring at either single synapses or groups of closely adjacent synapses. For sufficiently large groups of synapses spatial summation can occur with the result that an action potential is fired.

stimulation of the **presynaptic nerve** occurs randomly, leading to small depolarizations known as **miniature end plate potentials** (MEPPs). We find that some nerve terminals have synapses that connect directly to other terminals, which they can partially depolarize. The partially depolarized nerve terminal will release less neurotransmitter when it is depolarized by the invading action potential. The resultant synaptic potential will therefore be smaller; this grading process is termed **presynaptic inhibition**. Other neurotransmitters produce a selective increase in conductance of potassium or chloride only. In this case the resulting synaptic potential is called an **inhibitory postsynaptic potential** (IPSP), as it moves the potential away from threshold and makes the membrane less excitable.

The general picture of the operation of a chemical synapse reveals a complex and subtle mechanism in which we see that synaptic transmission has three characteristics unique to it. First, conduction at a synapse is delayed because of the brief interval of time required for the chemical events to occur. Second, because the two sides of the synapse are specialized to perform only one function, transmission can only occur in one direction across the synapse (this polarizes the nervous system for unidirectional impulse transmission). Third, temporal and spatial current summation occurs, allowing a neuron to integrate activity from many sources.

There is, in addition, a second type of synaptic junction, the *electrical synapse*, which does not possess the asymmetry in function and structure of the chemical synapse. In the electrical synapse (Figure 2.20(c)) signals can move in both directions and each cell may be pre- or postsynaptic at different times. Instead of the synaptic cleft which separates the two elements of the chemical synapse, electrical synapses are characterized by an area of very close opposition or juxtaposition of the membranes in the pre- and postsynaptic cells. Within these membranes are found *gap-junctions*, cell-to-cell pores that allow ions and small molecules to pass freely from the cytoplasm of one cell to that of the next. In this context a small molecule implies a molecular weight up to around 1500.

At the electrical synapse intercellular signalling occurs, mediated by the passage of small ions from one cell to another, through the cell-to-cell gap-junction channels. These channels are short and provide a low resistance pathway for ion flow (without leakage) to the extracellular space; signals can be transmitted without significant attenuation. Information transfer in the electrical synapse can be bidirectional, so matching the structural symmetry. However some examples exist where the efficiency of transmission in one direction is higher than that in the opposing direction, or even where the junction is markedly rectifying. A further contrast to the chemical synapse is that the delay which characterizes the chemical synapse is absent in the electrical synapse; electrical synapses are fast. Despite this contrast, electrical and chemical synapses may operate in concert so that, for example, a pair of neurons may be connected by both electrical and chemical synapses with the operation of each type of synapse being modified by the other. The speed of the electrical synapse tempts the suggestion that it operates mainly as a synchronizing entity for groups of cells so that they operate as one functional unit. However, electrical synapses are also found between pairs of neurons that do not always fire in synchrony so that their role may include allowing the synaptic inputs into one neuron to be registered in an adjacent neuron. The biological role of the gap-junction or electrical synapse is not yet well understood.

2.16 INITIATION OF MUSCLE CONTRACTION

All bodily movement, or for that matter just stable posture, relies on muscular control. In turn, muscular control is carried out by the nervous system, so that we need to understand how this is achieved. In the case of muscles, control is maintained by the motor unit which comprises the **motor neuron** together with the muscle fibres innervated by it, the motor neuron being connected to the muscle fibres by the **motor axon** (Figure 2.23).

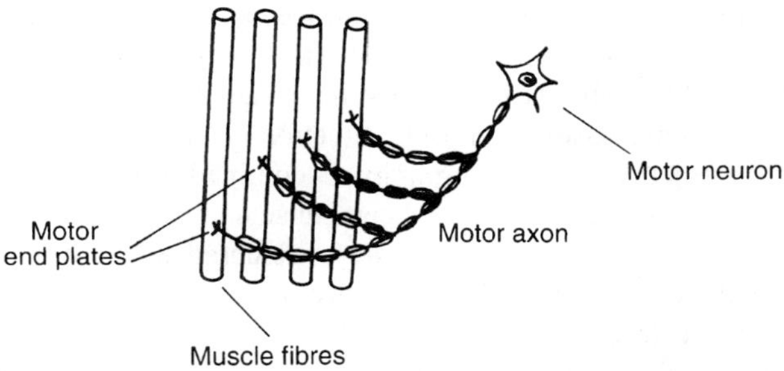

Figure 2.23 The structure of the motor unit which initiates muscular control. It comprises a motor neuron, which initiates action, and a motor axon that connects the motor neuron to the muscle fibres via an enlarged synapse, the motor end plate. Every motor axon branches at the nodes of Ranvier to provide innervation of many muscle fibres. The size of a motor unit is determined by the number of muscle fibres connected to each motor neuron (the innervation ratio).

Motor axons are large myelinated fibres, 6–20 μm in diameter, located in the peripheral nerves. Branching of motor nerve fibres occurs along the course of muscular nerves at the nodes of Ranvier. As they branch they become smaller in diameter and at their termination they are unmyelinated. Each terminal of a motor axon is coupled to the muscle by a modified synapse, actually an enormously enlarged synapse, known as the **motor end plate**.

The terminals of a single motor axon are on muscle fibres which may be distributed widely throughout the muscle, intermingling with muscle fibres innervated by other neurons. A muscle may contain 50–1000 motor units where a motor unit is the collection of muscle fibres innervated by a single motor neuron. The size of the motor unit is characterized by the number of muscle fibres per axon (motor neuron); the innervation ratio. The speed and strength of a movement is controlled by the number of motor units active, their rate of firing and the characteristics of the motor units activated; a firing rate of 5–20 per second is not unusual.

The motor unit is the physiological unit of reflex and voluntary contraction. Usually, the motor unit behaves in an all-or-none manner, which means that an impulse in the motor nerve fibre produces an action potential in, and synchronous contraction of, all the muscle fibres it supplies. The resulting contraction of the motor unit is thus the sum of the mechanical responses of the component muscle fibres. Activation of a motor neuron in the spinal cord or brain stem produces an action potential which spreads to each of the terminal branches of the motor axon in the muscle. As a result we can sense the performance of the nervous system from the electrical activity in the muscles themselves. This leads us to two distinct diagnostic approaches: one, electromyography, assesses muscle function from its electrical 'signature'; the other, electrocardiography, is confined to the measurement of the signals associated with the heart muscle. The characteristics of, and basis for, these measurements are considered in the next section together with the information gained from signals originating in the brain itself (electroencephalography).

The current flows accompanying nervous activity must, of course, have an associated magnetic field. Since the currents which flow are extremely small, so also are the resulting magnetic fields. Consequently, if we measure the magnetic fields, we obtain information on the current flows, but now with excellent spatial and temporal resolution. These techniques are magnetoencephalography (MEG), magnetomyography (MMG) and magnetocardiography (MCG); they embrace some new physics and are the subject of Chapter 3. By a curious coincidence we see that while the squid (*Loligo*) was central to enabling us to understand the electrical characteristics of the neuron and axon, it has a modern homonym which is the basis of our ability to measure the magnetic fields associated with this activity. This SQUID however is actually an acronym standing for 'superconducting quantum interference device'.

2.17 ELECTROENCEPHALOGRAPHY

The electrical activity of the nervous system has been described in some detail on the basis of the function of an isolated neuron/axon dendrite/synapse combination. In practice, of course, the situation is much more complicated since neurons do not function in isolation, but in aggregation, with richly endowed synaptic interconnections. The overall electrical activity reflects the summation of all the dendritic and action potentials and is necessarily very complex.

The electrical activity of the cortex, that is the layer of cells immediately below the skull and scalp, can be detected and recorded externally by means of electrodes placed on the scalp. They measure the dendritic potentials in the upper layers of the cortex as they appear on the scalp. These potentials comprise continuous, rhythmic or arrythmic oscillating waveforms that vary in frequency, amplitude, and shape. These potentials have a magnitude in the range 20–50 μV, and result from the summation of synaptic potentials at the many dendrites lying near the surface of the cortex. The fluctuation of the waveform is due to excitatory and inhibitory synaptic potentials impinging on the dendritic membranes. This electrical activity of the cortex is recorded as the electroencephalogram (EEG). Figure 2.24(a) shows the electrical origin of the EEG while Figure 2.24(b) shows the four basic EEG waveforms. The EEG is taken as a recording of the potential difference between five/six pairs of electrodes, usually laid down in the positions indicated in Figure 2.24(c), so that we may well see beta activity over the anterior regions of the skull, and alpha activity over the posterior regions. The theta and delta rhythms are not usually seen in an awake adult; if they are then this is an indication of abnormality. Indeed the principal use of the EEG is in the detection of abnormalities of cerebral function.

In order to record the electrical signals arising in an EEG measurement, or any other electrical measurement on the body for that matter, it is necessary to have electrodes which can be easily attached to the skin surface without penetrating it. In addition, these electrodes must

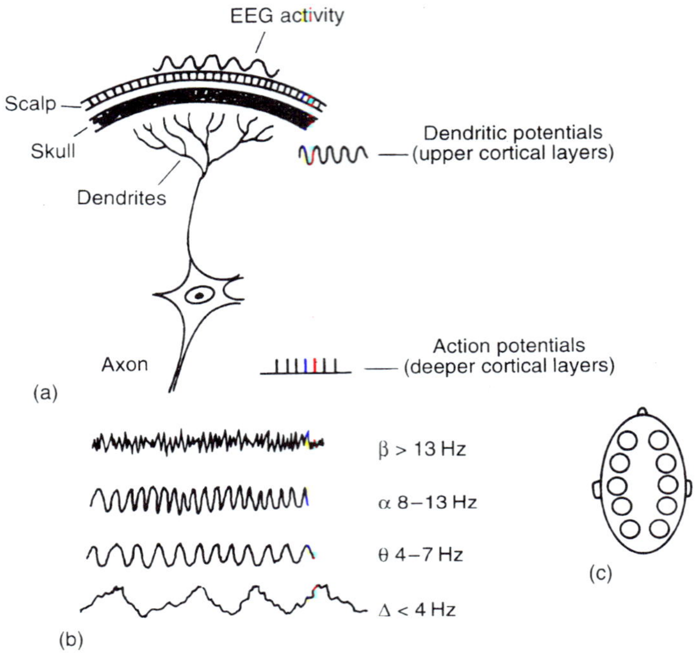

Figure 2.24 (a) The origin of potentials measured in an electroencephalogram (EEG). These potentials are actually due to the dendritic potentials in the upper cortical layers and lie in the range 20–50 μV. They are, in fact, a summation of the synaptic potentials of many dendrites lying near the surface of the cortex. (b) Electrical activity recorded by the EEG is classified into four main types according to frequency, alpha (α), beta (β), theta (θ) and delta (Δ). Typical waveforms for these types are shown. (c) Measuring points on the skull for EEG recordings. Potentials are taken between adjacent pairs in each column of measuring points.

present a low contact resistance. This latter requirement is met by coating the electrodes with a gel containing a strong electrolyte, such as sodium chloride, which penetrates the **epidermis** to some extent and yields an overall contact resistance of about 1 kΩ. The usual combination uses a silver/silver chloride electrode.

The electrical requirements for amplification of signals from the cortex are not particularly severe, in one sense, since the frequency range encountered is quite limited. However, artefact signals, which may be produced by the electrical signals from muscular activity, have to be reduced to an acceptable level, so also must mains hum pick-up for example and noise. The solution to these problems is to use an amplifier with differential and common-line inputs and high common-mode rejection.

2.18 ELECTROMYOGRAPHY

Another important aspect of the electrical activity of the body is that associated with muscle function since nerve disorders produce secondary changes in muscle function. Muscle function is evaluated by measuring its electrical activity by electromyography (EMG). Electromyography is valuable not only in primary muscle diseases, but also in differentiating between diseases of neuromuscular transmission and diseases of the central nervous system.

The natural activity in individual axons of a peripheral nerve is random and cannot readily be recorded. However, with the application of an electrical stimulus, all the large myelinated fibres can be discharged simultaneously, and the resulting compound action potential recorded and measured. If an electrical stimulus (20–100 V for 0.1 ms) is applied to a mixed peripheral nerve, action potentials will be initiated which travel in both directions along the nerve. Those travelling centrally will be perceived by the patient as a shock. Those travelling peripherally will reach each of the terminal branches of the nerve where they can be recorded, either from **cutaneous** sensory branches or from muscles innervated by motor branches. In this way either the motor or the sensory components of the peripheral nerves can be studied selectively, as the sensory nerve action potential or the compound muscle action potential. These two types of potential are characterized by their *amplitude* (related to number of axons involved), *latency* (time in ms taken by the largest fibres in the nerve to propagate action potentials down the axon) and *conduction speed* θ of equation 2.24 (θ depends on the axon diameter and extent of demyelination).

In EMG a fine hypodermic, coaxial, needle electrode is inserted into a muscle, and the electrical activity is recorded on an oscilloscope. Recordings are made in multiple muscles in three states: at rest, with needle movement, and with voluntary activity. At rest there is electrical silence in a normal muscle, except in the region of motor end plates, where end plate potentials or other small potentials may be recorded. If the needle is moved in the muscle, the mechanical stimulus produces a discharge of the muscle fibres. With voluntary contraction of the muscle, the motor units fire repetitively, in an orderly fashion, with a frequency and number proportional to the effort expended. The motor unit potentials are measured to determine their amplitude, duration, and firing rate.

In cases of primary **myopathy**, with random degeneration of muscle fibres, each motor unit will have a reduced innervation ratio and the motor unit potential will be of short duration and low amplitude. As a consequence of the reduction in power of each motor unit, many more will fire for a given strength than in a normal muscle. If there is partial destruction of some muscle fibres, those still viable may fire spontaneously.

The muscle action potential which initiates the contraction of a muscle has the waveform shown in Figure 2.25(a). The potential has an

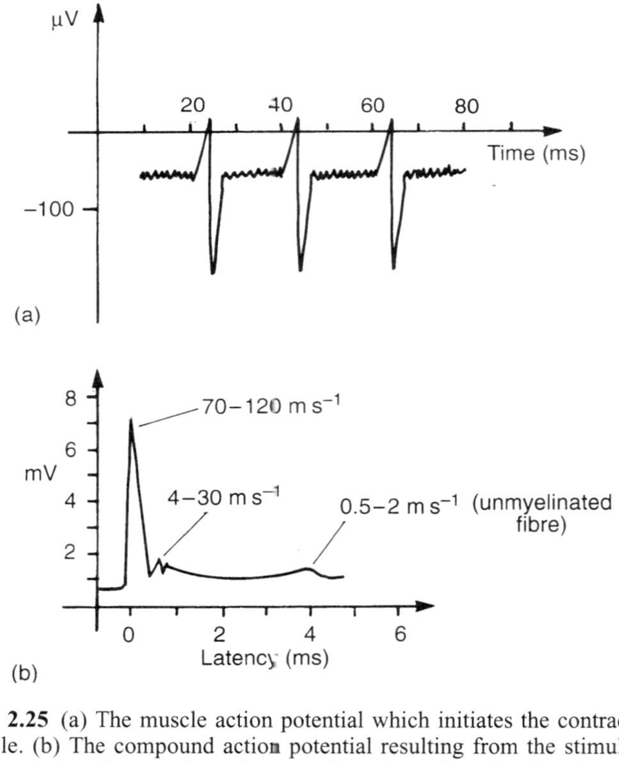

Figure 2.25 (a) The muscle action potential which initiates the contraction of a muscle. (b) The compound action potential resulting from the stimulus of a nerve trunk at a distance from the recording electrodes, showing the different components arising from the dispersion of the potentials from nerve fibres of different diameters. Large fibres having high conduction speeds are recorded first while smaller fibres produce later signals. In each case the typical spread of speed is indicated.

amplitude of around 100 μV and a period of 6–12 ms. In a normal muscle, voluntary contraction of the muscle merely results in this potential repeating in an orderly fashion. The signal seen from electrical excitation of a nerve trunk, the compound action potential, is somewhat different in magnitude (Figure 2.25(b)) and has separate components arising from the different transmission velocities of the various axons. The conduction speed, in $m\ s^{-1}$, is approximately five times the numerical value of the axon diameter in micrometres, so that the impulses in the large axons reach the recording electrode first, and the components of the compound action potential reveal and distinguish activity in groups of nerve fibres whose diameters are within certain size ranges. Since the compound action potential is the summation of action potentials from many axons, its amplitude can be graded by varying the strength of the stimulus. As more fibres are excited the amplitude of the action potential increases and reaches a maximum when all of the nerve fibres are excited. The action potentials in single axons are not graded but fire in an all-or-nothing fashion. Table 2.2 summarizes some conduction speeds for commonly tested pathways

2.19 ELECTROCARDIOGRAPHY

The heart is a compact and efficient pump, operating at a rate of around 70 beats per minute, which depends for its pumping action on the contraction of the heart muscles under the influence of an initiating pulse at the **sino-atrial node** (S-A node) in the right atrium (Figure 2.26(a)). The sino-atrial node sends out regular impulses which cause contraction of the atria. Having passed through the atrial muscle the

Table 2.2 *Nerve conduction speeds in healthy adults*

Peripheral nerve (in italics)	*Number of nerves*	*Speed range* ($m\ s^{-1}$)
Ulnar (motor) elbow to wrist	47	46.5–72.6
Medium (motor) elbow to wrist	45	46.1–72.1
Ulnar (sensory) fingers to wrist	52	41.7–59.4
Ulnar (sensory) finger to elbow	38	45.3–60.2
Median (sensory) finger to wrist	48	36.4–65.4
Median (sensory) finger to elbow	36	42.4–61.7
Personal (motor) knee to ankle	49	42.1–63.5
Posterior tibial (motor) knee to ankle	30	39.8–66.9

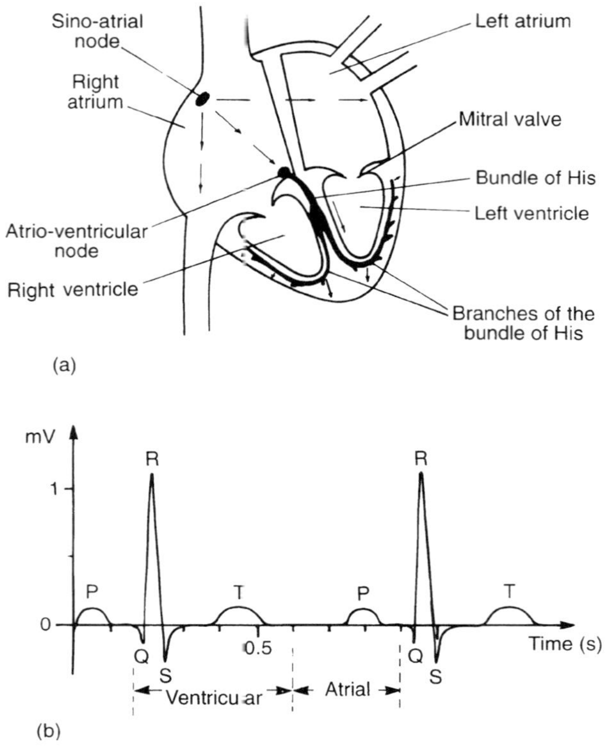

Figure 2.26 (a) Pathway of the electrical impulse in the heart. The spread of the impulse wavefront from the sino-atrial node is indicated by arrows. (b) Typical ECG waveform showing the five peaks P, Q, R, S and T.

impulse reaches the **atrio-ventricular node** (A-V node) near the ventricles. The atrial muscle is separated from the ventricular muscle by a non-conductive band which prevents the direct transmission of impulses from atria to ventricles, except at the A-V node. The A-V node is the beginning of special conducting tissue, the **bundle of His**, which divides itself into left and right branches, and conducts the impulse rapidly through the ventricles causing ventricular contractions. Both atria contract together, followed immediately by the contraction of both ventricles. On its own the S-A node sends out 60 impulses a minute, but it is under the control of the autonomic nervous system which alters this rate to suit the body's needs. The A-V node normally follows this rate, but can beat on its own at 45 times a minute.

The initiating pulse (pacemaker) conducts through the muscles of the heart and can be detected by electrodes on the heart itself, or more conveniently, by placing electrodes at specially selected, standardized sites, on the body surface. The resulting electrical signal is the electrocardiogram (ECG), which shows five waves designated (by Einthoven originally) P, Q, R, S and T (Figure 2.26(b)). The P-wave is due to the atrial impulse, Q, R and S are due to the ventricular impulse, and the T-wave is due to the recovery (repolarization) of the ventricular muscle. Changes in the shape or amplitude of any parts of this curve are of diagnostic significance.

Figure 2.27 shows the heart section and function, greatly simplified, together with a simplified view of the ECG signal. We can think of the heart as a two-stroke four-chamber pump which, in the **diastole** phase, contracts the right and left atria to pump blood into the right and left ventricles. The R-wave signifies the beginning of the **systole** phase in

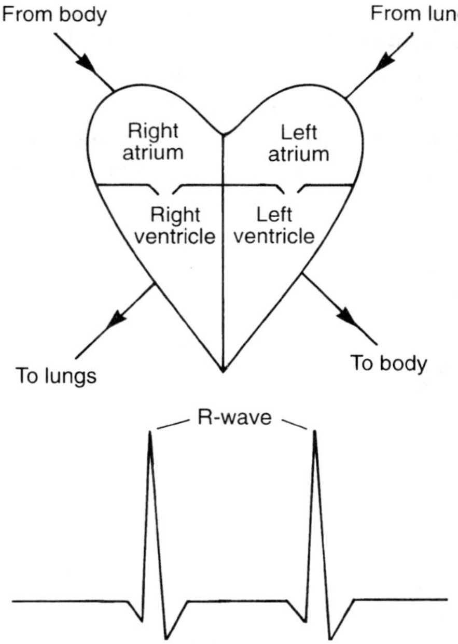

Figure 2.27 Greatly simplified section through the human heart which may be thought of as a two-stroke, four-chamber pump. The pumping cycle is divided into two phases, the diastole phase when the right and left atria contract, and the systole phase when the right and left ventricles contract; the division between these two phases is marked by the R-wave of the heart beat.

which the ventricles contract, dispersing blood to the lungs and body. The left ventricle has the largest muscle as it has the hardest job, that of pumping blood around the body.

Cardiac muscle is made up from a large number of individual cells with dimensions of the order of 15 μm in diameter and 100 μm in length, in contrast to peripheral nerves which may be up to 1 m in length. The intracellular spaces of adjacent cardiac cells are interconnected by a structure termed an **intercalated disc** containing proteins which provide channels for the movement of ions between cells. Because of the electrical interconnections provided by the intercalated discs, cardiac tissue functions as if it were a single large cell, a **syncytium**. Skeletal muscle cells, on the other hand, are not interconnected and can be stimulated individually. The shape of the cardiac action potential (Figure 2.28) is quite different from the nerve action potential of Figure 2.27, primarily because of a secondary inward current which brings Ca^{2+} ions into the cell and maintains a long depolarized plateau. Because the repolarization phase of the cardiac action potential is a hundred times longer than the depolarization phase, the repolarization currents are one hundredth of the depolarization currents. As a result, the depolarization dipole is much more prominent than the repolarization dipole. For this reason, and because of the large temporal and spatial separation of depolarization and repolarization, we

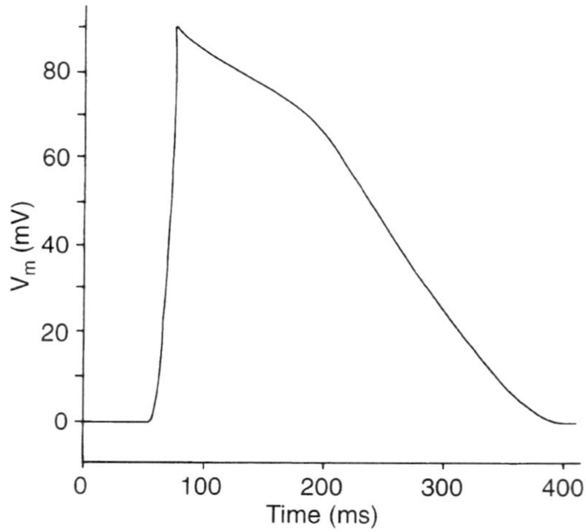

Figure 2.28 The transmembrane potential (action potential) for cardiac tissue showing the long depolarized plateau due to an inward current of Ca^{2+} ions.

might expect depolarization dipoles to be dominant. This proves to be of importance in our consideration of the magnetic effects of nerve action. In Chapter 3 we shall see that it is possible to gain a great deal more information about the electrical activity of the human body by monitoring the associated magnetic fields arising from the current dipoles and quadrupoles pictured in Figure 2.10(c). In this fashion the scope and range of the information available from EEG, EMG and ECG measurements may be significantly enhanced, particularly with regard to spatial disposition.

2.20 DISORDERS OF THE NERVOUS SYSTEM

We have examined the way in which the electrical activity of the human body arises and how it may be sampled through the specific techniques of EEG, EMG and ECG. It is not possible to give any sort of exhaustive overview of neurological problems within the confines of this book, rather it must suffice to outline the principal ways in which the nervous system is vulnerable and, where appropriate, show how this might manifest itself, either physically or electrically.

The human nervous system may be thought of as vulnerable on four fronts:

1. alteration to or disturbance of the contents of the axons, that is the axoplasm;
2. damage to or alteration of the function of the axonal membrane, or **axolemma**;
3. interference with, or cessation of, synaptic transmission;
4. damage to the myelin sheath.

In axoplasmic disorders the axon may lose continuity or blood supply or both. In either case the distal axon is deprived of axoplasmic flow from the neuron and it undergoes dissolution in a process known as **Wallerian degeneration**; peripheral muscle atrophy will accompany this degeneration and recovery can only occur through the growth of new axons. If an axon becomes completely severed, reinnervation is poor because the axon has no pathway to follow for regrowth.

In **chronic** axoplasmic disorder, the axons may degenerate just as they do in **acute** disorders, but now with intermediate stages in which the axon 'dies back' from the distal region, accompanied by narrowing, so-called **axonal dystrophy**. Nerve conduction, as determined by electromyographic measurements, becomes mildly slowed during degeneration, as a result of conduction failure of large fibres; there may be little functional impairment.

In many respects the **axolemma** may be considered a more sensitive region of the axon, since its function can be altered or destroyed not only by physical damage but also by chemical action. It is this latter property which is of most interest since in many instances the process is reversible and clinically useful. The axon membrane may undergo physicochemical alterations which block conduction without destruction or histologic alteration of the axon. The most important example of such an alteration is that produced by pharmacologic agents such as local anaesthetics, when applied to a nerve. Typical examples are compounds such as novocaine, or procaine. Local anaesthetics interfere with nerve conduction by preventing the membrane permeability changes which accompany depolarization (block sodium ions). The membrane is said to be 'stabilized' by local anaesthetics. The smallest axons are the most readily affected. A transient, reversible, conduction block can also be obtained simply by lowering the temperature of an axon, and hence the propagation speed of the action potentials. This method of blocking nerve impulse transmission is accomplished by application of ice or an ethyl chloride spray and produces superficial anaesthesia.

In addition to attack on the transmission properties of the axolemma we must include some details of those chemicals which attack the synapse specifically. Normal synapse operation involves the release of neurotransmitter chemicals, for example acetylcholine, from storage vesicles inside the nerve terminal, and their interaction with specific receptor molecules on neurons. This interaction can be blocked by chemicals such as the paralytic ingredient D-tubocurarine of the arrow poison curare which produces prolonged depolarization at striated muscles by competing with acetylcholine and binding with receptor sites; the result is paralysis. Although D-tubocurarine is destroyed in the tissues, this may occur too late to save the victim, since the muscles of the diaphragm are also affected. Strychnine, another well-known poison, operates by competing with glycine, another neurotransmitter.

Calcium acts primarily as a membrane stabilizer, but the binding of calcium in the membrane is also necessary for the release of neurotransmitter so that a low calcium level may impair synaptic transmission and produce spontaneous firing of a neuron or an axon. An excess of calcium tends to block action potentials and enhance synaptic transmission.

Abnormalities in the sodium and/or potassium levels must have significant clinical effects. Since the resting potential is primarily dependent on potassium, alterations to the extracellular potassium levels will alter the resting potential. A *reduction* in extracellular potassium increases the concentration gradient and also the resting potential (equation 2.8); fewer or no action potentials may be produced. An *increase* in extracellular potassium lowers the concentration gradient and also the resting potential; making the cell more excitable, it may even fire spontaneously. The other important ion in action potential production is the sodium ion and, as we have seen, local anaesthetics block its transmission through the membrane. Alterations to the extracellular sodium ion concentration alter the size of the action potential and its rate of rise; the latter increases with increasing sodium concentration and decreases with decreasing concentration. Action potentials may be blocked if the sodium concentration falls too low.

The final avenue of attack on an axon is through destruction of the myelin sheath. The most common and important example of this is in **multiple sclerosis**. The cause of this condition is unknown, but it is believed to involve both environmental and genetic factors. The condition is characterized by localized areas of demyelimation in the white matter of the central nervous system. The lengths of demyelination may range from a few millimetres to several centimetres. In some instances the demyelination may be demonstrated by determining nerve conduction velocities; demyelination leads to reduced conduction velocity.

PROBLEMS

1. The equilibrium membrane potential for a Ca^{2+} ion at 37 °C is measured as 129 mV. Calculate the concentration ratio $[Ca^{2+}]_o/[Ca^{2+}]_i$ which is required to produce this potential.
2. In establishing the equilibrium membrane potential, potassium ions must be transferred from inside the neuron to outside. This process actually changes the potassium ion concentration by a negligible amount. Calculate the number of moles of potassium ions which are transferred, per unit area, to establish an equilibrium membrane potential of −98 mV.
3. A cell membrane, permeable to both potassium and sodium ions, is in the resting state at a temperature of 30 °C. If $[K^+]_i = 38 \times 10^{25}\ m^{-3}$, $[K^+]_o = 1 \times 10^{25}\ m^{-3}$, the potassium ion permeability $p_K = 6 \times 10^{-9}\ ms^{-1}$ and $V_r = 80$ mV, calculate the power density which must be provided by the sodium–potassium pump in order to maintain the internal K^+ and Na^+ concentrations.
4. Myelination of an axon greatly increases the rate at which an action potential can be transmitted. The extent of myelination varies with

axon diameter. A myelinated axon has an external diameter d_o and an internal diameter d_i in the myelinated region. Use the cable model of an axon to show that the speed of action potential transmission θ is a maximum if the myelin sheath thickness is 32% of d_i. Assume that the capacitance per unit length is proportional to $1/\log_e(d_o/d_i)$. *Hint*: assume d_o is constant.

5. A small neuron may have 10^6 sodium–potassium pumps, each able to work at a maximum rate corresponding to the net transport of 60 Na^+ ions per second out of the neuron. The energy for this process is obtained by hydrolysing ATP to ADP yielding 29.3 kJ mol^{-1}. What is the maximum power that this neuron can provide?

3

Neuromagnetic signals

3.1 INTRODUCTION

At the beginning of the twentieth century the English physiologist, Charles Sherrington, described the brain as 'an enchanted loom where millions of flashing shuttles weave a dissolving pattern, always a meaningful pattern, but never an abiding one'. This description well conveys the mystery and complexity of brain function, so that the prospect of a new, and totally non-invasive, way of determining brain function is potentially very important. Although modern imaging techniques, such as computerized X-ray tomography (CT), positron emission tomography (PET) or magnetic resonance imaging (MRI), all provide high resolution 'windows' through which details of the anatomical structure or metabolism of the human brain may be obtained *in vivo*, none of these techniques are linked in any way to the electrical activity of the brain. The brain's, or for that matter the body's, innate electrical activity must always be accompanied by matching magnetic activity so that in principle if we can monitor this activity we are presented with not only another 'window' onto the 'enchanted loom', but also a 'window' onto the more general neuroelectrical scene.

The first attempts to detect the magnetic fields associated with neuroelectrical activity were not in fact connected with the brain and were made in 1963 by G. Baule and R. McFee. They used a two-coil magnetometer, in a gradiometer arrangement, where each coil comprised two million turns of wire on a dumb-bell shaped core of ferrite, approximately 300 mm long. They applied their equipment to the measurement of the magnetic fields associated with the heart, magnetocardiography (MCG), and recorded signals from their coils with a magnitude of some 30 μV, as a consequence of the changing magnetic

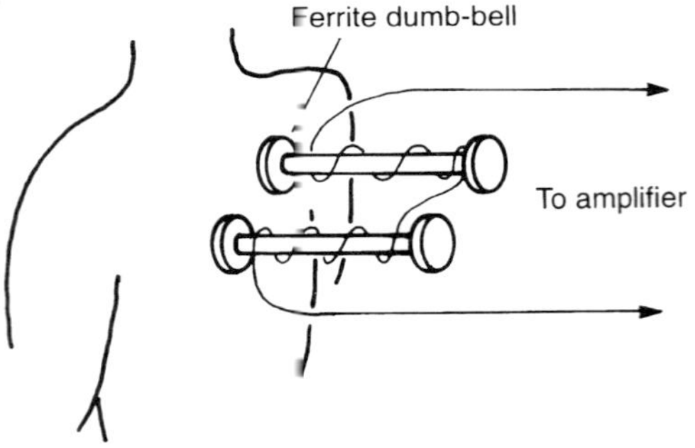

Figure 3.1 Outline of the experimental arrangement used in the earliest measurement of the magnetic field associated with nerve activity. Each ferrite dumb-bell was wound with two million turns of wire. (Adapted from Baule and McFee.)

field resulting from heart action. Figure 3.1 shows, in outline, the experimental arrangement used.

In 1968 the first attempt to measure the magnetic activity associated with the brain, magnetoencephalography (MEG), was made by D. Cohen using an induction coil and signal averaging techniques. Subsequently (since 1970) all attempts at measuring neuromagnetic signals have been made with equipment using a 'superconducting quantum interference device' (SQUID); the first commercially built neuromagnetometers became available in 1992.

The difficulties inherent in neuromagnetic measurements are made clear by Figure 3.2, which shows the neuromagnetic field strength on a spectrum of other common magnetic field sources. Although the total current flowing in a nerve bundle (possibly comprising thousands of individual nerve fibres) may be as much as 1 μA, the fall-off with distance of the magnetic field ensures that the field strength above the nerve bundle, measured outside the body, will typically be only 50–500 fT (roughly 10^8 times weaker than the geomagnetic field). This very low value of magnetic field places severe restrictions on the type of equipment which may be employed for neuromagnetic measurements; generally speaking conventional magnetometers will not be sensitive enough unless the extreme measures of Baule and McFee are employed. The SQUID, however, is the most sensitive magnetic flux detector known today and, at the time of writing, has a limiting noise level of around 2 fT $Hz^{-1/2}$. In consequence, all modern biomagnetic or neuromagnetic measurements are made using SQUID-based magnetometers or gradiometers. In some instrumentation the magnetic field

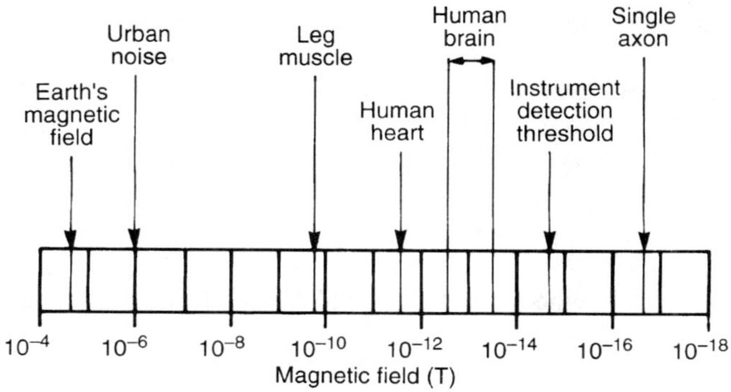

Figure 3.2 'Spectrum' of magnetic fields associated with biological signals. Those produced by background noise and the earth's magnetic field are presented for contrast. The instrumental detection threshold is also shown.

may be measured at many sites simultaneously; equipment measuring at as many as 61 sites has been described.

In order to understand the basis for neuromagnetic measurements we must return to a consideration of the action potential described in Chapter 2, only now from the point of view of the magnetic fields associated with it. We shall then proceed to outline the principles embodied in a SQUID, before looking at the characteristics of MEG, magnetomyography (MMG) and MCG. Finally, a brief examination of the new information gathered by the application of these techniques will be presented. It is useful, however, to begin by reviewing the basic principles of magnetic field production.

3.2 THE MAGNETIC EFFECTS OF CURRENTS

A magnet or a current-carrying coil are said to produce a magnetic field or magnetic induction $\boldsymbol{B}$ which exerts a force on other magnets or coils. $\boldsymbol{B}$ is a vector quantity and lines of $\boldsymbol{B}$ can be drawn whose direction at any point is that of $\boldsymbol{B}$, in the same fashion as lines of force are drawn in an electric field. The strength of $\boldsymbol{B}$ is shown by making the number of lines per unit area normal to $\boldsymbol{B}$ numerically equal to the value of $\boldsymbol{B}$.

The incremental force $\mathrm{d}\boldsymbol{F}$ exerted on an element of wire $\mathrm{d}\boldsymbol{s}$ carrying a current i, at a point where the magnetic field is $\boldsymbol{B}$, can be expressed in the form

$$\mathrm{d}\boldsymbol{F} = i\,\mathrm{d}\boldsymbol{s} \times \boldsymbol{B} \tag{3.1}$$

This equation defines the unit of magnetic field as that which exerts unit force on unit length of wire carrying unit current. The unit of $\boldsymbol{B}$ is thus N A^{-1} m^{-1}, or more customarily tesla, T. We can write the magnetic field increment d$\boldsymbol{B}$ produced at a distance $\boldsymbol{r}$ by an element of conductor d$\boldsymbol{s}$ carrying a current i, using the equation

$$\mathrm{d}\boldsymbol{B} = \left(\frac{\mu_0}{4\pi r^3}\right) i \, \mathrm{d}\boldsymbol{s} \times \boldsymbol{r} \tag{3.2}$$

Further we define the magnetizing field $\boldsymbol{H}$ such that, in free space,

$$\boldsymbol{B} = \mu_0 \boldsymbol{H} \tag{3.3}$$

where μ_0 is a constant known as the permeability of free space and is defined as $4\pi \times 10^{-7}$ henrys per metre (H m^{-1}) or N A^{-2}; the units of $\boldsymbol{H}$ are A m^{-1}. In consequence we may rewrite equation 3.2

$$\mathrm{d}\boldsymbol{H} = \frac{i}{4\pi r^3} \mathrm{d}\boldsymbol{s} \times \boldsymbol{r} \tag{3.4}$$

Equation 3.2 is generally known as the Biot–Savart law and its geometry is sketched in Figure 3.3(a).

Since the vector $\boldsymbol{r}$ has the dimensions of length, we can see that $\boldsymbol{H}$ falls off with the inverse square of the distance, for distances much greater than the length of the element of conductor. If we have an element of magnitude ds metres carrying a current of 1 A, then the magnetizing field $\boldsymbol{H}$ at a perpendicular distance of 1 m will be d$s/4\pi$ A m^{-1}, under the same approximation. The associated magnetic field in free space is μ_0 times greater

More conveniently we may think of a plane circular coil comprising just one turn. In this case $\boldsymbol{r}$ is the radius of the coil and the element d$\boldsymbol{s}$ of the coil is always perpendicular to $\boldsymbol{r}$. Integration of equation 3.4 gives a value for the magnetizing field at the centre of the coil as $H = i/2r$. Hence, unit magnetizing field strength is set up at the centre of a single turn coil carrying a current of 1 A and having a diameter of 1 m. The field is normal to the plane of the coil, and directed according to the right-hand rule (Figure 3.3(b)).

A final point concerns the concept of magnetic flux. If we consider an element of area δA at an angle Φ to a magnetic field, then the component of magnetic field $\boldsymbol{B}$ perpendicular to this area is $B \sin \Phi$ and the magnetic flux δN passing through the area δA is $B \sin \Phi \, \delta A$: magnetic flux has the units N m A^{-1} or J A^{-1} or T m^2 or H A, otherwise known as the weber (Wb). Consequently we may refer to $\boldsymbol{B}$ not only as the magnetic field, but also as the flux density; the usual symbol for the magnetic flux is Φ.

A sign is allocated to Φ using the following rules. The self-flux of a current-carrying circuit, across the area bounded by the circuit, is always considered positive. Any other magnetic flux over this area is considered positive if it has the same direction as the self-flux.

3.3 BIOMAGNETIC FIELD SOURCES

The propagating action potential is, of course, just one manifestation of the currents flowing into, and out of, the axon as the processes of depolarization and repolarization take place. We may think of these currents as flowing in loops which intersect the axon sheath, rather in the manner shown in Figure 3.4(a) for the depolarization front. From any point external to the axon sheath these current loops will appear as current dipoles accompanied always by an opposing current dipole due to the repolarization process; the whole constitutes a current quadrupole, as pointed out in section 2.7 and shown more completely in Figure 2.10(c). Associated with each set of current loops there will be

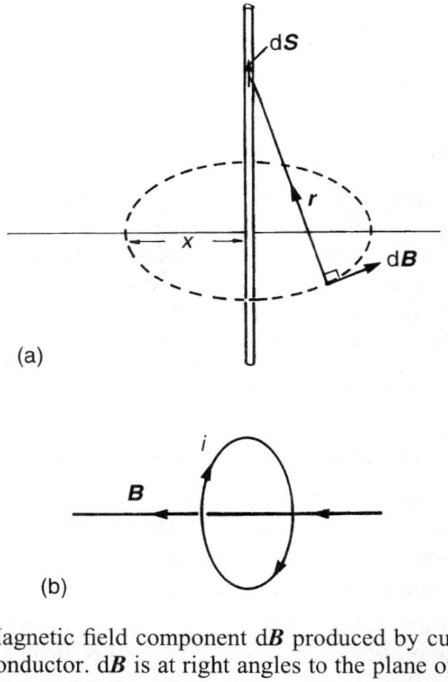

Figure 3.3 (a) Magnetic field component d***B*** produced by current *i* flowing in element d*s* of a conductor. d***B*** is at right angles to the plane of the diagram and directed into the page. (b) Direction of the magnetic field associated with a single turn current-carrying coil. This obeys the right-hand rule.

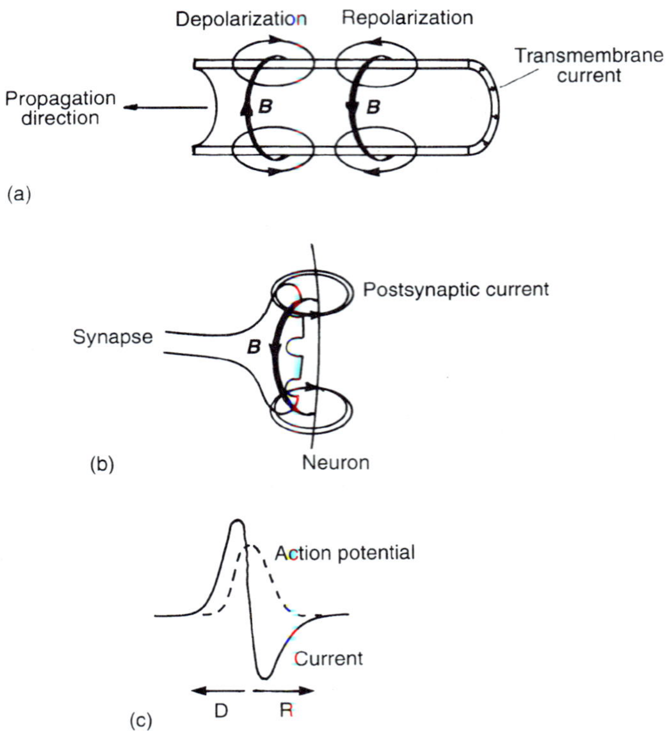

Figure 3.4 (a) Simplified view of the depolarization and repolarization current flow paths in an axon, together with their associated magnetic field loops ***B***. (b) Current flow loops at a synapse, as a consequence of the postsynaptic potential (PSP), together with the corresponding induction loop. The situation in (a) represents a current quadrupole while that in (b) is a current dipole; the latter has a less marked decrease in ***B*** with separation distance. (c) Action potential and dual-polarity action current, with accompanying depolarization current dipole D and repolarization current dipole R.

a corresponding magnetic field, whose direction is indicated by the broad arrows in Figure 3.4(a).

The model we adopt is based upon an axon in an infinite medium with homogeneous conductivity. We divide the current pattern into two parts, the primary or intracellular current and the volume or extracellular currents. It transpires that the net contribution to the field at any point in space from volume currents is zero. This results from the symmetry of the paths of the volume currents in the medium, the contribution from one region of space being cancelled by the contribution from another region. The waveform describing the axial intracellular current closely resembles the time derivative (slope) of the transmembrane action potential waveform. Thus the single-polarity action potential will produce a dual-polarity action current, such as is shown in Figure 3.4(c), and this will be accompanied by a corresponding magnetic field. The two polarities of the action current may be represented by opposing current dipoles (Figure 3.4(c)). The current dipole $\boldsymbol{D}$ is a vector quantity, and for a current i flowing in a dipole of length $\boldsymbol{s}$ is just $\boldsymbol{D} = i\boldsymbol{s}$, with units of A m. If we assume an axon of diameter d and length l having a conductivity σ, then the resistance $R = 4l/\pi d^2\sigma$ and from Ohm's law $i = V/R$, so that the current dipole moment D is given by

$$D = \pi d^2 \sigma V/4 \tag{3.5}$$

The resultant magnetising field is obtained by integrating the Biot–Savart law in the form

$$\mathrm{d}\boldsymbol{H} = \frac{1}{4\pi r^3}\,\mathrm{d}\boldsymbol{D} \times \boldsymbol{r} \tag{3.6}$$

Alternatively, the magnetic field may be obtained by integrating

$$\mathrm{d}\boldsymbol{B} = \frac{\mu_0}{4\pi r^3}\,\mathrm{d}\boldsymbol{D} \times \boldsymbol{r} \tag{3.7}$$

We assume that the permeability of biological tissues is equal to that of free space. If we make some reasonable assumptions for the values of the quantities in equation 3.5 namely, that $d \simeq 4\ \mu\mathrm{m}$, $\sigma \simeq 0.25\ \Omega^{-1}\ \mathrm{m}^{-1}$ and $V \simeq 10$ mV, then we estimate that $D \simeq 3 \times 10^{-14}$ A m. We are now in a position to estimate the magnetic field at a point some distance from the axon, due to the presence of the current dipole. For a distance of say 10 mm, $\boldsymbol{B}$ is calculated to be 3.2×10^{-2} fT and far too small to be detected by *any* magnetic sensor.

This analysis is an oversimplification of the actual situation since, if we were able to measure the magnetic field very close to the axon, that is at a distance much less than the length of the dipole, the situation would be akin to current flow in a conductor of infinite extent and the magnetic field would fall off as $1/x$, where x is the perpendicular distance from the current flow. At a greater separation distance, such as has been assumed for the above calculation, the finite length of the dipole becomes significant and the magnetic field falls off as $1/x^2$. Even further away from the axon the effect of the opposing current dipole becomes important and the magnetic field falls off even more rapidly with distance, as $1/x^3$. Figure 3.5 shows a plot of the dependence of $\boldsymbol{B}$

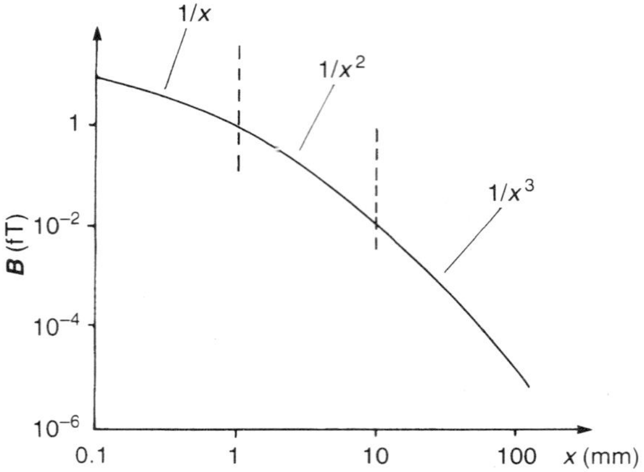

Figure 3.5 Variation of magnetic field ***B*** with distance, for an action potential in an axon, showing how the dependence on separation distance x changes. The magnetic field associated with a *single* neuron is far too small to be detectable. In practice tens of thousands of neurons fire with temporal coherence.

on x over a range of values for x. The reader who wishes for a more extended treatment of magnetic topics will find one in *Basic Electromagnetism* by E.R. Dobbs, a companion volume in this series.

On the basis of the above description of the magnetic effects of currents, we see that the transmembrane current, which gives rise to the action potential in the first place, can play no appreciable part in producing the external magnetic field, simply on symmetry arguments; only the intracellular and extracellular currents produce an external magnetic field. In addition, the external magnetic field should be far too weak to be detectable. However, actual measurements outside the scalp, in response say to auditory stimulation, yield magnetic fields of 100 fT (or more). This implies that a very large number of neurons (>10 000) act with temporal coherence to generate the magnetic field, adding their individual responses to form the overall signal. Individual action potentials last of the order of 1 ms, so they must occur in adjacent axons within about 1 ms.

The postsynaptic currents (Figure 3.4(b)) are not accompanied by an opposing repolarization current dipole and last for about 10 ms so that there is a greater opportunity for the individual magnetic fields to sum; the absence of an opposing current dipole ensures that the magnetic signal drops off with a $1/x^2$ dependence rather than the more severe $1/x^3$ dependence which would be seen for the action potential. This means that postsynaptic current dipoles should be much more readily recorded. We turn now to a consideration of the manner in which these very weak magnetic signals are detected, by examining the basis for the operation of a SQUID.

3.4 NEUROMAGNETIC MEASUREMENTS

All biomagnetic or neuromagnetic measurements are conducted with superconducting systems. Superconductivity itself was discovered by Kamerlingh Onnes in 1911 and, prior to 1986, was confined to operating temperatures around that of liquid helium, namely 4.2 K. In 1986 a new class of superconducting materials was discovered by J.G. Bednorz and K.A. Müller which permitted operation at liquid nitrogen temperature, 77 K. The first example of the application of these new materials in neuromagnetic measurements (actually magnetocardiography) was described in 1993 by Y.A.Tavrin and co-workers.

The term superconductivity refers to a specific physical property, namely that the electrical resistance of a material is zero below a critical temperature T_c. This perfect conductivity is obtained, however, only below a critical frequency, a critical field and a critical current density. In the superconducting state, a magnetic flux does not penetrate the material and it exhibits perfect diamagnetism, known as the Meissner effect. The superconducting state is important, not only as the basis for neuromagnetic measurements, but also as the means by which high magnetic fields are attained for use in magnetic resonance imaging (MRI), the subject of Chapter 8. We will therefore find it useful to consider the subject in some detail.

According to the microscopic theory of J. Bardeen, L. Cooper and J.R. Schrieffer (the BCS theory), the zero resistance of superconductors arises from the existence of Cooper pairs of electrons. The electrons of a Cooper pair have a weak attraction for each other through interaction with the surrounding lattice of positive ions. This may be thought of as a two-step process in which the passage of one electron distorts the lattice which, in turn, attracts the second electron. The diameter of a Cooper pair is large on an atomic scale (a few hundred nanometres), far larger than the diameter of any atom. The Cooper pairs in a superconductor all have the same drift speed and their formation causes the appearance of a gap in the electron energy distribution. This energy gap prevents Cooper pairs from moving to higher energy levels. The states that occupied the energies where the gap has appeared are shifted away from the Fermi energy to make a sharp increase in the density of states just above and just below the energy gap. As a result of this energy gap Cooper pairs are unable to scatter to higher energy levels, and the current is not limited by scattering as it would be in a normal conductor. We can say that the energy gap will be zero above the critical temperature and constant below it. The Cooper pairs are in a macroscopic quantum state which can be described by a single wavefunction. In a closed loop superconductor the requirement that this wavefunction be single-valued leads to a further restriction, which is that the magnetic flux enclosed by the loop can only take on values given by

$$\Phi = n\Phi_0 \qquad (n = 0, 1, 2, 3, \ldots) \tag{3.8}$$

where Φ_0, the flux quantum, equals $h/2e$ (Planck's constant divided by the Cooper pair charge) and has the value 2.07×10^{-15} Wb.

The quantization of magnetic flux is illustrated in Figure 3.6 for a closed superconducting loop, with an open loop for comparison. The magnetic flux inside the superconducting loops Φ_i is plotted as a function of the externally applied flux Φ_e, both normalized to the flux quantum Φ_0. The closed superconducting ring exhibits flux values

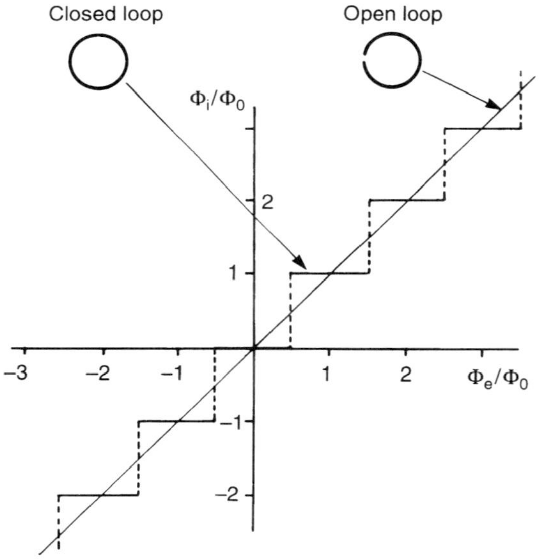

Figure 3.6 Comparison of the relationship between internal flux Φ_i and external flux Φ_e, both normalized to the flux quantum Φ_0, for an open and closed loop superconductor, showing the flux quantization which occurs for the closed loop superconductor.

which are quantized and appear only as multiples of the flux quantum Φ_0, the integer steps of Figure 3.6. For the open loop $\Phi_i \simeq \Phi_e$.

3.5* THE FLUX TRANSFORMER

While a loop or ring remains superconducting, the loop not only encloses an integer number of magnetic flux quanta $n\Phi_0$, but it will also resist any attempt to change the internal magnetic flux Φ_i, that is

$$\delta\Phi_i = 0 \tag{3.9}$$

Equation 3.9 embodies the principle of the flux transformer upon which superconducting magnetometers and gradiometers are based. A magnetic flux transformer contains two or more coils, normally coupled in series (Figure 3.7). A change of magnetic flux through one of the coils causes a change in magnetic flux in the other coil, or coils.

For the sake of clarity we will consider initially a two-coil system which is typical of a conventional magnetometer (Figure 3.7). The coil L_1 is the main sensing coil and any external magnetic field should act upon it. The coil L_s is the signal coil which is coupled to the magnetic flux sensor, in this instance, the SQUID. Application of equation 3.9 to this coil combination gives

$$\Delta\Phi_1 + \Delta\Phi_s = 0 \tag{3.10}$$

Hence the change in magnetic flux $\Delta\Phi_1$ through the sensing coil L_1, is transferred to the signal coil L_s and the optimum configuration is achieved when $L_1 = L_s$ (neglecting the influence of the sensing

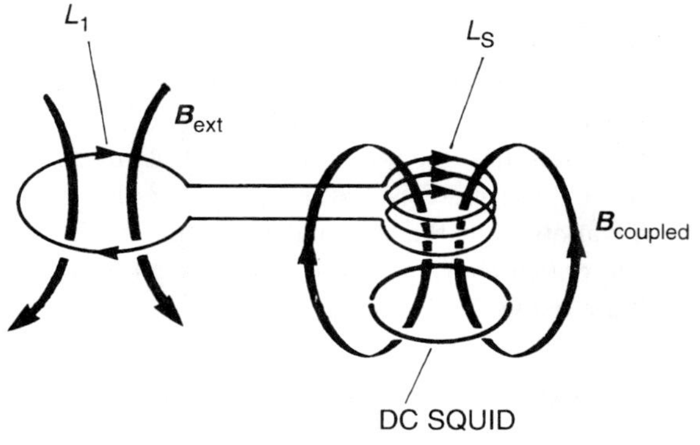

Figure 3.7 Two-coil flux transformer comprising a sensing coil L_1 coupled to a signal coil L_s. The signal coil interacts with the DC SQUID loop, immediately beneath it, transferring flux from the sensing coil.

element). In practical terms the flux transfer is caused by a shielding supercurrent I_s maintaining the magnetic flux constant inside the superconducting loop, where the loop here consists of the coils L_1 and L_s, together with any interconnecting wires; the inductance of these wires is neglected for our purposes. Any external magnetic flux $\Delta\Phi_e$ introduced at the sensing coil is compensated for by the change in the shielding current i_s, so that

$$\Delta\Phi_e = (L_1 + L_s)\,\Delta i_s \tag{3.11}$$

When $L_1 = L_s$ about half the magnetic flux introduced into the sensing coil is transferred to the SQUID sensing element attached to the coil L_s. The simple configuration described above for a magnetometer is seldom used in practice; a more complex structure known as a gradiometer is preferred, since it confers more effective external magnetic noise rejection.

Gradiometers exist in first-, second- and third-order configurations. The first-order gradiometer configuration is shown in Figure 3.8(a). It comprises two sensing coils L_1 and L_2 together with a signal coil L_s (not shown) which would, in each case, be in series with the conductor linking the top and bottom coils. The second-order gradiometer has three sensing coils L_1, L_2 and L_3, while the third-order gradiometer has four sensing coils (Figure 3.8(c)). In each case the coils are uniaxial and the gradiometers are asymmetric, that is the diameter of the main sensing coil L_1 is smaller than that of the compensating coils L_2, L_3 and L_4. Typical diameters of the main sensing coils are between 10 and 30 mm with a coil separation distance, or baseline, in the range of 30–150 mm. The coils are made from niobium wire wound on precision-ground quartz formers. It should be noted that the coils are wound with opposing sense so that all the gradiometers are insensitive to a uniform, external magnetic field. In addition, the second-order gradiometer configuration has the property that it is insensitive to the external magnetic field gradient while the third-order gradiometer is insensitive to variation in the external magnetic field gradient. The gradiometers described above are axial; it is now more common to find a planar structure being used (Figure 3.9).

3.6* THE JOSEPHSON EFFECT

The Josephson effect and the Josephson junction form the basis of a SQUID. In 1962 B.C. Josephson derived a relationship connecting the supercurrent through a weak link, a thin oxide layer separating two

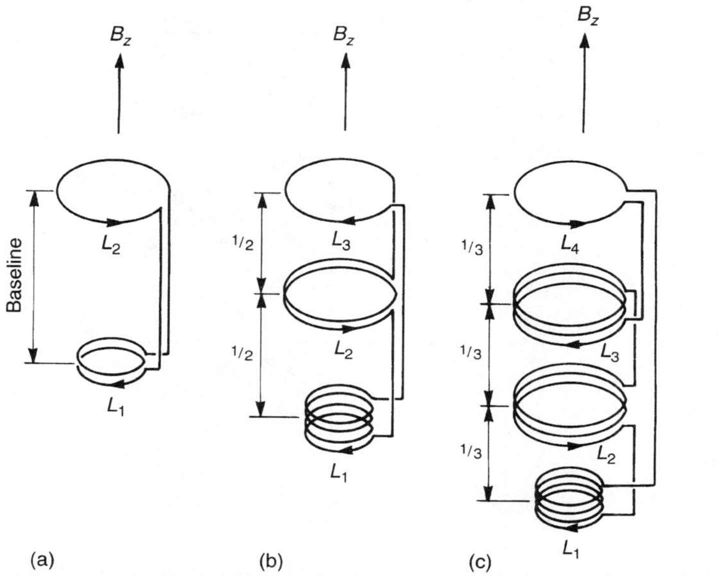

Figure 3.8 Axial gradiometer configurations for biomagnetic measurements: (a) first-order gradiometer; (b) second-order gradiometer; (c) third-order gradiometer. In each case the signal coil which couples to the SQUID, as in Figure 3.6, has been omitted, but would be connected into the return conductor joining the top and bottom coils. The positioning of coils in the second- and third-order gradiometer is shown as fractions of the first-order gradiometer baseline.

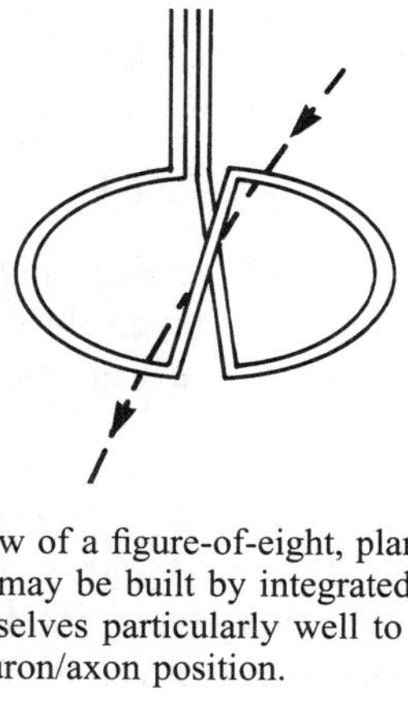

Figure 3.9 Simplified view of a figure-of-eight, planar gradiometer configuration. Planar gradiometers may be built by integrated circuit, photolithographic techniques and lend themselves particularly well to MEG measurements. The dashed line shows the neuron/axon position.

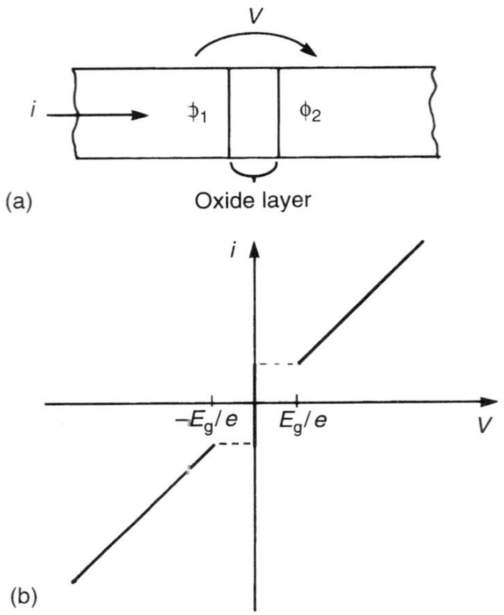

Figure 3.10 (a) An oxide weak link joining superconductors. (b) i–V characteristic for the oxide weak link.

superconductors (Figure 3.10(a)). Josephson predicted that a Cooper pair could tunnel between the superconductors with the same probability as a single electron, rather than with the square of the single electron tunnelling probability. The supercurrent (tunnelling of Cooper pairs) through the weak link is a periodic function of the phase difference $\Delta\phi$ between the phases of the time dependent wavefunctions representing the superconducting electron pairs in the two superconductors, thus

$$i_s = i_c \sin \Delta\phi \tag{3.12}$$

where i_c is the critical current of the weak link (the maximum supercurrent through the weak link). For a typical weak link with a resistance of approximately 1 Ω, i_c is <1 mA and in practical devices may range between 1 μA and 1 mA.

According to equation 3.12, a supercurrent will flow through the weak link even in the absence of a voltage difference across it; this is the DC Josephson effect (Figure 3.10(b)). However a new effect appears in the presence of a voltage V across the junction. The relative energy difference of Cooper pairs between the two sides of the junction will then be $2eV$, and since the rate of change of phase of a wavefunction varies as its energy divided by $\hbar$, we may write

$$\frac{\partial \Delta\Phi}{\partial t} = \frac{2eV}{\hbar} \tag{3.13}$$

If we integrate equation 3.13 and put $\Delta\phi = \delta_0$ at $t = 0$, we can rewrite equation 3.12 as

$$i_s = i_c \sin\left(\delta_0 - \frac{2eVt}{\hbar}\right) \tag{3.14}$$

Equation 3.14 tells us that with a voltage V across the junction, the supercurrent oscillates at a frequency ν given by

$$\nu = \frac{2eV}{\hbar} = \frac{2\pi V}{\Phi_0} \tag{3.15}$$

The phenomenon that a voltage across a Josephson junction produces current oscillations is known as the AC Josephson effect. Equations 3.12 and 3.14 represent the basic Josephson relations. We may summarise their implications as follows:

1. A finite supercurrent, of maximum value i_c, can flow through the insulating junction, with zero voltage drop, making the junction behave as if it were not there. This behaviour is found to involve lower values of the critical current and field than before and we have so-called 'weak superconductivity'; this is the DC Josephson effect.
2. A constant applied voltage will lead to an alternating current of frequency ν where $\nu/V = 486.6$ MHz μV^{-1}; this is the AC Josephson effect.

3.7* RF AND DC SQUIDS

The SQUID exists in two forms, the RF SQUID and the DC SQUID. The RF SQUID is essentially a superconducting ring with one weak link, while the DC SQUID has two weak links. In the superconducting state, the current i_s contributes a shielding term to the flux inside the ring

$$\Phi_i = \Phi_e + L_{sq} i_s \tag{3.16}$$

where L_{sq} is the inductance of the ring. The dependence of Φ_i on Φ_e at low frequencies now takes the form shown in Figure 3.6. The RF

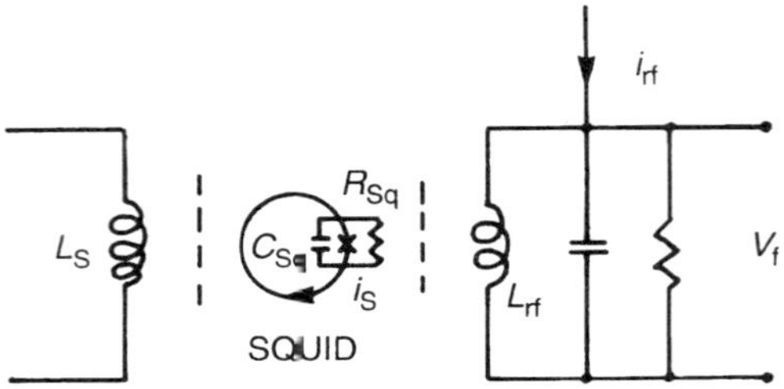

Figure 3.11 The RF SQUID is coupled to a signal coil L_s (part of a flux transformer) and to the RF resonant circuit L_{rf}. The RF circuit introduces an alternating magnetic flux, shifting the working point back and forth. The consequent flux jumps cause hysteretic losses in the RF tuned circuit.

SQUID is coupled inductively to the signal coil L_s and to the inductance of an RF resonant circuit L_{rf} shown in Figure 3.11. The flux transformer couples part of the slowly varying external magnetic flux, affecting the main sensing coil L_1, through to the signal coil L_s and thence to the SQUID. This shifts the working point from A, say, in Figure 3.12, to the range A* to A**, as the RF circuit introduces an alternating flux Φ_c. When the external flux increases, the internal flux Φ_i changes value, represented by the path B to C. However, when the external flux decreases, Φ_i follows the path from C to D before experiencing a flux jump from D to E. In other words the flux Φ_i follows a hysteretic path BCDEB, causing a hysteretic loss in the resonant circuit. If the working point in Figure 3.11 were at the point P, then a change in magnetic flux introduced by the RF circuit would not produce a corresponding hysteretic path, unless the amplitude of the RF change were made very large. To summarize, the hysteresis enclosed depends on the RF amplitude as well as the working point. This analysis ignores the inherent capacitance C_{sq} and resistance R_{sq} of the Josephson junction which contribute additional terms to equation 3.12. Along a closed path inside the superconductor, including the weak link, we have the relationship

$$2\pi \frac{\Phi_i}{\Phi_0} = \Delta\phi = 2n\pi \qquad \text{with } n = 0, \pm 1, \pm 2, \ldots \tag{3.17}$$

The periodicity of the curve in Figure 3.12 is that of the flux quantum Φ_0, as a consequence the output voltage V_f of the RF circuit in the RF SQUID is also periodic in Φ_0. The form of this voltage is sketched in Figure 3.13 and has an approximately triangular shape. The ratio

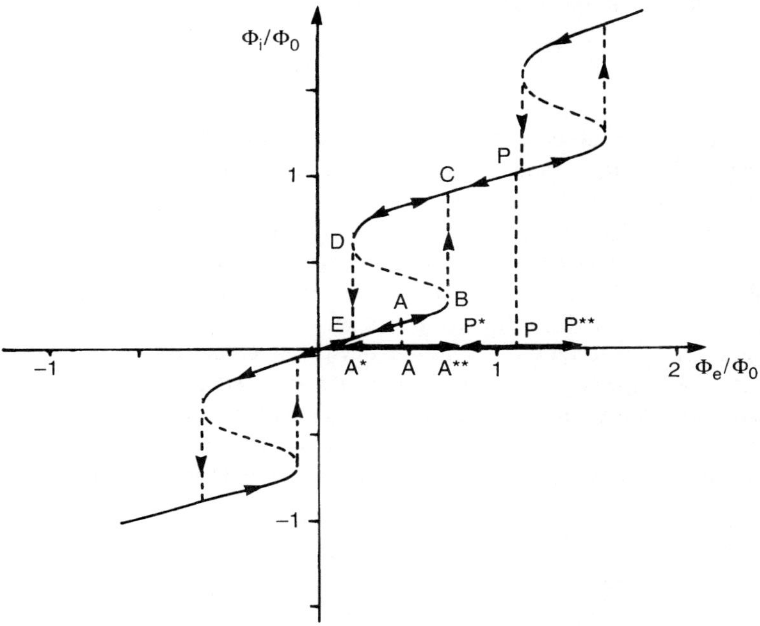

Figure 3.12 The magnetic flux Φ_i inside a superconducting loop with one weak link (RF SQUID), as a function of the externally applied flux Φ_e, with the fluxes normalized to the flux quantum Φ_0.

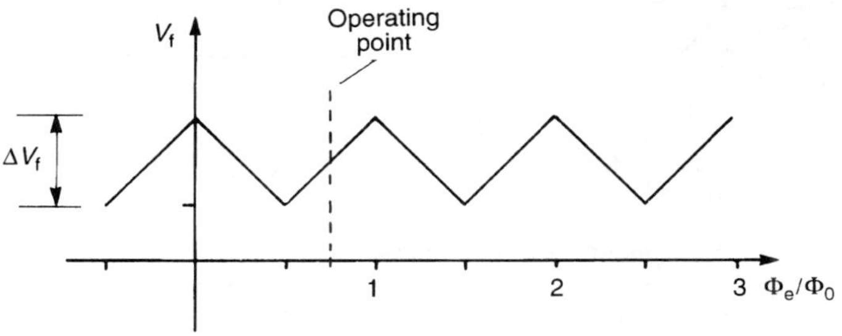

Figure 3.13 Approximate form of the average output voltage $\langle V_f \rangle$ of an RF SQUID amplifier as a function of the externally applied magnetic flux Φ_e, normalized to the flux quantum Φ_0. The dashed line indicates a typical operating bias.

between the time averaged output voltage and the external magnetic flux is the SQUID sensitivity S, and is given by the equation

$$S \simeq \frac{\omega_0}{k}\sqrt{\frac{L_{\mathrm{rf}}}{L_{\mathrm{sq}}}} \tag{3.18}$$

where k a dimensionless coupling coefficient between the RF coil and the SQUID loop, and ω_0 is the RF angular frequency. S has the dimensions of V Wb^{-1} or frequency.

The DC SQUID operates with two weak links in the superconducting ring, and a constant current i_0, from an external source, is driven through it. The average value of the voltage across the SQUID is measured; the circuit arrangement is shown in Figure 3.14. The equation for the current through the device is obtained by modifying equation 3.14, neglecting junction capacitance and assuming two identical weak links. In the presence of two junctions, however, the current depends on the magnetic flux through the loop formed by the two junctions, because of quantum mechanical interference between the two

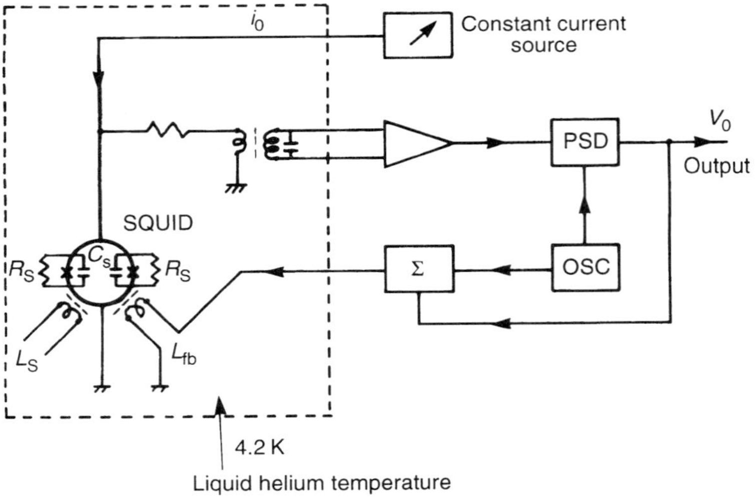

Figure 3.14 Simplified circuit arrangement for a DC SQUID. The inductance L_s (part of a flux transformer) couples the signal into a DC SQUID with two weak links, fed from a constant current source. The resulting output signal undergoes phase-sensitive detection and is fed back via the inductance L_{fb}. The DC SQUID functions as a null detector. All circuitry within the dashed line operates at liquid helium temperature.

currents through the separate junctions. The phase difference around a loop which encloses a magnetic flux Φ_i is

$$\phi_2 - \phi_1 = \frac{e\Phi_i}{\hbar} \tag{3.19}$$

The total current in the loop which contains the two junctions is

$$i_s = i_c \sin\left(\delta_0 + \frac{e\Phi_i}{\hbar}\right) + i_c \sin\left(\delta_0 - \frac{e\Phi_i}{\hbar}\right) + \frac{2V}{R_s} \tag{3.20}$$

$$i_s = 2i_c \sin\delta_0 \cos\left(\frac{e\Phi_i}{\hbar}\right) + \frac{2V}{R_s} \tag{3.21}$$

and the voltage is given by

$$V = \frac{R_s}{2}\left[i_s - 2i_c \sin\delta_0 \cos\left(\frac{e\Phi_i}{\hbar}\right)\right] \tag{3.22}$$

$$= \frac{R_s}{2}\left[i_s - 2i_c \sin\delta_0 \cos\left(\pi\frac{\Phi_i}{\Phi_0}\right)\right] \tag{3.23}$$

where R_s is the junction resistance.

In practice the oscillatory behaviour of V is not measured but only the time averaged voltage $\langle V\rangle$ after amplification; the value of the bias current i_0 is used as a parameter, with the strongest signal being produced when $i_0 = 2i_c$. The output is periodic in much the same way as that of the RF SQUID (Figure 3.13). The device is usually operated with a flux bias for which the output voltage is almost linear in applied flux. A typical operating point is indicated in Figure 3.13.

3.8* SQUID MAGNETOMETERS

For the purposes of explanation we will concentrate on the operation of a DC SQUID, since both RF and DC SQUIDs have comparable sensitivities and similar peak-to-peak output voltages (of the order of 10 μV). The DC SQUID, however, can exhibit a lower noise level than an RF SQUID. The function of the SQUID is simply that of a zero point detector, that is, it operates as a flux-to-voltage converter. In use the SQUID output voltage is coupled via a transformer and amplifier to a phase-sensitive detector (PSD) (Figure 3.14), the output from which is proportional to the in-phase and on-frequency component of $\langle V\rangle$. A constant current from a room temperature, constant current, source is passed through the DC SQUID, and the voltage across the device is sensed by the PSD and fed back via the feedback coil L_{fb} to bring the SQUID voltage close to zero. In order to do this the DC feedback signal, which is also the PSD output V_0, is modulated at the PSD reference frequency (generally between 10 and 100 kHz) before being supplied to the feedback coil. All of the coupling coils are operated at 4.2 K, and the coil L_{fb} couples a signal into the SQUID loop sufficient to compensate for the signal introduced by the signal coil L_s, which in turn has transformed flux from the pick-up coil. Any low frequency flux change due to the signal coil is compensated for by the corresponding feedback signal V_0, which is measured.

The practical embodiment of a SQUID will consist of Nb or NbN electrodes separated by artificial tunnel barriers of Al_2O_3 or MgO. The problems associated with wire-wound gradiometers (which are expensive to manufacture) may be avoided by using planar integrated sensors which offer an elegant solution, being compact, having good intrinsic balance and excellent precision. Recent devices have been constructed on silicon (Si) wafers, each wafer containing two SQUIDs with integrated, inductance-matching transformers for easy coupling to the pick-up coils. The coil disposition for a planar gradiometer configuration is shown in Figure 3.15 where we see a planar figure-of-eight detector coil, which couples most strongly to a current dipole directly underneath it or, put another way, the measured signal is strongest immediately above the source. The two independent gradient components obtained at each site give both the magnitude and direction of the dipole. This type of gradiometer is particularly suited to MEG measurements where its compact nature permits the use of many gradiometers arranged to cover the whole head in helmet-like fashion. The typical separation distance between skin and sensing coil is around 10 mm. The diameter of SQUID pick-up coils should be no less than the source-to-coil spacing for optimum sensitivity and this latter distance determines the spatial resolution. Each pair of gradiometers measures the tangential derivatives, $\partial B_z/\partial x$ and $\partial B_z/\partial y$, of the magnetic field B_z perpendicular to the gradiometer. Since the gradiometer has four-fold rotational symmetry the lead fields are orthogonal and strictly independent information is measured by each member of the pair.

As stated earlier, when high reliability and excellent balance are required, the pick-up coils should be made by lithographic techniques applied to thin superconductive films on a flat substrate. It is difficult to fabricate three-dimensional axial gradiometers using thin film pick-up

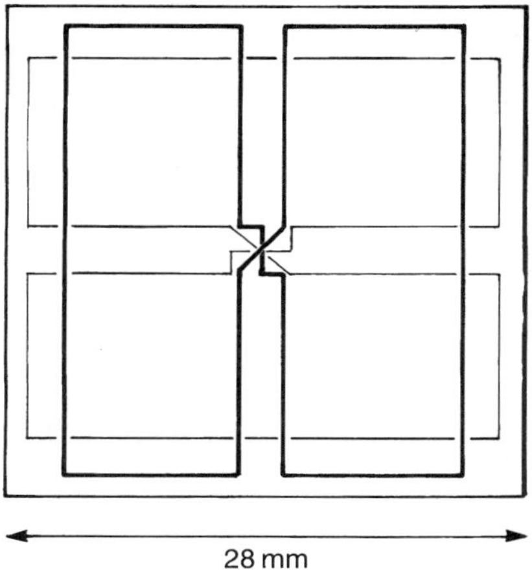

Figure 3.15 Gradiometer chip with two, orthogonal, figure-of-eight, superconducting coils, shown as bold lines and fine lines. The coils are produced photolithographically on a 28 mm × 28 mm silicon wafer. Each pair of radiometers measures the orthogonal tangential derivatives, $\partial B_z/\partial x$ and $\partial B_z/\partial y$ of the magnetic field B_z, normal to the wafer.

coils because a superconducting connection is needed between the sensing coil and the cancellation coil. This problem is avoided in the planar-type gradiometer referred to previously, but it is not easy to make a direct comparison between the mapping data obtained from the two types. In 1994 a way round this problem was proposed by K. Yokosawa and S. Kuriki, who described a new type of thin film axial gradiometer, a direct feedback gradiometer which requires no superconducting connections. This type of gradiometer is suited to SQUID magnetometers for multichannel systems, especially those made of high T_c superconductors.

The conventional axial first-order gradiometer has a pick-up coil consisting of a sensing coil and a cancellation coil connected in series, each wound in the opposite sense. The new type of axial gradiometer, the direct feedback gradiometer, is actually composed of two magnetometers, a cancellation magnetometer (CM) and a sensing magnetometer (SM). Each of these magnetometers consists of a pick-up coil, a SQUID and a flux-locked-loop (FLL) circuit. The pick-up coils of these two magnetometers are parallel and have the same area as in a conventional gradiometer. The pick-up coil of the sensing magnetometer is placed near the signal source, for example the brain; Figure 3.16 shows the circuit arrangement. Although both magnetometers are driven by conventional flux-locked-loop (FLL) circuits, the feedback flux of the cancellation magnetometer is applied to both the CM and sensing magnetometer SM pick-up coils by connecting the feedback coils of the CM and SM in series, using normal conductive wire. The noise signal generated by the sensing magnetometer is cancelled by the feedback flux from the cancellation magnetometer. A signal representing the spatial difference in magnetic field between the two coils is obtained at the output of the sensing magnetometer. Unlike conventional electronic gradiometers, exact matching of the response of the control circuits is unnecessary here and noise cancellation by a feedback current coupled directly to the pick-up coil of the sensing magnetometer ensures that the gradiometer does not distort the magnetic field around itself. The layout of the sensing magnetometer and SQUID is shown in Figure 3.17 while the pick-up coil and gradiometer device is shown in Figure 3.18. These two figures serve to establish the scale of practical devices. The SQUID here is fabricated with a Nb/AlO_x/Nb junction while the

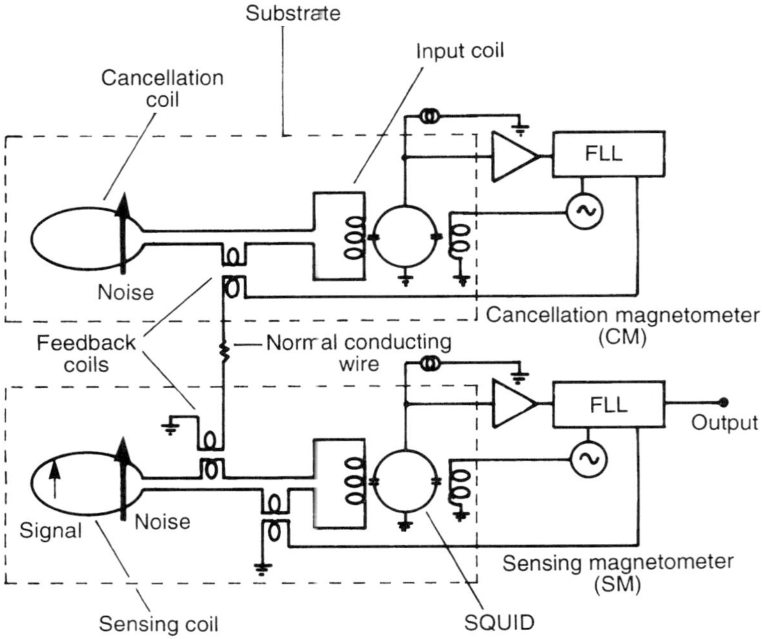

Figure 3.16 Configuration of the direct feedback gradiometer. (Adapted from K. Yokosawa and S. Kuriki, *Rev. Sci. Instrum.* **65** (1994) 3814–18.)

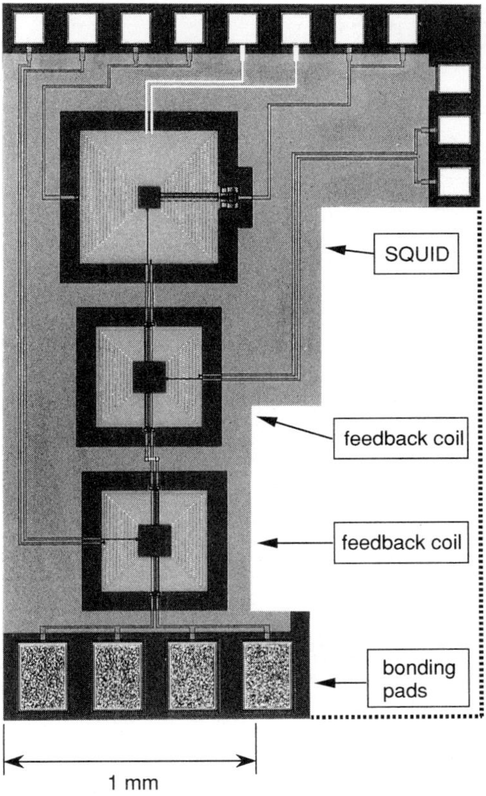

Figure 3.17 Sensing magnetometer (SM) device with a SQUID. (K. Yokosawa and S. Kuriki, *Rev. Sci. Instrum.* **65** (1994) 3814–18, with permission.)

pick-up coil is formed from a Nb film. It is, of course, a low temperature SQUID.

3.9 THE INVERSE PROBLEM

The central problem in interpreting neuromagnetic data, or for that matter neuroelectric data, is that of locating the active area in the brain, or heart, or muscle, from the details of the magnetic field distribution. In other words, how to determine the source currents responsible for the externally measured magnetic field; this is the inverse problem. In 1853 Hermann von Helmholtz showed that this problem has no unique solution for either magnetic or electric fields. In consequence, it is necessary to use source models, such as current dipoles, to interpret the

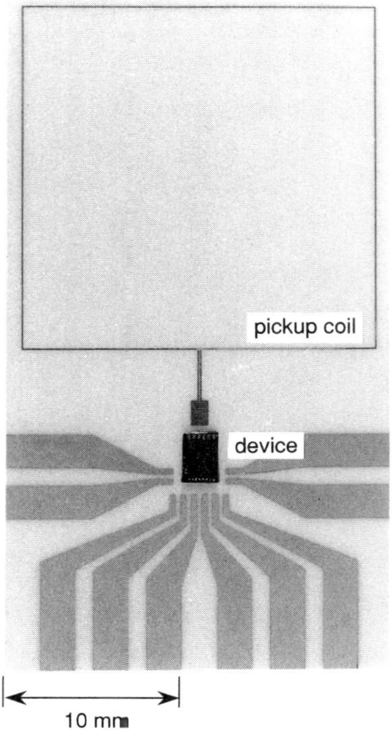

Figure 3.18 A pick-up coil and mounted gradiometer device. (K. Yokosawa and S. Kuriki, *Rev. Sci. Instrum.* **65** (1994) 3814–18, with permission.)

data. If the dipole assumption is correct, then the active area of tissue or muscle can be defined unambiguously. A least-squares, iterative method is used to find the parameters of the dipole so that the sum of the squared differences between the recorded and predicted data are a minimum. The error in dipole location by this method can be as little as 3.5 mm.

For an assumed dipole source, the distance between the minima and maxima of the magnetic field, that is the magnetic field leaving and entering the body, will be related to the depth of the dipole beneath the tissue surface. A deep dipole will yield a large separation between extrema. Since the dipole orientation is orthogonal to a line joining the minimum and maximum of the magnetic field, its direction can be determined by which of the field extrema represent field emerging from the body. Figure 3.19(a) shows the position of an 'active' dipole and

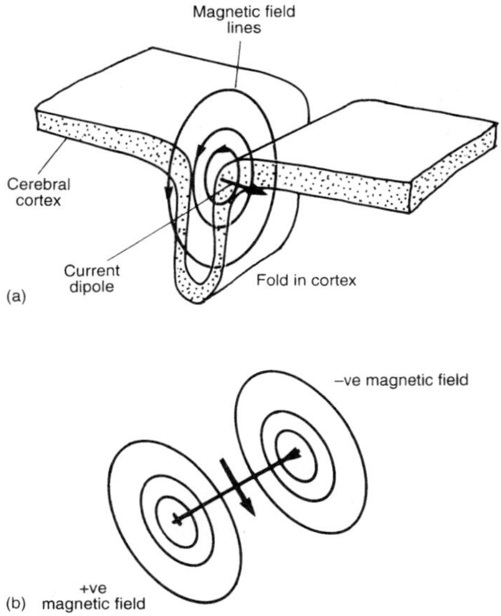

Figure 3.19 (a) Fold in cerebral cortex showing position of dipole which will be active in MEG measurements. (b) Dipolar field pattern showing the field contours. The current dipole (arrow) lies at right angles to the line joining the centres of the field contours. The left-hand isocontours represent magnetic flux entering the head.

Figure 3.19(b) shows the relative orientation of field contours and a source dipole. Clearly, the above procedure is likely to work well only if the field really is due to a dipole; in practice the measured fields may well not be dipolar and/or contain several maxima and minima. For example, before MEG data can be analysed we need some physical model of the brain with which to associate the measured dipoles. An excellent approximation for the conductor geometry in MEG data analysis is a sphere with a conductivity profile $\sigma(r)$. The magnetic field outside such a spherically symmetric conductor has several special properties:

1. radial dipoles and their volume currents cancel each other and do not generate any magnetic field outside the conducting body;
2. the radial component of the magnetic field is produced solely by primary intracellular currents sources;
3. both radial and tangential field components are independent of the conductivity profile.

The fact that radial dipoles cannot be seen in MEG is not a serious drawback since most of the cortical surface is in the fissures (Figure 3.20(a), (b)). Pyramidal cells (Figure 3.21(a)) are orientated perpendicular to the cortex so that the flow of primary current in these fissures is tangential to the surface of the head. Magnetoencephalographic data are plotted as isocontour maps of the magnetic field distribution, as a function of time. A simple field pattern over one hemisphere can often

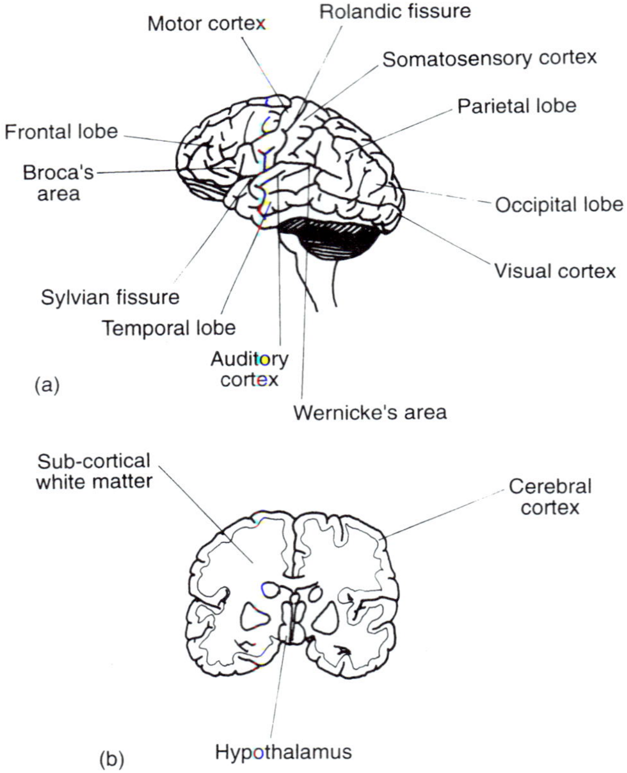

Figure 3.20 (a) Human brain viewed from left-hand side (left hemisphere) showing the creased nature of the cerebral cortex and some of the important structural features. The area just in front of the Rolandic fissure, in the frontal lobe, is the motor cortex associated with muscular activity. Behind this fissure, in the parietal lobe, is the primary somatosensory cortex which deals with the reception and analysis of information from the skin, joints and muscles. The primary visual cortex is in the occipital lobe at the back of the head. (b) Coronal section through the cerebral hemispheres. The outer band is the heavily creased cerebral cortex (grey matter).

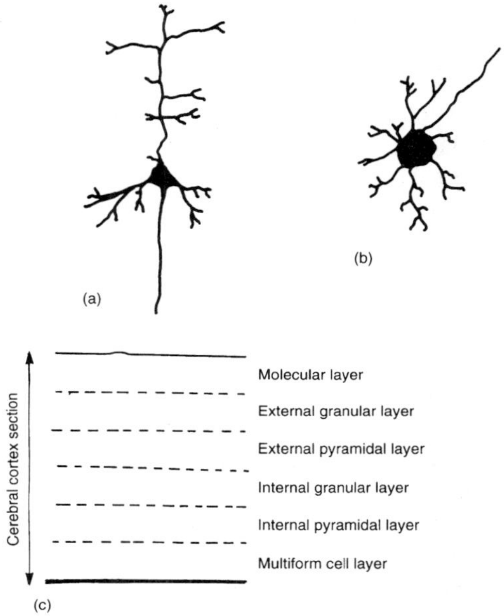

Figure 3.21 Cells of the cerebral cortex and the six-layer structure of the cortex. (a) Pyramidal cell; (b) stellate or granule cell; (c) disposition of cells within the cerebral cortex; neurons in each layer are interconnected both horizontally and vertically. The molecular layer is principally dendrites of pyramidal cells, while the multiform cell layer comprises various cell types.

be accounted for by a single current dipole, the parameters of which are found by a least-squares fit. The equivalent source location may be determined with millisecond time resolution.

The electric and magnetic fields produced by cardiac muscle are, in many respects, easier to understand than the fields from nerves or skeletal muscle. We have seen in section 2.19 that cardiac tissue functions as if it were a single large cell or syncytium. The electrical resistance of the intercalated discs results in the axial resistance of a strand of cardiac tissue being higher than that of a corresponding strand of peripheral nerve and the propagation velocity is lower. Because the repolarization phase of the cardiac action potential is a hundred times longer than the depolarization phase, the repolarization currents are one hundredth of the depolarization currents and the depolarization dipole is the more prominent. We have, in addition, large temporal and spatial separation of depolarization and repolarization so that we can think of a single dipole current source, in contrast to the pair of dipoles for a peripheral nerve.

The vast majority of cardiac muscle, however, is a three-dimensional structure in which adjacent cells in all directions are interconnected. As a result the depolarization wavefront in cardiac muscle is an extended surface that moves through the heart with a velocity of approximately $1\ \mathrm{m\ s^{-1}}$. The isolated current dipole model is no longer appropriate and we need to divide the depolarization wavefront into many small segments, each with its own dipole. The magnetocardiogram (MCG) is determined by the sum of the magnetic fields from each segment of the wavefront.

The lack of unique solutions to the inverse problem would appear to reduce the usefulness of neuromagnetic techniques. However, it turns out that, in practice, new and exciting information may be derived from these techniques; in the next sections we shall review some of the results obtained from neuromagnetic measurements, chiefly for the brain and heart. The range of data from magnetomyography is at present quite limited and we can only outline how this field might develop. Neuromagnetic measurements pose additional experimental problems since the environmental noise in a normal laboratory is likely to be many orders of magnitude greater than the neuromagnetic signals of interest (Figure 3.2). Magnetic field protection is afforded by carrying out the measurements within a special, magnetically shielded room constructed with two to three shielding layers made from mu-metal and a high conductivity material such as aluminium. In the absence of shielding, carefully balanced second-order gradiometers have been used successfully.

3.10 MAGNETOENCEPHALOGRAPHY

There are only two non-invasive windows into the brain functions actually associated with information processing, namely electroencephalography (EEG) and magnetoencephalography (MEG). In EEG the potential distribution caused by electrical activity in the brain is measured by contacts on the head, whereas in MEG, the weak magnetic fields generated by the same activity are measured with SQUID magnetometers. Magnetoencephalography offers good spatial and temporal resolution, on a millimetre and millisecond scale, so providing the possibility of determining the actual position of sources which are sensed by EEG; the magnetic fields are not distorted by the skin, skull and brain as are electrical fields associated with the EEG signals, and there is no need to attach contacts to the skull. It is now possible to exploit the spatial–temporal resolving power of MEG with a SQUID based magnetometer which covers the whole head of the patient in the manner of a helmet. Figure 3.22 shows the typical arrangement of sensor chips, and a section through the Dewar magnetometer arrangement. This illustration is based loosely on the 122 channel DC SQUID magnetometer built by Neuromag Ltd.

As we have seen, information is conveyed from one neuron to another via an axon and its terminal synapse. The action potential propagates along the axon manifesting itself as a current quadrupole (Figure 3.4(a)). At the synapse the action potential initiates the post-synaptic potential (PSP) which constitutes a current dipole (Figure 3.4(b)). If the spatial or temporal sum of the PSPs exceeds a certain threshold, an action potential is initiated in the receiving neuron. From the point of view of an external measuring system, the magnetic induction associated with these dipoles and quadrupoles in isolation is far too weak to be detectable by any known measuring system, as explained in section 3.3. However, if a large number of neurons act in concert, then we are able to measure a resultant signal. The observed magnetic fields and the potential differences over the scalp are largely due to PSPs in pyramidal cells of the cerebral cortex which, like the net

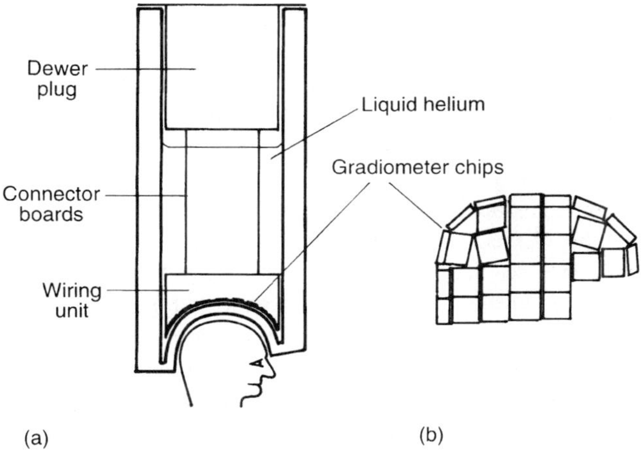

Figure 3.22 (a) Schematic diagram of the magnetometer probe and liquid helium dewar used for multichannel MEG measurements. (b) Disposition of the sensor array which is situated at the helmet-like base of the liquid helium Dewar. (This depiction is based loosely on the 122 channel DC SQUID neuromagnetometer built by Neuromag Ltd.)

current, are perpendicular to the surface of the cerebral cortex. The cells of the cerebral cortex and their disposition in the layered structure of the cerebral cortex are shown in Figure 3.21.

Magnetoencephalographic measurements are carried out on signals generated under two different sets of conditions, namely spontaneous brain signals and rhythms (section 2.17), or those generated or 'evoked' by an external stimulus. In practice, the strength of a typical equivalent dipole in an evoked response experiment is around 10^{8} A m compared with the 3×10^{-14} A m we calculate for a single dipole. We must conclude, therefore, that approximately 10^{5}–10^{6} neurons are activated at the same time! The dipoles associated with spontaneous brain activity are measured to be roughly an order of magnitude greater than this at 10^{-7} A m. The spontaneous rhythms are continually changing and reflect activity of the brain that is not necessarily locked to immediate stimulus input. Spontaneous alpha MEG activity, for example, has been localized to the **parieto-occipital** areas (Figure 3.20(a)). Estimates of the generator location are difficult because signal averaging is required to improve the signal-to-noise ratio. In order to carry out the averaging, the signal has to be locked to some part of the alpha rhythm, and this presupposes that the same underlying processes are occurring during each phase of the selected activity.

It has been proposed by Grey Walter that the spontaneous EEG recorded from electrodes outside the brain results from the summation of brain signals having different amplitudes, frequencies and phases, so that Fourier analysis of the spontaneous activity should reveal the component signals. In practice the distribution of magnetic fields over the head is first recorded and Fourier analysed for each position. A particular frequency of interest, say 8 Hz from an alpha rhythm, is then selected and the inverse computation carried out for that frequency, retaining the phase information in the data from different locations. This procedure results in sine waves of the selected frequency and returns the data to the time domain. The next step is to select a point in the time domain for estimating the dipole position. This selection is, in principle, unrestricted, but should include the point that results in the greatest number of phase reversals in the time domain. This point is then mapped over the head, just as a point in an evoked field would be mapped, and an estimate made of the location of a theoretical dipole which could be responsible for the map. The result is that the dipole is the theoretical location of the generator for that frequency of the magnetic field. The process is repeated to locate generators of different frequencies.

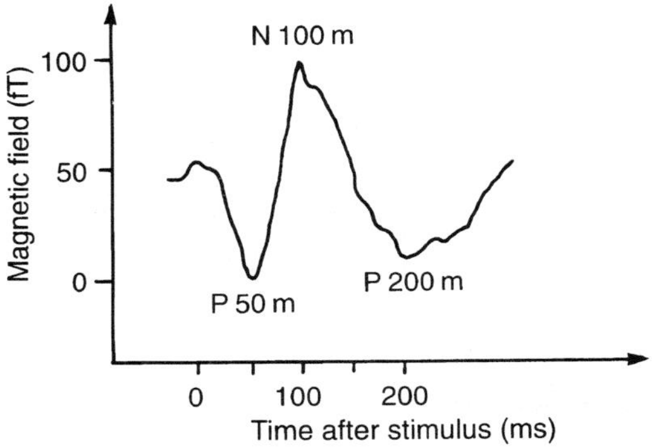

Figure 3.23 Typical evoked response obtained close to the auditory cortex. The signal was evoked by short tones. The numbers refer to the latency (in ms) after the stimulus onset, the letters show the sign of the associated magnetic field P (positive), N (negative).

In sensory evoked response studies we examine the magnetic response as a function of time after the stimulus. Figure 3.23 shows the sort of signal which is obtained close to the auditory cortex following short tones. The customary method of labelling the peaks is shown here also; the letter indicates an associated positive (P) or negative (N) magnetic field (that is fields entering or exiting from the skull, respectively) while the numbers refer to the latency (in ms) after the stimulus onset.

In these studies we are able to separate five processes in the brain associated with the sensory input namely:

1. the registration of sensory input
2. selective and generalized attention
3. memory update and storage
4. the interaction of the input with previously stored information
5. the processes associated with retrieval and preparation for the motor output.

These processes form part of an information processing line which will have associated magnetic fields arising from different parts of the brain. During these processes some feedback loops will be established.

The time scale of the response to external stimuli suggests that early brain stem electrical activity occurs within about 10 ms of the stimulus. These early components index function in different parts of the brain stem, depending on the kind of sensory input. In the interval from approximately 10 ms to 80 ms the information is transferred from brain

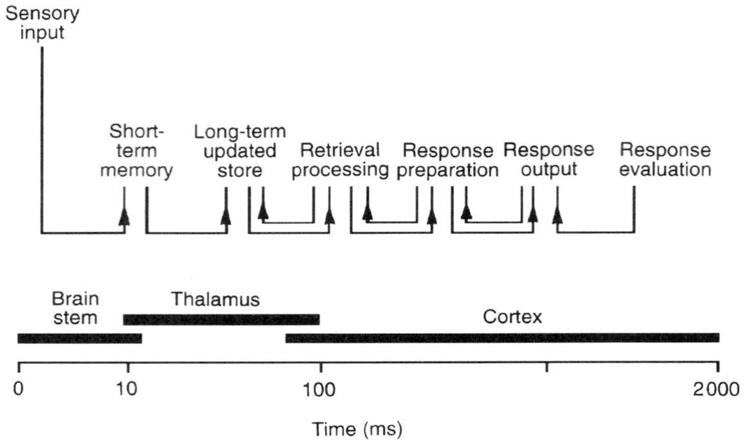

Figure 3.24 Information processing spectrum for the brain showing the time scale and function for the various processing sectors.

stem to thalamus and other midbrain structures. During the timespan of approximately 80 ms to 300 ms, complex cortical processes occur, for example memory update. These functions, and their likely time scale, are shown in Figure 3.24. In principle, we can identify function from the signal latency (time after stimulus) and sense (positive or negative peak), so that if we are able to locate a generator responsible for changes in the amplitude and latency of N100, then this generator must be responsible for, or involved in, processing, storing and retrieving information.

The concept of a generator (e.g. a dipole) is an important one and we can distinguish between generators associated with cognitive function (information processing, memory access) and generators associated with the first stages of sensory processing, and indeed the last stages of motor output. In addition we may think of generators as 'localized', that is confined to some small site or space, or 'distributed', spread out three-dimensionally and simultaneously active. A final generator classification is 'stationary' or 'non-stationary', where by stationary we imply that the generator does not change its location.

As a consequence of the above categorization we can distinguish four classes of generator, namely:

1. localized and stationary
2. localized and non-stationary
3. distributed and stationary
4. distributed and non-stationary.

If the period of activity associated with a generator is relatively long, it is unlikely that a single localized generator would be responsible.

Reference has already been made both in Chapter 2 (section 2.17) and earlier in this chapter to the alpha rhythm, which represents electrical fluctuations between 8 and 13 Hz and can be detected on the **occipital** scalp; it is attenuated by visual stimuli. Studies indicate that there are time series of the rhythm lasting for up to 1 s, during which the oscillation period is constant. These time series are called *spindles*, whether or not they occur in the sleeping state (where they were originally observed). The magnetic field pattern during a spindle is relatively stable, implying that its source is a specific configuration of neurons. We can conclude that the human alpha rhythm represented by the spindles is generated by a large number, or possibly a continuum, of different source configurations. The neuronal excitation producing a spindle has been called an *alphon*. Strong alpha rhythm generally exhibits field extrema of opposite polarity over the left and right hemispheres, indicating that the corresponding alphons lie near the midline (the signals are very weak more than 90 mm to either side of the midline), probably in the **bilateral striate cortex**. Each alphon can be modelled by a current dipole so that its centre of activity can be specified.

In order to identify the neuronal structure responsible for alphons we need a preferentially aligned population of cortical neurons. The most likely candidates are the pyramidal cells depicted in Figure 3.21(a). The measured distributions of current dipole moment for alphons show that the most probable value is approximately 40 nA m, which suggests that the alphon is a characteristic excitation involving about the same number of neurons, regardless of where it occurs. In contrast, a dipole moment of 40 nA m is only an order of magnitude stronger than a typical sensory evoked response. If we model the dipole on post-synaptic currents in pyramidal cells, then an area of only a few square millimetres will contain sufficient neurons (approximately 10^5), if coherently active, to account for an alphon.

To summarize, the helmet-like assembly of gradiometers used in the most recent MEG studies permits the recording of magnetic fields over many areas of the brain simultaneously and provides the opportunity to investigate the interaction between separate brain structures.

3.11 MAGNETOCARDIOGRAPHY AND MAGNETOMYOGRAPHY

The typical MCG, Figure 3.25, exhibits essentially the same features as an ECG, although it is notable for the magnitude of the magnetic field

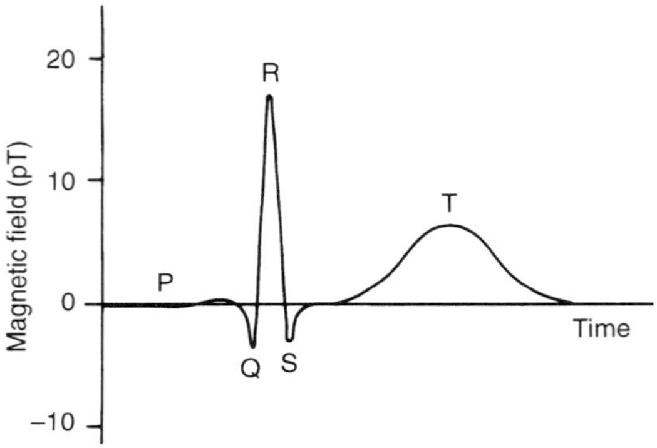

Figure 3.25 A typical MCG revealing the formal similarity to the standard ECG and the relatively substantial magnetic field associated with an MCG, which is measured in pT rather than fT.

compared with that seen in MEG measurements; data is recorded in pT rather than fT. We have seen in section 2.19 that heart muscle operates as a three-dimensional structure in which adjacent cells, in all directions, are interconnected. As a result, the depolarization wavefront in cardiac muscle is an extended surface that moves through the heart with a velocity of approximately 1 m s^{-1}. The isolated current dipole model we have considered thus far is not appropriate in this instance, instead the depolarization wavefront must be divided into many small segments, each of which has its own dipole. The MCG is determined by the sum of the magnetic fields of each segment of the wavefront. If these segments are made small enough it becomes possible to think of the current dipole associated with each segment in terms of a current dipole density, usually known as the 'impressed current density', which is dispersed uniformly over the wavefront. Figure 3.26(a) shows the result of the measurement of a magnetic field map on a normal (healthy) patient integrated over the QRST-wave of a cardiac cycle. Figure 3.26(b) shows the analogous measurement for a patient with **inferior myocardial infarction.**

In order to record magnetic signals from muscles elsewhere in the body (the basis of magnetomyography), it is necessary to stimulate the muscle electrically in the manner described in section 2.18. The resulting neuromagnetic field may be evaluated in terms of latency, intensity, composition, characteristic behaviour and site of generation of the signal. Figure 3.27 shows the magnetic signal obtained by stimulating the **tibial nerve** in the hollow of the knee, the measurement being

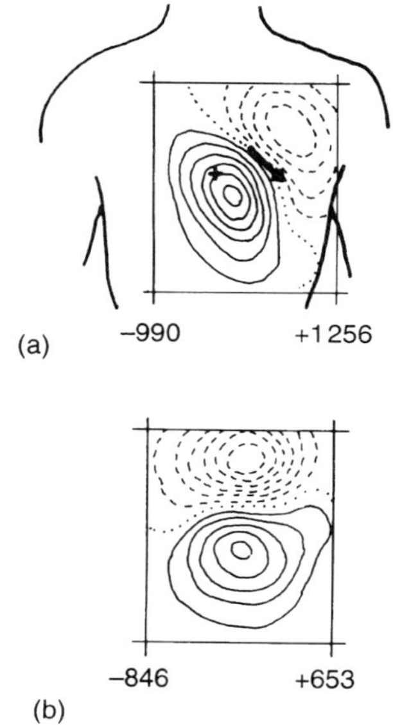

Figure 3.26 (a) Cardiac magnetic field plot (240 mm × 280 mm) for a normal subject. This is an iso-integral plot (∫ QRST-wave). The black cross locates the junction of the **sternum** and the fourth **intercostal space**. Solid contours indicate a magnetic field directed inwards (positive), dashed lines indicate magnetic field directed outwards (negative). The dotted line is the zero magnetic field contour. The arrow indicates the direction of the time averaged equivalent current dipole. (b) The same measurements repeated for a patient exhibiting **inferior myocardial infarction**. The extrema values for both sets of contours are displayed, in pT ms, below each figure. (Adapted from the work of Stroink *et al.*)

conducted over the **soleus muscle** in the back of the leg. Magnetomyographic measurements provide information outside that provided by EMG; this technique provides diagnostic information on the state of the neuronal reflex pathways, including muscle function. The spread of excitation through the muscle can also be followed, non-invasively, a facility unique to MMG.

The measurement of neuromagnetic fields is a relatively new technique and it has still to reach maturity. Nevertheless, it is already possible to obtain diagnostic information which is unobtainable by any other means. The chief point of interest in MEG, for example, is those situations in which differences arise between the neuroelectric and neuromagnetic measurements. Thus MEG is good for studying activity located in a cortical fissure since this produces a tangential dipole,

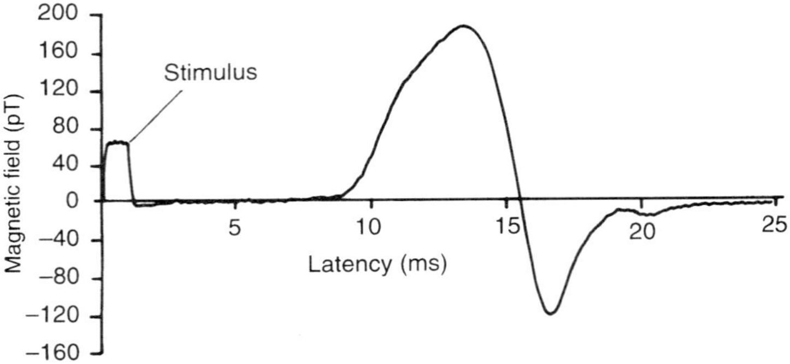

Figure 3.37 Magnetomyographic signal obtained by stimulating the tibial nerve in the hollow of the knee. The data is obtained with a SQUID and second-order gradiometer over the **soleus** muscle in the back of the leg. (After M. Reincke, *Biomed. Technik* **38** (1993) 276–81.)

whereas a radial, cortical current dipole will have no magnetic signal, but only a simple electric pattern. Again MEG requires no surface electrodes or reference electrodes so that as the size of SQUID magnetometers diminishes it will be easier to map the MEG than to prepare an equivalent number of skin electrodes. That situation has already been reached. Neuromagnetic signals are more closely related to the intracellular currents than are the extracellular potentials. In all situations where we now measure potentials at the skin surface it may therefore be preferable to measure the magnetic field, since this eliminates the effect of tissue resistance associated with anisotropic tissue. In the case of the heart tissue, anisotropy is important, but poorly understood. Neuromagnetic fields are unaffected by tissue, but this is not the case for neuroelectric measurements. Localization of neuroelectric sources may ultimately prove easier to achieve by magnetic methods than by electrical ones.

PROBLEMS

1. Calculate the magnetic field of the current dipole associated with an axon of 4 μm diameter, having conductivity $0.25\ \Omega^{-1}\ \text{m}^{-1}$, where the potential drop across the depolarized region is 10 mV.
2. A magnetometer is used to measure the field arising from the depolarization current dipole of a nerve fibre comprising 10^3 axons operating in synchrony. Assuming that each axon has the same characteristics as the axon described in question 1, calculate the maximum magnetic field 5 mm away from the nerve axis.

4

X-rays and projection radiography

4.1 INTRODUCTION

The interaction of physics with medicine spans some 400 years, yet, during the whole of that period only one discovery in physics has seen more or less instant adoption in medical practice, and that is the use of X-rays in projection radiography. On 8 November 1895, W.C. Roentgen demonstrated the existence of 'a new kind of rays', X-rays, using a Crookes tube; by February of 1896 applications of the new rays in medicine were being described in the *British Medical Journal*. Roentgen had actually been looking for the 'invisible high-frequency rays' which Hermann von Helmholtz had predicted from the, then new, Maxwell's equations. It had been proposed by Helmholtz that these high frequency radiations might be present in cathode-ray discharges. The immediate outcome of Roentgen's work was one of the most famous scientific photographs ever published (Figure 4.1).

Roentgen's discovery of X-rays brought him instant fame, but took him away from his laboratory at what was perhaps the most productive time of his life. With hindsight, it is clear that he was ideally placed to discover radioactivity as well as X-rays. Had he done so he might, in addition to winning the first Nobel prize for physics (1901), have joined A.H. Becquerel and Pierre and Marie Curie as a recipient of the third Nobel prize for physics (1903).

The discovery of X-rays actually rested on developments in many fields of science and technology; all of these developments had to be in place before X-rays could be discovered. We begin with the discovery of vacuum in the middle of the seventeenth century by Otto von Guericke. The improvements in vacuum technique introduced by R. Boyle, and later by F. Hauksbee, resulted in pumps which could pump

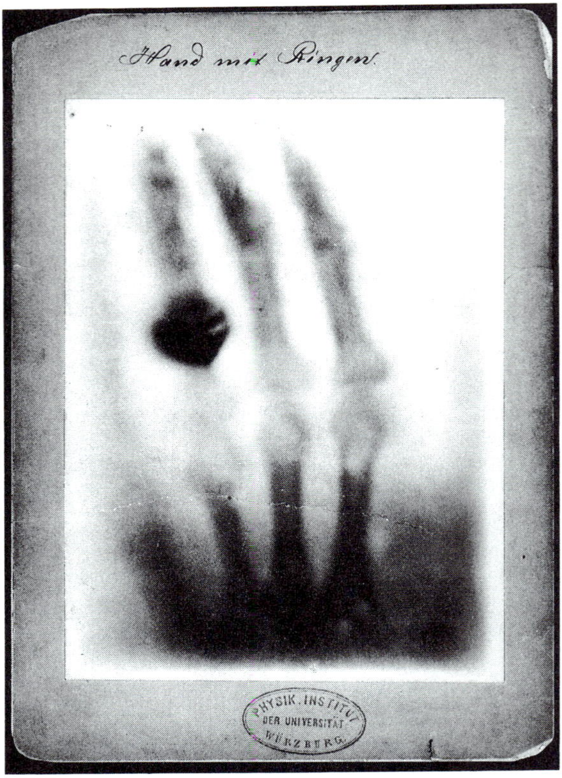

Figure 4.1 Original X-ray image taken by Roentgen to demonstrate the peculiar properties of the 'new kind of X-rays'. The image is of his wife's hand! (Courtesy of the Wellcome Institute Library, London.)

down to about 10 mbar, or roughly 10 mm of mercury. Next came the work of M. Faraday, which laid the basis for the induction coil (of Rühmkorff) and later the transformer; these were the means by which a large electric potential could be applied to a vacuum tube. Finally, in 1838 a glassblower, H. Geissler, showed how platinum wire could be sealed into a soda-glass, vacuum envelope. This was the first glass–metal seal, and low-vacuum tubes came to be called Geissler tubes. Improvements in vacuum technique led to the introduction of high-vacuum tubes and the discovery of cathode rays by J.W. Hittorf.

It is ironic that at least four pioneering workers, namely W. Crookes, J.W. Hittorf, P.E.A. Lenard (winner of the Nobel prize for physics in 1905) and H. Jackson, undoubtedly discovered X-rays without recognizing that they had done so. Probably the discovery of X-rays should have been made by Sir William Crookes since, using high-vacuum tubes which he designed himself (Crookes tubes), he gave a complete description of an X-ray tube in action. Nevertheless, he failed to recognize the presence of X-rays despite, on occasion, complaining to his supplier of photographic plates that plates had been delivered fogged, that is partially exposed. This was the unrecognized effect of exposure to X-rays. The development of sensitive photographic recording emulsions was yet another prerequisite for the introduction of X-ray radiography.

The early X-ray tubes (prior to 1913), were in fact cold-cathode tubes powered by one of four possible types of voltage generator, either:

1. an induction coil (Rühmkorff coil);
2. an influence machine (Wimshurst machine);
3. a high voltage transformer; or
4. a dynamo.

Of these, the most popular choice was the induction coil (this was actually used by Roentgen in his work), large Wimshurst machines were not uncommon, and it should be remembered that all of these voltage generators could produce voltages as high as 100 kV, even in 1895.

Figure 4.2 shows sections through typical examples of early X-ray tubes, in this instance a Crookes–Hittorf tube (Figure 4.2(a)) and a Jackson focus tube (Figure 4.2(b)). The Jackson focus tube contained a dished or concave aluminium cathode which focused the secondary electron beam onto a platinum anode, which, angled at 45° to the electron beam, projected the resulting X-rays in a more useable way than other tube designs. Both tubes were fitted with appendages or side arms. By heating or cooling the side arms the pressure in the X-ray tube could be increased or reduced, respectively. The Jackson focus tube was very important since within it, X-ray emission was confined to a spot on the anode, thus yielding X-ray images of much greater sharpness than before. This tube can be said to have launched clinical radiography.

Gradual improvements in basic tube design, and particularly power sources, permitted increases in tube voltage which highlighted the principal drawback of the cold-cathode tube, namely, that the gas pressure in the tube decreased during use, due to 'gettering' action; the tube was said to 'harden'. The tube's ability to sustain any sort of discharge gradually disappeared, despite the higher tube voltage, and

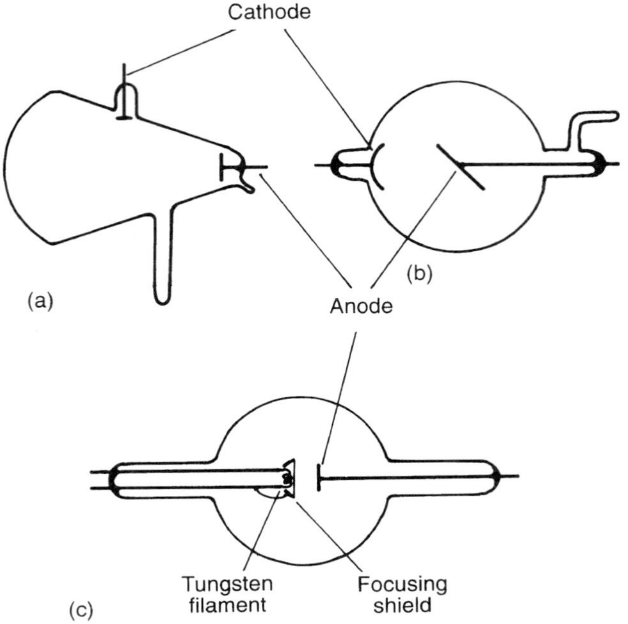

Figure 4.2 Outlines of the earliest forms of X-ray tube showing (a) the Crookes-Hittorf tube, (b) the Jackson focus tube and (c) the Coolidge tube. Tubes (a) and (b) are cold-cathode tubes. Tube (c), which revolutionized clinical X-ray practice, is a hot-cathode tube.

with it went the secondary electron production upon which its operation depended. Put another way, the current flowing in the tube could not be varied independently of the applied voltage and ultimately, with continuous use, the current reduced to zero despite increases in applied voltage; X-ray production then ceased. It should be appreciated that in order to produce X-rays the cold-cathode tube had to retain some gas filling. The reason for this is that the discharge in the tube is initiated by the random ionization of a gas atom. The resulting positive ion is accelerated towards, and collides with, the cathode, where it is neutralized with high efficiency producing a secondary electron. The secondary electron is, in turn, accelerated in the electric field of the tube until it has sufficient energy to ionize additional gas atoms by collision. This process repeats and multiplies to yield a significant electron flux between cathode and anode, with the electrons acquiring enough energy to yield X-rays on impact with the anode target.

The solution to the problems associated with the cold-cathode X-ray tube was the introduction in 1913, by W.D. Coolidge, of the hot-cathode, high vacuum, X-ray tube. This tube was really a consequence of the application of technical advances made in the electric lamp industry, then growing rapidly in the USA. Tungsten, the most refractory metal known, is also very brittle and difficult to fabricate, but Coolidge showed how to make tungsten ductile which, in turn, permitted its ready conversion into delicate filaments for light bulbs. Irving Langmuir had shown that temperature is the sole criterion for the emission of electrons from a metal, so that the stage was set for the introduction of hot filament cathodes into X-ray tubes. This change, together with improved vacuum technique (due to Langmuir), permitted the production of an X-ray tube in which the electron current could be varied independently of the applied voltage and, with the use of a tungsten anode, permitted very high power dissipation. Figure 4.2(c) shows a section through a Coolidge tube. The hot-cathode tube has remained in use since that time, little altered in basic design. It is worth noting though, that had vacuum technology developed ahead of the other technologies associated with the production of the cold-cathode X-ray tube, the introduction of X-rays into medical use might well have been delayed by 20 years, or perhaps even longer.

4.2 INTERACTION OF ELECTRONS AND X-RAYS WITH MATTER

Before we can fully comprehend an X-ray imaging system we must understand both the interaction of energetic electrons with matter, which forms the basis of X-ray production, and also the interactions of X-ray photons with matter, since these latter interactions dominate the imaging process. Figure 4.3 summarizes all of the physical processes involved in X-ray production and their medical application, from the emission of electrons at the cathode of the X-ray tube, to the reception of X-rays by the X-ray detector or receptor.

There are two basic processes involved in X-ray production and these manifest themselves in the spectrum of X-rays produced by electron impact on a metal target. The spectrum from a molybdenum anode is shown in Figure 4.4; the tungsten spectrum appears as an inset to Figure 4.3. These spectra are characterized by a relatively smooth background radiation (the same for all materials), superimposed upon which are discrete lines, characteristic of the anode material. The smooth background is bremsstrahlung or ‘braking radiation’, while the

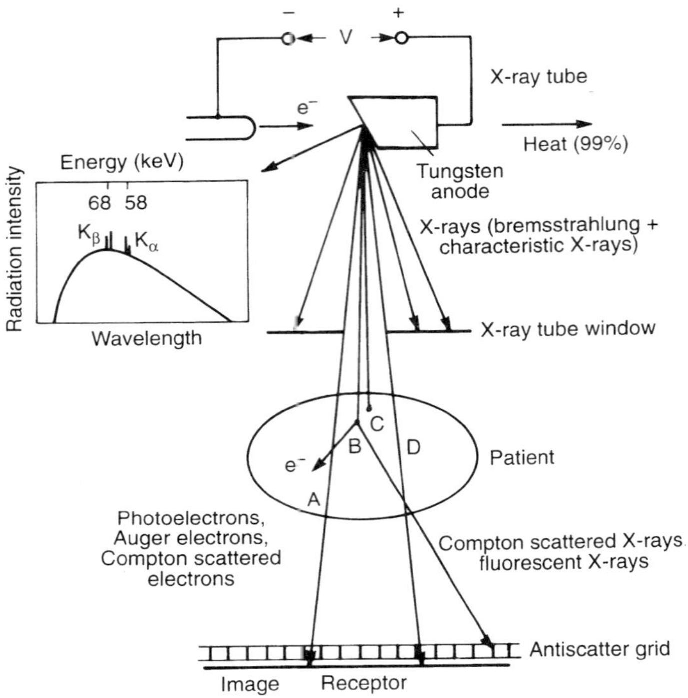

Figure 4.3 Summary of the phenomena involved in clinical X-ray usage showing the processes which result in X-ray production and (inset) the typical X-ray spectrum from a tungsten anode. The secondary electron products, that is photoelectrons, Auger electrons or Compton scattered electrons, consequent upon X-ray impact, are indicated together with the secondary X-ray beams (Compton scattered and characteristic fluorescent X-rays). Of the four X-ray beams shown impinging on the target, beams A and D contribute to the final image, beam B is scattered and intercepted by the antiscatter grid, whilst beam C is absorbed and contributes to patient dose.

discrete lines are characteristic X-rays. The mechanisms responsible for the two types of X-ray emission are shown schematically in Figure 4.5(a), (b).

Bremsstrahlung radiation is produced when electrons pass through matter and is caused by an electromagnetic interaction with an atomic nucleus in which a photon is radiated by the accelerated electron as it is deflected, or when it collides with nuclei. We can set a lower limit on the wavelength of the X-rays produced in this process, given by

$$\lambda_{min} = hc/eV \tag{4.1}$$

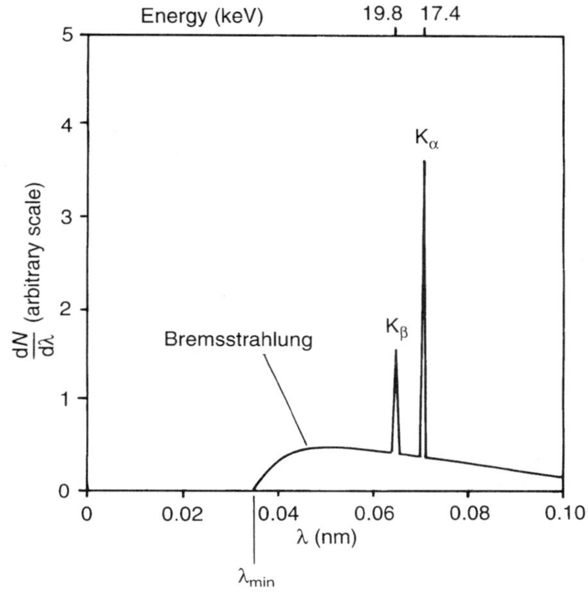

Figure 4.4 X-ray spectrum from a molybdenum anode showing the typical structure of background bremsstrahlung radiation superimposed upon which are the K_α and K_β peaks for molybdenum at 17.4 and 19.8 keV, respectively. The tube excitation energy is 35 keV.

where V is the electron accelerating potential, e the electron charge, h Planck's constant and c the velocity of light.

The characteristic X-ray is emitted when an electron changes from a stationary state at $n = 2$ say, to $n = 1$ (n is the principal quantum number), the original occupying electron having been ejected by electron impact. The possible transitions are shown in Figure 4.5(b) and are grouped into K series, L series, M series, etc. The transitions within each group are labelled α, β, γ and so on.

In the energy range applicable to medical usage, the interaction of X-rays with matter is complex since it can lead to the production of photoelectrons, Auger electrons and Compton-scattered electrons, together with scattered X-rays and characteristic fluorescent X-rays. Characteristic fluorescent X-rays are produced in a manner analogous to the production of a characteristic X-ray, though now initiated by incident X-rays rather than incident electrons. In the photoelectric interaction an atom absorbs an X-ray photon which has sufficient energy to free an electron from the atom. The electron emerges with a

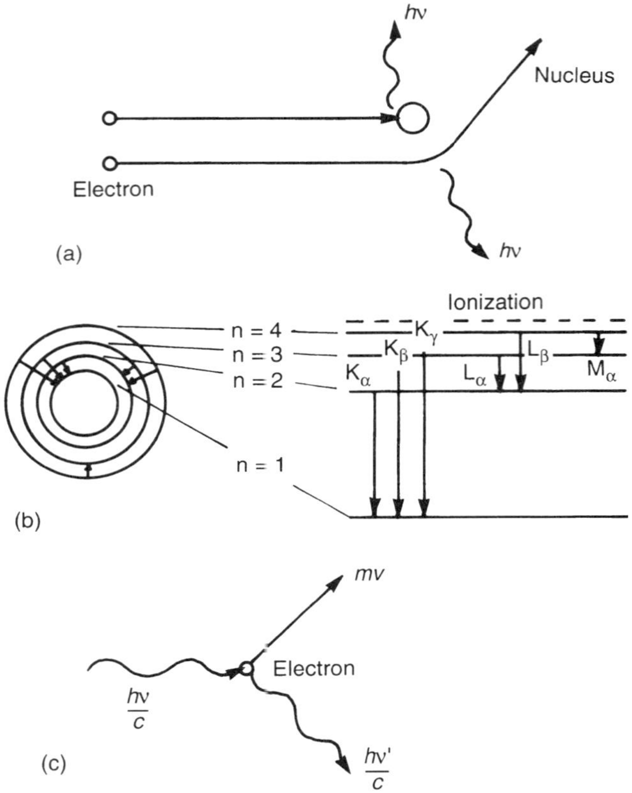

Figure 4.5 Models for different types of X-ray production. The production of bremsstrahlung (braking radiation) in (a) is a consequence of the interaction of an energetic electron with an atomic nucleus, either through collision or deflection. In (b) we see the transitions between energy levels responsible for the characteristic X-ray lines, whilst (c) shows the interaction responsible for a Compton scattered X-ray and its associated scattered electron.

kinetic energy equal to the incident photon energy minus the electron binding energy. Alternatively, the X-ray photon can excite an inner electron from an atom. The atom may then return to the ground state with the emission of an electron carrying the excess energy; this is Auger emission. Compton scattering (Figure 4.5(c)) involves the scattering of an energetic X-ray photon by an outer electron in an atom. Kinetic energy is transferred from the X-ray photon to the electron. In this case the photon energy is so large that the electron may be considered a free particle and the remainder of the atom plays no role in the scattering process. The outcome is a photon of longer wavelength than the incident photon, plus a scattered electron. X-ray photons with energies greater than 30 keV interact with body soft tissue primarily by Compton scattering. Photoelectric absorption occurs primarily in substances with a high *Z*, such as positive contrast media and bone.

To begin with it is convenient to think of the X-ray source as being monoenergetic, although in practice this will never be the case, as we have seen. The transmission of monoenergetic X-ray photons through tissue varies with photon energy as well as tissue thickness. If the fraction of photons transmitted is very low, then very few photons will reach the detector or receptor. Instead they will be absorbed by the tissue and the radiation dose to the patient will be very high. If, on the other hand, the fraction transmitted is very high, that is close to unity, then there will be very little difference in transmission for different types of tissue, and image contrast will be very poor. The variation in transmission of monoenergetic X-ray photons through tissue, as a function of photon energy, is shown in Figure 4.6.

In practice, therefore, we have to establish a balance between image contrast and patient dose and, as a consequence, the energy

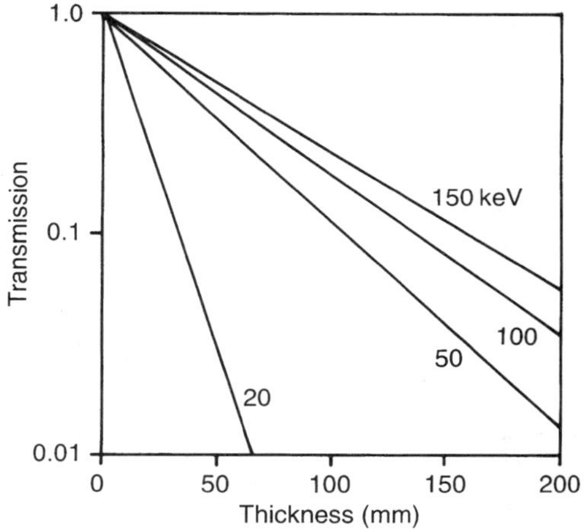

Figure 4.6 Variation in the transmission probability of X-rays against material thickness, with X-ray energy as a parameter. When the X-ray energy is decreased, the transmission probability diminishes also, and increases the patient dose. A polychromatic X-ray beam will thus 'harden' (lose its low energy constituents) on passage through matter.

range usually chosen is between 20 and 150 keV. In this energy range the important photon interactions involve the photoelectric effect, Compton scattering, characteristic fluorescent X-radiation and Auger electron production. The output from these processes is divisible into four classes:

1. X-rays which have passed through the patient without deflection;
2. X-rays which have undergone deflection, or do not lie on the axis of the incident beam;
3. X-rays which have been absorbed by the patient; and
4. electrons produced during the X-ray interaction.

The highest energy that electrons are likely to have in diagnostic X-ray work is 150 keV. At this energy, however, the electron range will be of the order of 0.3 mm and all of the electrons produced may be considered to be absorbed locally, contributing substantially to the patient dose or, if the electrons are produced in the image receptor, to the energy deposited therein.

The X-ray, or radiographic, image is formed by the interaction of X-ray photons with a photon detector of some sort, and records the distribution of those photons which have been transmitted through the patient. It is only the primary X-ray photons, those which have reached the image receptor without deflection, that is A and D of Figure 4.3, which carry useful structural information; they give a measure of the probability that a photon will pass through the patient without interaction. The important parameter describing the attenuating property of the tissue along the line of the X-ray beam is the linear attenuation coefficient. The X-ray image itself is simply a shadowgraph, a projection of the attenuating properties of all the tissues along the path of the X-ray beam. Before examining the basis of X-ray image formation it is important to define the terms which will be encountered in the imaging process. In particular we must distinguish between absorption and attenuation.

The effect of X-radiation on a material is determined by the amount of energy absorbed per unit mass. The rate of removal of energy from the X-ray beam is given by $I(\mu/\rho)_e$, where I is the X-ray intensity, $(\mu/\rho)_e$ is the mass energy absorption coefficient for a medium of density ρ at a beam energy e, and μ is the linear attenuation coefficient. As we have seen, this energy may be given up in a photoelectric interaction or a Compton scattering interaction. In the latter case only part of the photon energy is transferred to an electron, while the scattered photon may not be absorbed at all. The photoelectric interaction largely determines the energy absorbed.

Usually however, when considering the effect of a medium on radiation, it is the reduction in intensity which is of interest and this is characterized by the mass attenuation coefficient μ/ρ (at low energies <30 keV this is almost equal to the mass energy absorption coefficient) (Figure 4.7). Consequently, the attenuation of intensity may be written as $I_t = I_0 \exp[-(\mu/\rho)\rho t]$, where t is the linear thickness of the medium, I_0 is the incident intensity and I_t the transmitted intensity.The units for both the mass energy coefficient and the mass attenuation coefficient are $m^2\ kg^{-1}$, while 'thicknesses' are referred to in $kg\ m^{-2}$. Generally speaking, in describing the problem of image formation, it is just the linear dimensions which are relevant and here we use the linear attenuation coefficient μ which is the mass attenuation coefficient times density. The linear attenuation coefficient has the units m^{-1}. The linear attenuation coefficient μ is just the fractional loss of intensity in travelling *unit distance* through the medium. In contrast, the ratio μ/ρ,

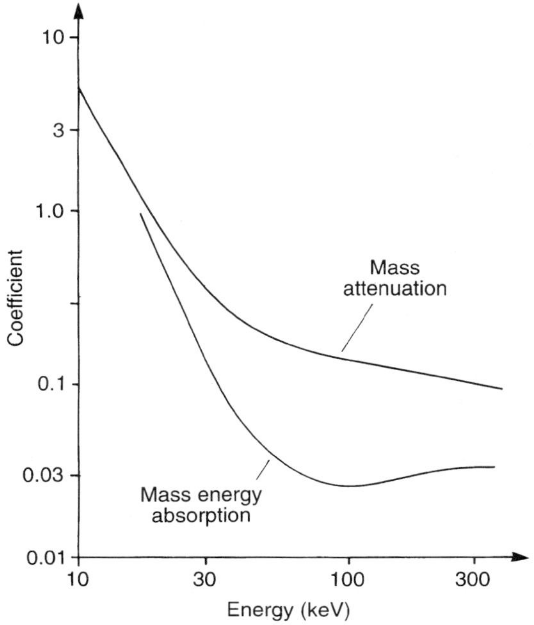

Figure 4.7 Variation of the mass energy absorption coefficient and mass attenuation coefficient of water, in the diagnostic X-ray energy range.

or mass attenuation coefficient, described above, represents the fractional loss of intensity for a beam of *unit cross-section* traversing *unit mass* of material. For a homogeneous beam of X-rays this quantity is a constant for any given material and increases with the atomic weight of the material. It is worth noting that if μ were *exactly* proportional to density, then μ/ρ would be a universal constant.

4.3 X-RAY IMAGE FORMATION

If the human body were entirely homogeneous, then we could describe the outcome of the X-ray imaging process by a simple equation which relates the transmitted intensity I_t to the incident intensity I_0. For an X-ray beam traversing a medium of thickness t we have the relationship

$$I_t = I_0 \exp(-\mu t) \tag{4.2}$$

where μ is a constant for a given wavelength and is the linear attenuation coefficient described above. Equation 4.2 is Lambert's law of absorption, found also in optics. To a large extent μ is dependent on the electron density and is therefore roughly proportional to the density of the material, so that we can expect the X-ray image to reflect this latter property.

The linear attenuation coefficient μ, is in fact, the sum of the individual attenuation coefficients for several types of X-ray interaction. In the energy range embraced by medical imaging, two kinds of interaction predominate: the photoelectric effect and Compton scattering. The photoelectric coefficient increases with atomic number Z (actually as Z^3) because X-ray photons interact photoelectrically with the inner, tightly bound electrons of an atom. The Compton coefficient, on the other hand, is relatively independent of atomic number, since the interaction is with the outer, loosely bound electrons. Both coefficients increase linearly with tissue density. Figure 4.8 shows that if we wish to achieve satisfactory contrast between tissues which vary only slightly in Z, then we must use low energy X-rays where the dominant interaction is by the photoelectric effect, and the Z^3 dependence may be used with advantage.

The human body is, of course, inhomogeneous, so that equation 4.2 must be modified to take account of this fact. The attenuation coefficient along any ray path is now no longer constant, but varies from point to point along the path. We shall denote this by the density function $\mu(x, y, z)$ so that equation 4.2 must now include a line integral

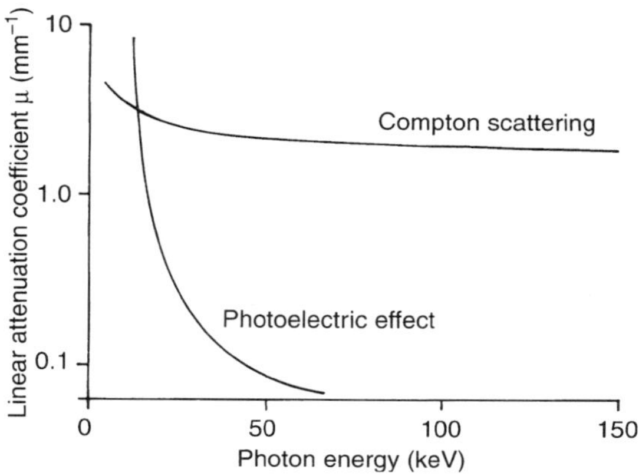

Figure 4.8 Plot of the principal components of the linear attenuation coefficient for the diagnostic X-ray energy range.

along the path of the X-ray beam (assumed here to be the z axis) in place of the attenuation coefficient, thus

$$I_t = I_0 \exp\left[-\int_{\text{detector}}^{\text{source}} \mu(x, y, z)\, \mathrm{d}z \right] \tag{4.3}$$

Although the path is here taken in the z direction and is fixed, we shall see in Chapter 6 that this need not always be so.

Equations 4.2 and 4.3 only take account of those photons which form the primary beam; the scattered photons are ignored. From the point of view of our photon detector, or receptor, this is unfortunate since the scattered photons are likely to be recorded or absorbed more efficiently than the primary photons, largely because their pathlength in the receptor is greater, that is they are not incident normally. It is possible to reduce the effect of secondary electrons and X-ray photons by means of an antiscatter grid, or by reducing the beam size, or simply by increasing the distance between the patient and the receptor (Figure 4.9). Alternatively, we can assume that the ratio R of the scattered to the primary radiation is constant and may be measured or calculated. We then replace equation 4.3 by

$$I_t = I_0 \exp\left[-\int_{\text{detector}}^{\text{source}} \mu(x, y, z)\, \mathrm{d}z \right](1 + R) \tag{4.4}$$

We have so far assumed a monoenergetic or monochromatic X-ray energy distribution. In practice this is not the case and, as we have seen, the X-ray beam comprises a continuous spectrum of energies due to

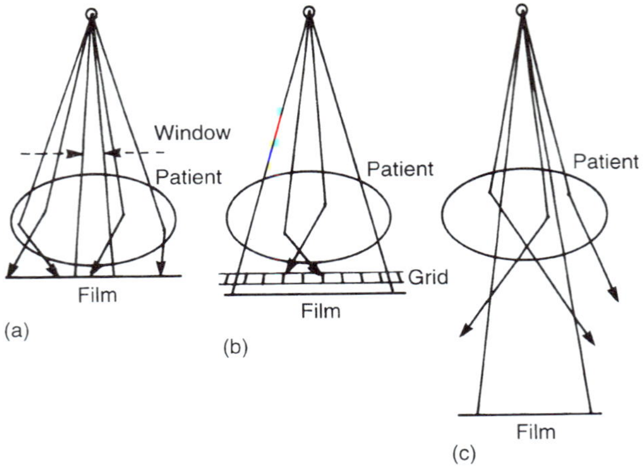

Figure 4.9 Usual means by which scatter is reduced in X-ray imaging: (a) reduction of beam size by means of a window (dashed lines); (b) use of an antiscatter grid which intercepts most scattered X-rays; (c) increase in distance between patient and X-ray image receptor. Scattered beams pass by rather than through the film.

bremsstrahlung (braking radiation) superimposed upon which are specific peaks due to the characteristic K lines of the X-ray anode material. The attenuation of the beam at any point depends on the material at that point and on the energy distribution of the beam so, as reference to Figure 4.6 shows, the beam spectrum must change during passage. In fact it is said to 'harden', that is the lower energy components of the beam are progressively removed.

4.4 X-RAY PARAMETERS

The performance of an X-ray imaging system for diagnostic use may be characterized by four quantities: contrast, blur or lack of focus, noise and radiation dose. Suppose we idealize the physical situation (Figure 4.10) and consider a cubic block of material having dimensions t_2 and linear attenuation coefficient μ_2, implanted in a substantial amount of a material having thickness t_1 and attenuation coefficient μ_1. The receptor response is assumed to be linear with the energy absorbed. If the energy absorbed in unit time per unit area of receptor outside the 'target' region is I_1 and that within is I_2, then we define the contrast C as

$$C = (I_1 - I_2)/I_1 \tag{4.5}$$

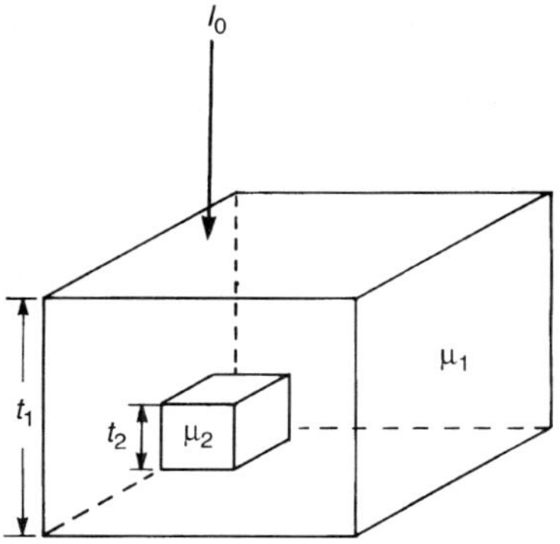

Figure 4.10 Tissue model for the calculation of contrast in an X-ray image. A cubic block of material with attenuation coefficient μ_2 and linear dimension t_2 is embedded in a slab of material with attenuation coefficient μ_1 and thickness t_1. The incident beam intensity is I_0.

If we assume for the moment that scattering processes may be ignored, then we can use equation 4.2 and write that

$$I_1 = I_0 \exp(-\mu_1 t_1) \tag{4.6}$$

and

$$I_2 = I_0 \exp[-\mu_1(t_1 - t_2) - \mu_2 t_2] \tag{4.7}$$

so that

$$C = 1 - \exp[-(\mu_2 - \mu_1)t_2] \tag{4.8}$$

If we include the effect of scattering, using equation 4.4, then we obtain

$$C = [1 - \exp(-(\mu_2 - \mu_1)t_2)]/(1 + R) \tag{4.9}$$

Equation 4.9 tells us that contrast must be determined by three factors of which two are very significant, namely the difference between the two attenuation coefficients, and the thickness of the implanted or 'target' material; contrast increases as both these terms increase. The ratio of scattered radiation to primary radiation will also have an effect, of course. The significance of equation 4.9 may be seen more clearly by examination of Figure 4.11, which displays the percentage by which the linear attenuation coefficient of body materials differs from that of water, and emphasizes the high contrast that we can expect, and indeed get for bony structures, against that which will be found for fat and tissue. Unfortunately, the position is made more complicated by the fact that

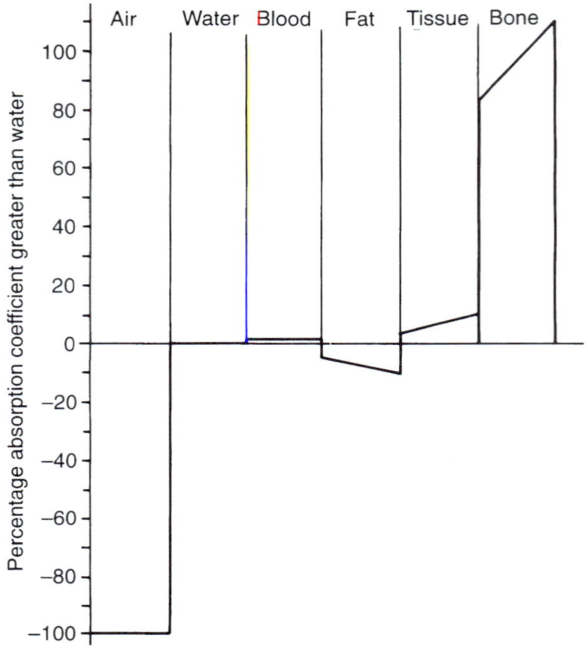

Figure 4.11 Variation of the absorption coefficient μ for various components of the human body, normalized to that of water (principal component of human body). In the cases of human tissue and bone there is a range of values for μ, and this is indicated by the slope at the top of these columns. An air space or a bone will offer the greatest contrast in the X-ray image, the differences between fat and tissue being more marginal.

contrast decreases rapidly with photon energy, roughly in the manner indicated in Figure 4.12, so that for best contrast we should use low photon energies. However, as has already been pointed out, low photon energies result in a high patient dose so that a compromise must always be struck between patient dose and contrast. The situation is further complicated by the fact that practical image receptors do not have the linear response assumed above, and the contrast given by equation 4.9 is incorrect; equation 4.9 has to be modified in practice. X-ray film usually allow contrasts of approximately 2% to be visualized.

It is possible in some situations to enhance the contrast by artificial means **(angiography)**. The methods by which this is achieved vary slightly, but usually depend on filling the organ in question with a high atomic number compound. The choices are between barium sulphate in

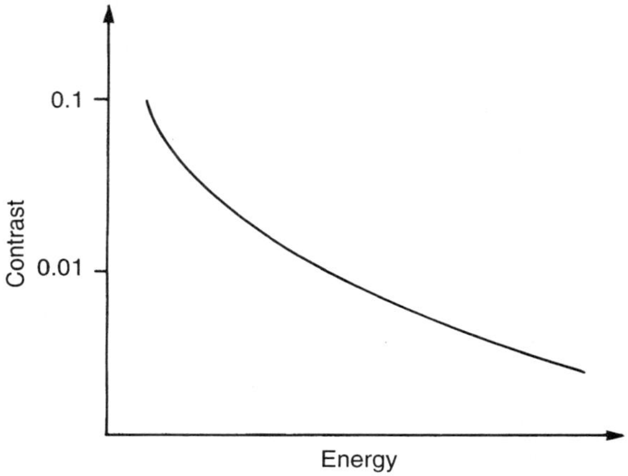

Figure 4.12 Trend in the variation of contrast for human tissue with X-ray energy. The energy axis is logarithmic.

aqueous suspension (barium sulphate is very insoluble in water) or an organic compound containing up to 50% iodine, the actual selection depending on the situation. A barium meal (barium sulphate) is used to outline the oesophagus, stomach, gastro-intestinal tract, etc., while iodine compounds may be used to outline small blood vessels. In some instances air is a useful contrast medium.

Blur or lack of focus in an X-ray image can best be understood by considering the problem of imaging a bar pattern. Image blur can then be defined in terms of the highest bar frequency (lines per mm) which can be visualized. There are, however, three easily identifiable contributions to image blur. The first and most obvious is the finite size of the X-ray emitting area on the X-ray tube anode. This yields a penumbra around every image and constitutes geometric image blur. Tubes with very small focal spot size, say 0.3 mm × 0.3 mm, reduce this problem. Second, the receptor itself may impose a limit on image resolution (receptor blur), while finally we must include patient movement (movement blur). Usually, movement blur can be neglected, but it is very significant when imaging parts of the body which are always in motion, for example the heart and its associated blood vessels.

Geometric blur may be understood more readily by reference to Figure 4.13. The image blur is just the width of the penumbra Δl and is given by

$$\Delta l = xd_2/d_1 \tag{4.10}$$

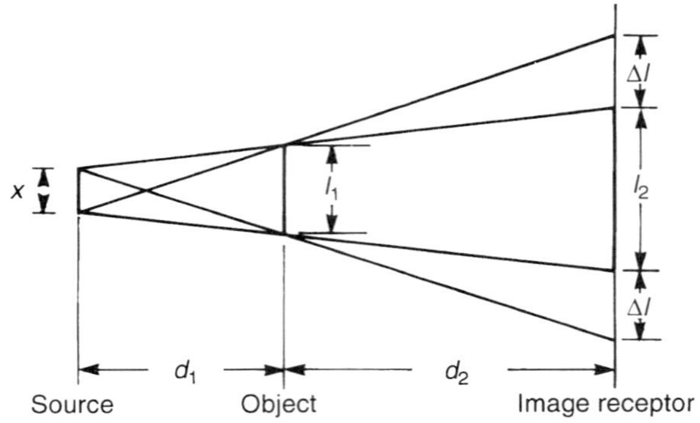

Figure 4.13 Geometry of X-ray source, object and image for the calculation of geometric blur or lack of focus. The source has a width x while the object being imaged has a width l_1; Δl is the penumbral width.

where x is the X-ray source width and d_1 and d_2 are the source–object and object–receptor distances, respectively. The penumbral blur is normalized by dividing by the system magnification m, where $m = (d_2 + d_1)/d_1$. The normalized blur ΔL is given by

$$\Delta L = x(1 - 1/m) \tag{4.11}$$

The overall image blur is, of course, not only that due to the source, but that due to the receptor as well. Receptor blur, which we will designate ΔR, may be combined with source blur to yield the overall blur U, where

$$U = (\Delta R^2 + \Delta L^2)^{1/2} \tag{4.12}$$

and we have assumed that the overall blur or lack of focus may be obtained by adding the two types of blur in quadrature.

The receptor blur ΔR is actually the intrinsic blur Δr, which would be measured if the object had zero thickness and was in contact with the receptor, normalized to the magnification m, i.e. it equals $\Delta r/m$. If we substitute the appropriate values then we obtain, for an X-ray source of width x

$$U = \left(\frac{\Delta r^2}{m^2} + x^2\frac{(m-1)^2}{m^2}\right)^{1/2} \tag{4.13}$$

Inspection of equations 4.10 and 4.13 shows that if ΔL is much greater than ΔR then the overall blur must increase with increasing magnification, whereas if ΔR is much greater than ΔL then the overall blur will be inversely proportional to the magnification. For general radiography the overall blur will increase with the magnification and the patient should be as close as possible to the image receptor; the source–receptor distance should be as large as possible. For magnification radiography, the overall blur or lack of focus will decrease with increasing magnification at modest magnifications, but thereafter will increase, so that careful choice of magnification is required.

Before considering the problem of patient dose we must define the units involved. Radiation dose is defined in terms of the energy deposited by the radiation, per unit mass of material. The unit is the gray (Gy): $1\ \text{Gy} = 1\ \text{J kg}^{-1} = 6.24 \times 10^{12}\ \text{MeV kg}^{-1}$. This measures the *radiation absorbed dose*. Some types of energy input do more biological damage than others, for example a 1 MeV α-particle (helium nucleus) does more damage than a 1 MeV electron. This is taken into account by defining the *radiation dose equivalent*, in which the radiation absorbed dose is multiplied by a factor Q which depends on the type of radiation. The more dangerous the radiation, the higher the value assigned to Q. The unit of the radiation dose equivalent is the sievert (Sv): $1\ \text{Sv} = 1(\text{Gy}) \times Q$. Table 4.1 gives the values assigned to Q for various kinds of radiation.

The problem of dose is related to the final imaging problem, that of 'noise'. An imaging system may have high contrast and good resolution, but this is of little consequence if the noise level in the image is too high, since the image will be obscured. The two major contributions to the noise level are statistical fluctuations in the number of X-ray photons detected per unit area (quantum noise) and fluctuations due to the properties of the image receptor and display system. The appearance of noise in an image will be affected by the spatial frequency response of the system, so that a poor spatial frequency

Table 4.1 *Radiation Q factors*

Radiation type	*Q value*
Gamma-ray	1
X-ray	1
Electron	1
Protons (10 MeV)	1
Protons (1 GeV)	2
Thermal neutrons	3
Fast neutrons	10
Alpha-particles	20

response will yield a blotchy image, described by the term 'quantum mottle' or 'radiographic mottle'.

The noise due to quantum mottle may be reduced by increasing the number of X-ray photons used to form the image, but this results in increased dose to the patient. Clearly we need some means by which we can quantify this dependence. If we return to the model from which we derived equation 4.9 we can ask the question: What is the surface dose required in order to establish a contrast C over an area A against a background of quantum mottle? Put another way, we need to establish the equation describing the signal-to-noise ratio (SNR).

The signal we are trying to detect, arising from an area A, is just

$$\text{Signal} = (I_1 - I_2)A = I_1 CA \tag{4.14}$$

If we substitute for I_1, using equation 4.6, only now including the effect of scattering as well as taking account of the number of photons incident per unit area N, the photon energy E, and the receptor efficiency α, then we can write

$$\text{Signal} = CA[\alpha NE \exp(-\mu_1 t_1)(1 + R)] \tag{4.15}$$

The noise in the image arises from fluctuations in the energy absorbed by the receptor and is described by Poisson statistics, so that the number of photons detected in an area A, adjacent to the target is $E(I_1 A/E)^{1/2}$, and the noise is given by

$$\text{Noise} = E[\alpha NA \exp(-\mu_1 t_1)(1 + R)]^{1/2} \tag{4.16}$$

The signal-to-noise ratio is then

$$\text{SNR} = C[\alpha NA \exp(-\mu_1 t_1)(1 + R)]^{1/2} \tag{4.17}$$

Substituting for C from equation 4.8 we obtain

$$\text{SNR} = [1 - \exp(\mu_1 - \mu_2)t_2][\alpha NA \exp(-\mu_1 t_1)/(1 + R)]^{1/2} \tag{4.18}$$

The signal-to-noise ratio is important in another way because once it exceeds a certain value, an object becomes detectable in the X-ray image; this condition is identically that for the minimum patient dose.

If we set this threshold signal-to-noise ratio equal to r, then we can obtain an expression for N_{min}, the minimum necessary number of photons per unit area, incident on the patient. Rearranging equation 4.18 and assuming that the difference $(\mu_1 - \mu_2)$ is small we can write that

$$N_{\text{min}} = \frac{(1 + R)r^2}{\alpha(\mu_2 - \mu_1)^2 t_2^4} \exp(\mu_1 t_1) \tag{4.19}$$

We have assumed in deriving this expression that the object being imaged is actually a cube of side t_2. The minimum surface dose is obtained as the product of the number of photons per unit area N_{min}, the mass energy attenuation coefficient $(\mu_1/\rho)_e$ and the photon energy E, so that

$$\text{Minimum dose} = (\mu_1/\rho)_e E r^2 (1 + R) \exp(\mu_1 t_1)/\alpha(\mu_2 - \mu_1)^2 t_2^4 \quad (4.20)$$

Equation 4.20 is very instructive in that it tells us that *the minimum dose required to visualize an object increases as the inverse fourth power of the size of the object, and the inverse second power of the difference in attenuation coefficient*. Of course the model for equation 4.16 is very idealized. In practice there will be other constraints so that if, for example, a film-based receptor is used, there will be a minimum dose required to produce appreciable blackening, and for that matter a maximum dose which saturates the film. The speed of the image receptor will have a critical effect on the noise in the image and hence on the resolution available at low contrast.

We can summarize, at this point, the outcome of our consideration of the minimum radiation dose and image contrast in the radiographic process, since it serves to emphasize the significance of differences in linear attenuation coefficient as they occur in the body. Equation 4.9 shows that contrast increases with the difference between the attenuation coefficients and the dimensions of the organ being viewed. The minimum dose, on the other hand, increases as the inverse second power of the difference in attenuation coefficient and the inverse fourth power of the object dimensions. For a fixed organ size the ratio between contrast and dose is increased by a large difference between the attenuation coefficients. Alternatively, if we relax the restriction on organ size, we should aim to be imaging large organs if we wish to maintain a favourable relationship between contrast and dose.

4.5 X-RAY PRODUCTION

The modern x-ray tube is based in its essentials upon the design introduced by Coolidge, although with many refinements. The actual X-ray tube is a borosilicate glass envelope within which are mounted a tungsten filament electron source and a tungsten anode (Figure 4.14). The filament is heated by the passage of an electric current and produces a narrow beam of electrons which are accelerated by a potential difference of between 20 and 150 kV before striking the anode. During their collision with the anode material the electrons slow

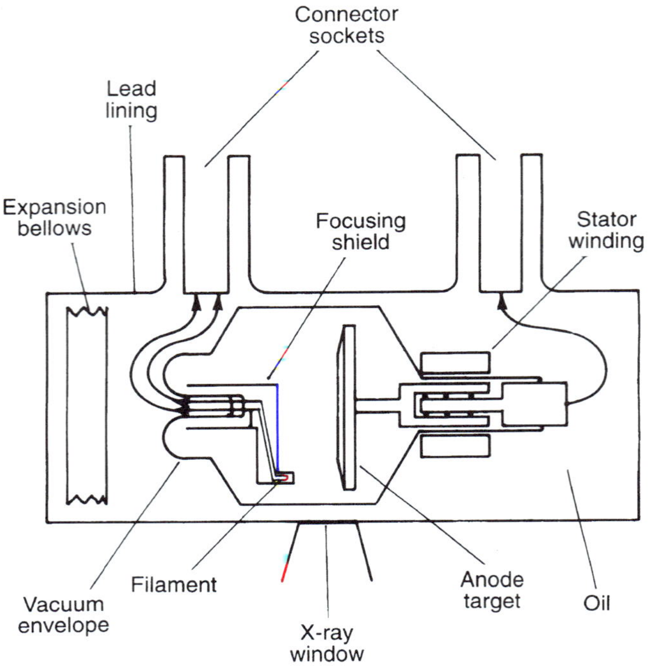

Figure 4.14 Section through a typical diagnostic X-ray tube with a rotating anode, indicating the essential features. The focusing shield is shown connected to the filament, although this will not always be the case.

down and stop, delivering their energy to the anode where it appears largely as heat (99%) and X-rays (1%). Some of these X-rays pass through the exit window in the oil-filled outer housing which encloses the X-ray tube; the remaining X-rays are absorbed by the lead lining of the housing.

The design of the filament and the electron optics which guide the electrons to the anode is of critical importance, since they control the size and shape of the electron spot on the anode and, in turn, the size of the X-ray source. The X-ray output from the tube is determined by the magnitude of the electron current striking the anode. The filament is constructed from a spiral of tungsten wire set in a nickel block. This block not only supports the filament, but is also shaped to create an electric field which focuses the electrons into a slit, or rectangular cross-section beam. The anode has a bevelled edge which is at a steep angle to the axis of the electron beam. The exit aperture accepts X-rays which emerge approximately at right angles to the electron beam axis so that the X-ray source, viewed from the receptor, appears to be approximately square; Figure 4.15 shows this diagrammatically. The choice of anode angle depends on the application. Generally a bevel of 17° is used, although the anode disc may have two bevels at different angles (and two filaments) so that either a coarse or a fine focus may be selected.

A central problem in the design of an X-ray tube is how to disperse the heat generated in the anode during electron beam impact. Although the focal spot on the anode should be as small as possible from the point of view of obtaining a sharp X-ray image (reduced penumbral blurring), this has the consequence that the power density in the spot will be very high and the local heating very intense. Two methods are adopted to overcome this problem. The first has already been described, and is the use of a rectangular section electron beam impinging on a bevelled anode, so that the impact area is higher than the apparent source area. A source projecting as 2 mm × 6 mm on the anode will appear as 2 mm × 2 mm owing to the bevel on the anode; the power

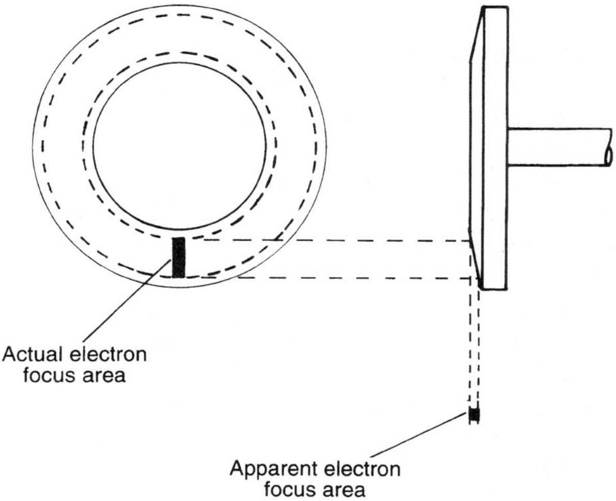

Figure 4.15 Two methods by which the power density delivered by an electron beam to an X-ray anode is reduced. The actual focus area is the rectangle shown on the plan view of the anode. The apparent X-ray source is a square (shown on the anode elevation). By rotation of the anode the power from the electron beam is delivered to an area represented by the annulus enclosed by the dashed lines on the plan view. The power density is thus dramatically reduced.

density is consequently reduced by a factor of three. A more dramatic reduction in power density is achieved by rotating the anode so that the electron beam actually impinges on an annular ring 6 mm wide. A rotation speed of 3000 rpm and an anode diameter of approximately 100 mm are typical, giving an effective area of approximately 1500 mm^2. Although rotation of the anode reduces the power density at the anode, the total power delivered to the anode is not altered. This power is dissipated as heat, which is removed by radiation through the vacuum of the X-ray tube and then by absorption in the surrounding oil. Heat from the oil is transferred by convection into the surrounding air. Rotation is imparted to the anode by mounting it on the squirrel-cage of an induction motor, the stator of which lies outside the vacuum envelope. The rotating magnetic field, supplied by the stator winding, produces rotation of the anode within the vacuum envelope.

The anode material usually chosen is tungsten (as in the early Coolidge tubes) although molybdenum is also used, particularly where a low energy X-ray beam is required. Tungsten has a high melting point (3410 °C), adequate thermal conductivity and thermal capacity, and a high atomic number (74). The high atomic number is important because the bremsstrahlung (braking radiation) yield from the target increases with atomic number and the X-ray spectrum from an element of high atomic number is well suited to imaging the thicker body sections. The continual heating and cooling of the X-ray anode during normal operation leads to crazing of the anode surface; this is overcome by the adoption of a 90/10 tungsten/rhenium alloy. A further important parameter is the thermal capacity of the anode, which should be high. A high thermal capacity permits a high power rating and hence short exposure times and, for that matter, a shorter interval between exposures. The thermal capacity of a tungsten anode can be increased artificially by backing it with molybdenum, which has a higher specific heat than tungsten. Reference to Figure 4.14 shows that the anode is mounted on a thin stem. This is usually made from molybdenum, and is intended to reduce conductive heat flow into the copper bearings of the squirrel-cage rotor. The total heat load on the anode of an X-ray tube can be very high so, for example, at an emission current of 10 mA and an accelerating voltage of 100 kV the total power delivered to the anode is 1 kW, of which 0.99 kW must be dispersed as heat. In practice, power dissipation as high as 75 kW can occur.

We have seen that the output from a conventional X-ray tube is a broad spread of wavelengths – bremsstrahlung – upon which are imposed the characteristic wavelengths corresponding to the anode material. As the cathode/anode potential is increased the short wavelength cutoff of the bremsstrahlung moves to ever shorter wavelengths. Our discussion of the problem of dose to the patient showed that if the X-ray spectrum is too soft then the lower energy photons will only contribute to patient dose and not to image contrast. It is important that these photons are removed from the X-ray beam before it reaches the patient. In principle we have at our disposal two basic X-ray spectra, that from tungsten and that from molybdenum. However, the details of these spectra may be 'tailored' either by using different accelerating potentials, or by adding filters, or both. Usually these filters are of copper or aluminium, although molybdenum filters are used for energies below 20 keV, particularly for **mammography.** The use of tungsten as an anode material, and the consequent tungsten X-ray spectrum, is most suitable for imaging the thicker body sections. Molybdenum anodes are more appropriate for imaging thinner body sections at high contrast. Filters of aluminium or copper, mentioned above, are selected according to tube potential.

4.6 IMAGE RECEPTORS

Historically the first image receptor was direct exposure photographic film. Nowadays, direct exposure photographic film is used only in special situations, because it has low absorption efficiency for X-ray photons in the diagnostic energy range. It is however used as the final display medium for many types of imaging system. A preferred alternative is the screen–film combination. This is faster than direct exposure film, but does not have equivalent spatial resolution. Here there is a trade-off between the dose to the patient (which is lower) and resolution. An even lower patient dose may be achieved using an image intensifier, but once again the trade-off is image quality. Nevertheless, in many applications there is no alternative. We begin our consideration of image receptors by examining the way in which X-ray film works.

Figure 4.16 shows a cross-section through a direct exposure X-ray film. The film comprises two sensitive photographic emulsions, some 20 μm thick, deposited on either side of a transparent sheet (usually polyester or acetate around 200 μm thick) which forms the film base. The photographic emulsions are attached to the film base by means of so called subbing layers, whilst their outer surfaces are protected from abrasion by a thin surface coating. The sensitive photographic emulsions are suspensions of silver bromide particles, some 1 μm in

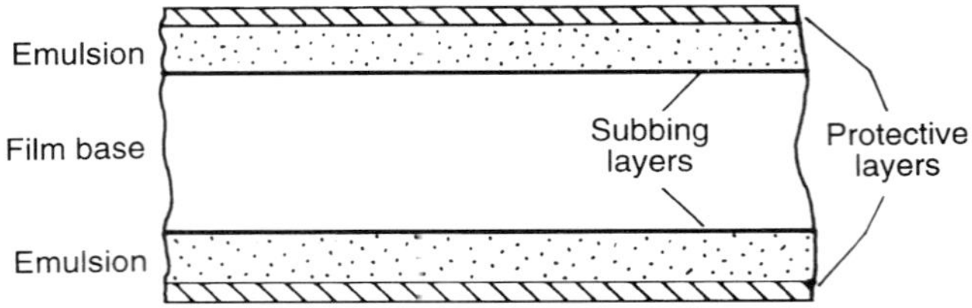

Figure 4.16 Cross-section through a typical X-ray film showing the manner in which the photographic emulsion is placed either side of a base layer.

diameter, in a gelatine base. Two emulsions are used to increase the X-ray absorption efficiency. When X-ray photons interact with the photographic emulsion, the silver and bromine atoms of the halide particles present a much larger effective absorption cross-section than the lighter elements, carbon, hydrogen, etc. forming the supporting gelatine. Consequently, most of the interaction will be with the halide grains. During this interaction energetic electrons are released, and as these electrons slow down, further electrons are released by ionization. Some of these electrons are captured at trapping centres within the silver bromide grains and a few silver bromide atoms are reduced to silver; they form a latent image. The few silver atoms in each grain act as sensitizers or catalysts so that when the film is developed the whole of the grain is reduced to silver. Light is absorbed most strongly in the regions of the film containing many silver grains, producing there a region of blackness. Where rather less radiation has fallen the film will appear grey, while in those areas where no X-ray radiation has fallen the film will appear almost transparent.

This blackening of the film is usually expressed in terms of the optical density D, which is defined by the equation

$$D = \log_{10} I_0/I_t \qquad (4.21)$$

where I_0 is the incident light intensity and I_t is the transmitted light intensity. The human eye is very sensitive to different shades of grey and can discern radiographic detail very accurately. The physiological response of the eye is logarithmic, so that if we require a quantitative measurement it should have a logarithmic basis such as we see in equation 4.21. In addition, we need some idea of the relationship between the optical density and the film exposure. A simple model of the exposure situation permits us to do this.

We begin by considering an emulsion containing k grains of silver bromide per unit area which, after irradiation with X-rays delivering N

photons per unit area, generates k^* sensitized grains per unit area each presenting a cross-section b. The rate of production of sensitized grains may then be assumed proportional to the unsensitized area times the area per grain. If we assume that a single photon hit ensures the production of a sensitized grain, we can eliminate the proportionality and write

$$\frac{dk^*}{dN} = (k - k^*)b \tag{4.22}$$

on integrating this becomes

$$k^* = k[1 - \exp(-bN)] \tag{4.23}$$

The product bN represents the number of X-ray photons hitting bromide grains and is related to the dose.

The connection with the optical density D can be made by assuming that every sensitized grain contributes a speck of silver having cross-section area σ. Then, for an emulsion of thickness t, the optical intensity $I(y)$ at a depth y is given by the equation

$$\frac{dI}{dy} = -I(y)\left(\frac{k^*}{t}\right)\sigma \tag{4.24}$$

Integration yields the relationship

$$I_t = I_0 \exp(-k^*\sigma) \tag{4.25}$$

where I_t is the intensity after passing through thickness t. Equation 4.25 gives us a further relationship between I_t and I_0 so that we can rewrite equation 4.21 to give

$$D = 0.434k^*\sigma \tag{4.26}$$

This equation represents Nutting's law. The maximum possible optical density D_{max} is achieved when $k = k^*$ so that

$$D_{max} = 0.434k\sigma \tag{4.27}$$

Equations 4.23, 4.26 and 4.27 can be combined to give

$$D = D_{max}[1 - \exp(-bN)] \tag{4.28}$$

Although this result has been derived by means of a rather simple model, it does reflect actual film properties quite well, as reference to Figure 4.17 shows. These plots are for film density against exposure on the one hand and of equation 4.28 on the other. The prediction of equation 4.28 does not reflect the fact that for a real X-ray film even

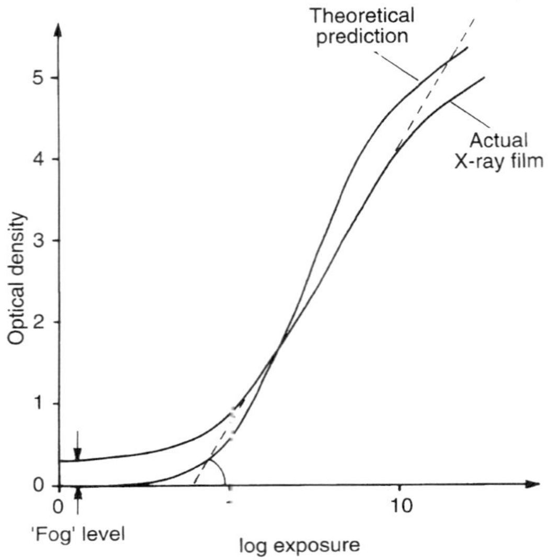

Figure 4.17 Graphical representation of actual film density against the prediction of equation 4.28. The off-set shown by actual X-ray film is due to background 'fog'. The slope of the dashed line represents the film 'gamma', or speed measure.

unexposed areas will contain some grains which reduce to metallic silver and thus provide a background 'fog' level of indeterminate magnitude. Nevertheless, it can be seen that the comparison is quite good.

Over a range of exposures, the plot of Figure 4.17 shows a more or less linear response to exposure as does the film itself. The slope of this straight section is known as the film 'gamma' (Γ) and is a measure of the film speed, a typical value would be in the range 2–3. Fast film has a large gamma. The whole curve is known as the characteristic curve of the film.

One way in which the speed of a film may be increased is by the use of an intensifying fluorescent screen. Figure 4.18(a) shows a cross-section through an image intensifying screen while Figure 4.8(b) shows a screen–film combination. As usual in these matters there is a trade-off, so that although apparent film speed may be increased by a factor of around 30, thus reducing patient dose, the resulting image does not possess such good resolution. Despite this latter drawback, a screen–film combination would usually be the image receptor of choice since,

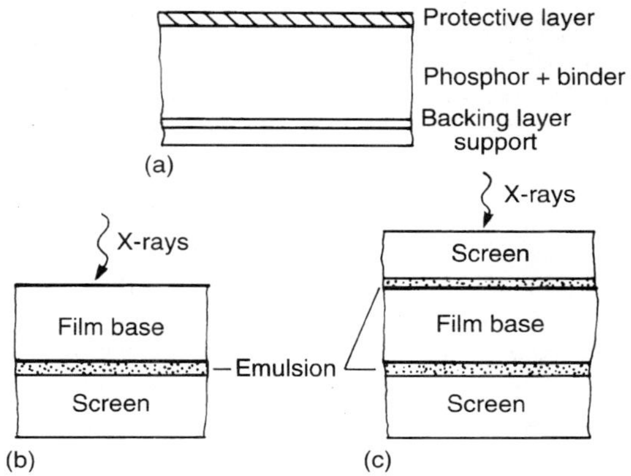

Figure 4.18 (a) Cross-section through a typical fluorescent screen structure used in conjunction with X-ray film. (b) Configuration for single screen–film combination. (c) Configuration for double screen–film combination.

in most applications, low dose is more important than the loss of very fine detail.

Screen–film receptors produce an image in a multistep process. Initially an X-ray photon is absorbed by the screen phosphor and some of its energy is re-emitted in the form of light fluorescent photons. These photons expose the photographic emulsion of a film in contact with the screen (Figure 4.18(b)). The photographic image is then developed in the normal way. The phosphors usually used are calcium tungstate, or terbium-activated, rare earth oxysulphides, where the rare earth is usually gadolinium, lanthanum or yttrium. These phosphors have a high mass energy absorption coefficient. The film-screen combination of Figure 4.18(b) is sometimes replaced by a double screen (Figure 4.18(c)), where the photographic emulsion is sandwiched between two screens.

4.7* IMAGE INTENSIFIERS

We have already noted the need to minimize patient dose whilst obtaining an X-ray image. An approach that reduces the dose to a very low level, at the cost of some deterioration in image quality is the use of an image intensifier. The image intensifier not only reduces the dose in routine radiological studies, but it may be used in an entirely different way, a fluoroscopy mode. In this manner of operation the X-ray tube runs continuously, but at a very low current, so that observations may be made of bodily processes which involve movement. An important example of such a situation is cardiac catheterization where the insertion of the catheter and its transport into the heart must be monitored.

The basic structure of an image intensifier is shown, in simplified form, in Figure 4.19. It comprises an evacuated glass tube with a fluorescent screen at each end, a photocathode in close contact with one of the fluorescent screens and simple electron optics. At the input end, the window of the tube must have very high transparency to X-rays. This is achieved by using thin glass or, in some cases, a thin metal window (aluminium) joined to the glass envelope by a glass–metal seal. When the X-rays have passed through the window they strike a fluorescent screen deposited on the vacuum side of the window. Light photons from this screen impinge on the adjacent photocathode which emits photoelectrons. These photoelectrons are then accelerated towards the viewing or output phosphors by means of a potential difference of some 20–30 kV established between the photocathode and the output phosphor, the photoelectrons being guided by appropriate electron optics. The diameter of the input screen may be anywhere in the range from 125 to 570 mm, while that of the viewing screen is typically 25 mm. This demagnification of the image, coupled with the electron acceleration, ensures that the image intensifier has a very high photon gain. As a guide, the increase in the light output intensity of the

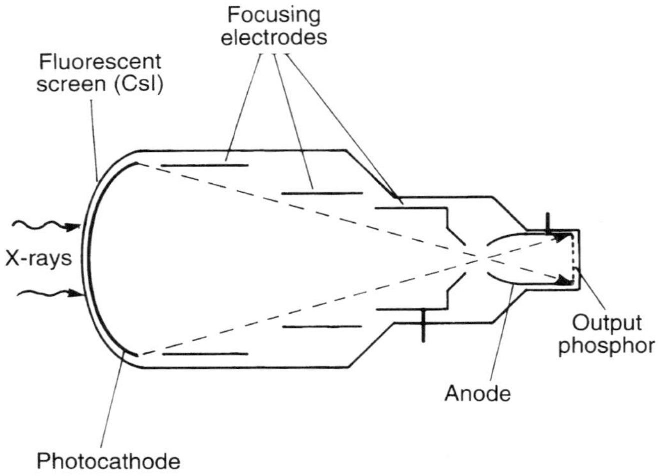

Figure 4.19 Simplified cross-section through an image intensifier.

image intensifier, compared with that from a standard fluorescent screen, can be as high as 10^4.

Modern image intensifiers depend upon caesium iodide input screens. This phosphor can be laid down with a packing density of 100%, as no binder is required. Caesium iodide produces about 2000 light photons per incident X-ray absorbed and since the photocathode efficiency is typically 0.1 we can expect 200 electrons per single incident X-ray photon absorbed. These photoelectrons are focused and accelerated towards the output screen to form the image. The output from an image intensifier is not really suitable for direct viewing by eye, instead it is usually recorded on film or viewed with a TV camera.

4.8* XERORADIOGRAPHY

A radiographic technique, which found particular application in **mammography**, is xeroradiography. As the name suggests, it is based on a principle similar to that of the xerox copier. Xeroradiography is a dry non-silver photographic system which produces images on paper which are viewed by reflected light. Although it is slower than systems using screen–film receptors, it has better resolution and produces images which are edge enhanced.

The xeroradiographic image process depends on the photoconductive properties of selenium. The image receptor comprises a metal plate (aluminium) evenly coated with a thin (125 μm) layer of amorphous selenium. Before exposure the selenium is given a uniform positive charge and this is maintained until exposure by keeping the receptor plate in a light-tight enclosure. In practice this situation can only be maintained for a limited amount of time since the surface charge leaks away with a half-life of about 1 h. When the plate is exposed to X-rays, the incident X-ray photons produce electron–hole pairs within the bulk of the selenium which migrate towards the surface of the selenium under the influence of the internal electric field. Those electrons which reach the surface of the plate form a latent image by virtue of cancellation of the original charge distribution.

The resulting charge image is developed by spraying the plate with an aerosol of fine blue powder particles. These particles are charged by friction and are attracted by the charge distribution on the plate. As a result a powder pattern builds up on the plate which replicates the charge pattern. This powder image is transferred to plastic coated paper by a contact process and fused into its surface by heating; the image is then viewed in reflected light. The powdering process is actually controlled by applying a bias to the selenium plate, and since the polarity of this bias is subject to choice, we can have either positive or negative images. A positive image has dark blue regions corresponding to the most highly charged areas of the plate, while the reverse is true for a negative image.

When the selenium plate is developed, it is held in a horizontal position so that, without the influence of the electric field, the charged particles fall perpendicularly towards the plate by gravitational attraction. Of course the electric field dominates and the charged particles follow the lines of electric force. Where a difference, or step change, in charge density exists, the electric field lines are perpendicular to the plate at a distance from the step change, and they impinge on the plate normally. However, at the step change in charge density, the situation is more complex and the electric field lines loop between the two regions, this results in powder piling up on one side of the step and being depleted on the other. Figure 4.20(a) shows the distribution of equipotentials which gives this effect, while figures 4.20(b) and (c) show the vertical and horizontal field components respectively. Figure 4.21 shows an image of the female breast obtained by this technique.

4.9* ADVANCED IMAGE RECEPTORS

Under the title of advanced image receptors we must include some of the more recent developments in receptor design. These have a special importance since they are ideally suited for use in digital radiology, the subject of the next section. We shall consider two advanced image receptor systems, namely photostimulated luminescence based receptors and ionographic receptors. The former employs a phosphor which exhibits stimulable luminescence when irradiated with high energy radiation (X-rays), while the latter employs a novel, gas-filled ion chamber for image production.

The phenomenon of stimulable luminescence of phosphors irradiated with high energy radiation has been known for a long time; luminescence was mentioned by Robert Boyle in 1663 in connection with certain kinds of diamond. Upon irradiation of a storage phosphor, part of the absorbed energy is stored in the phosphor by trapping of free charge carriers. The stored energy can be released in stimulated emission. If the phosphor is irradiated with X-rays a dose-proportional amount of energy is stored in the phosphor by the trapping of free electrons and holes. This stored energy can be released by scanning the phosphor screen with a focused helium–neon laser. For each spot on

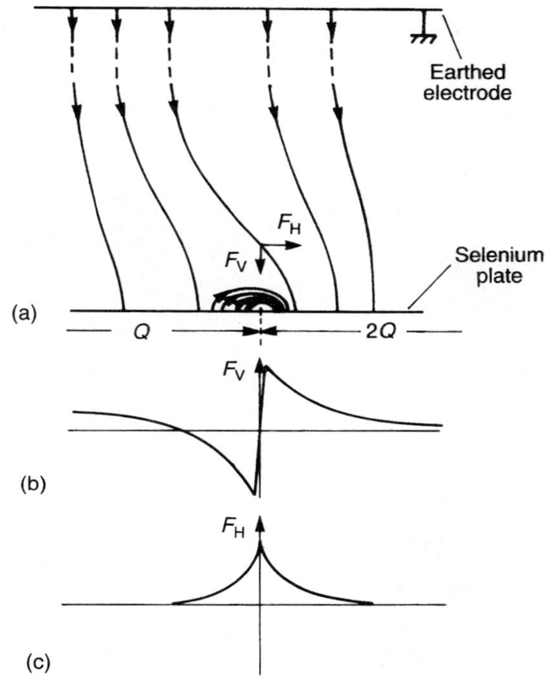

Figure 4.20 (a) Electric field distribution associated with a charge density step on a xeroradiograph (not to scale). The charge density Q changes to $2Q$ at the step. (b) Vertical component of electric field. (c) Horizontal component of electric field.

the screen the intensity of the stimulated emission is measured using a photomultiplier tube, and the signal is then digitized.

The photostimulable phosphor generally used for X-ray imaging systems is a europium-activated barium fluorobromide. The mechanism responsible for photostimulated luminescence is believed to involve a halide vacancy acting as a trap for electrons, while the europium ion Eu^{2+} acts as a hole trap. Electron–hole recombination then occurs via a direct recombination mechanism involving tunnelling, since the photoluminescence decay, under continuous stimulation, is temperature independent. This type of mechanism seems to apply to a range of phosphors in which halide vacancies act as electron traps and Eu^{2+} is involved in hole trapping. Phosphors introduced in 1991, based upon europium-activated barium bromoborate and barium bromogermanate, have the same mechanism.

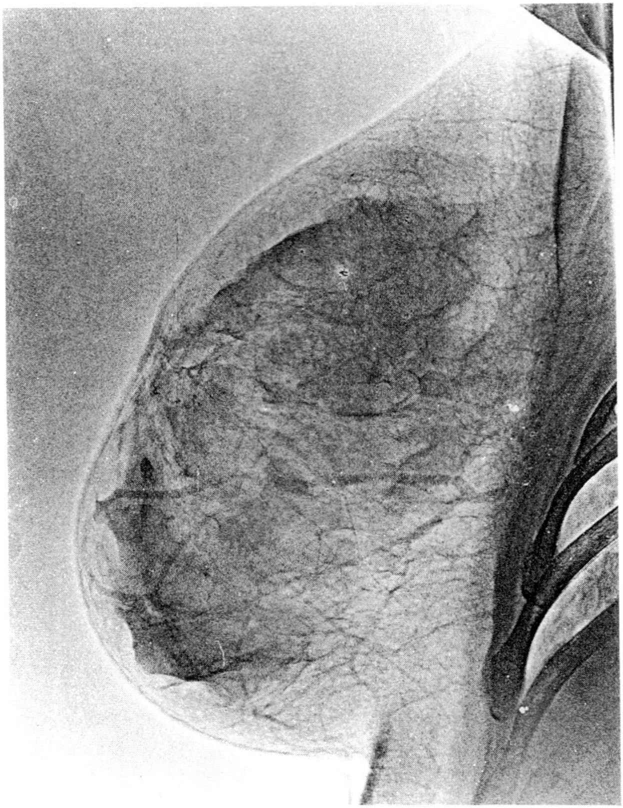

Figure 4.21 Xeroradiograph of the female breast. (Courtesy of Mr A. Ashley.)

The actual image receptor is a flexible plate coated with photostimulable phosphor, which can be scanned by a laser beam using an oscillating mirror (Figure 4.22). After each lateral scan the receptor plate is advanced by one pixel width. The laser spot size is 0.1 mm and the image is sampled at 5–10 pixels mm^{-1}. The radiation emitted by the plate is collected by a light guide, which is coupled to a photomultiplier tube. The resulting signal is digitized by means of an analogue-to-digital convertor and then stored. The earliest example of this type of imaging system (due to the Fuji Photo Film Company) has a dynamic range of more than 10 000 : 1 and can accept a wide range of radiographic exposures. The spatial resolution of such a system is

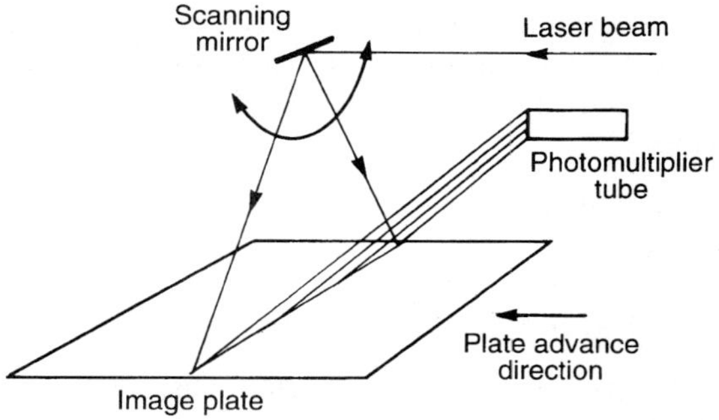

Figure 4.22 Laser scanning system for reading out the image stored on a photostimulable plate.

determined by the laser spot size. Phosphors of this type have a high photon efficiency (conversion efficiency); figures of approximately 70% are typical, depending on the phosphor. The amount of energy stored in the phosphor upon X-ray irradiation is usually in the range 1–4 pJ mm^{-2} mR^{-1}, where R is the unit of exposure, the roentgen. The roentgen is defined in terms of the charge deposited per kilogram of material: $1\ R \equiv 2.58 \times 10^{-4}\ C\ kg^{-1}$.

Ionography, or as it was initially called, electron radiography, was introduced in 1973 by Muntz and co-workers. Ionographic imaging systems use a special gas-filled chamber rather than a solid detector. A high atomic number gas, at high pressure, is used to absorb incident X-ray photons and generate charge carriers. Each incident X-ray photon strips an electron from a target atom to produce a positive ion. The gas is situated between two electrodes which supply an electric field to separate the charges. Either positive or negative charge can be collected on an aluminized, Melinex membrane stretched over the appropriate electrode. The charge image is developed by dusting with a powder cloud. High efficiency is achieved by using electrode separations and gas pressures such that the product of gas pressure and gas thickness is 50–100 atm mm, in a chamber depth of 10 mm. There are engineering problems associated with the use of gas at high pressures; equally there are X-ray absorption problems, particularly if low X-ray energies are used. Thus for mammography, where the X-ray energies are low (between 30 and 40 kV), it is desirable to operate with relatively low (1 atm) gas pressure.

The design of an ionographic chamber should maximize sensitivity, resolution and contrast. In this instance contrast depends to a very great extent on the development process and need not be considered further; we are left with a consideration of sensitivity and resolution. The sensitivity and resolving capability of an ionographic chamber depends on four factors:

1. the nature of the gas employed;
2. the gas pressure;
3. the electrode separation;
4. the electric field and therefore the applied voltage.

All of these factors interact to some extent. Thus, if a low gas pressure is chosen, for engineering and ease of use reasons, then a deep chamber is required to ensure adequate X-ray absorption. This has the consequence that high voltages must be employed to provide adequate electric fields. The choice of filling is also important since the gas should have a high atomic number, a high density and good resolving capability. The range of halogenated hydrocarbons (Freons), which are used as refrigeration agents, provide convenient and relatively cheap solutions to the gas filling problem. The choice of gas filling also has a bearing on the value of electric field selected; high electric fields minimize blur from lateral diffusion. If a deep chamber is chosen then it must possess spherical plate geometry to minimize blur due to obliquity of the X-ray beam. In addition, the X-ray source must be aligned with the centre of curvature of the chamber. With a deep chamber the electrostatic field lines at the edge of the chamber have to be corrected to remove radial distortion which would produce image blur.

Figure 4.23 shows a cross-section through an atmospheric ionographic chamber, displaying the features outlined above, and based upon the design of Moores *et al.* (1980). It was intended for use in mammography, where it can produce clinical images with a lower exposure than film–screen combinations. It comprises a cylindrical Perspex container closed at one end by a concave Perspex disc and at the other by a planar Perspex disc. The inside of the concave disc is lined with an aluminized, 50 μm thick, plastic (Melinex) sheet, the aluminized surface facing towards the Perspex disc. Mounted upon the bottom Perspex plate is an aluminium electrode with curvature concentric with that of the top electrode. The Perspex sleeve is lined with a series of guard rings to correct the edge electric field between the two concave electrodes. The potentials for the guard rings are obtained from a potential divider across the applied voltage (approximately 40 kV). The space between the electrodes is filled with gas, in this instance at

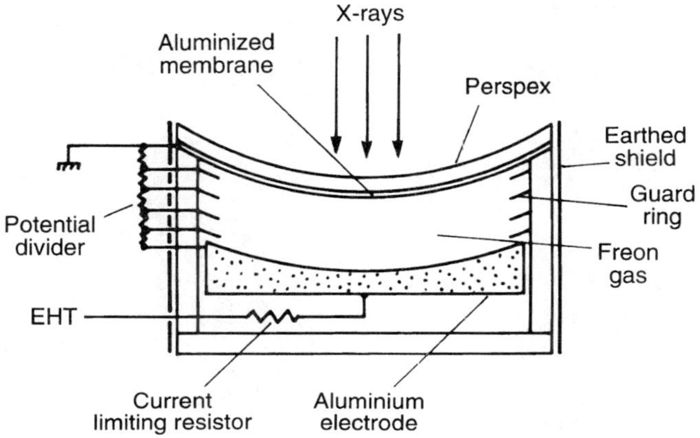

Figure 4.23 Simplified cross-section through a low pressure ionographic chamber. (Based on the design of B.M. Moores, J.A. Ramsden and D.L. Asbury, *Phys. Med. Biol.* (1980) **25**: 893–902.)

atmospheric pressure. The charge image is captured on the thin plastic sheet and may then be developed by a powder-cloud process, with a back-bias voltage applied to the aluminized coating.

Alternative designs, employing higher gas pressures, provide high ionization efficiency and need only 10 mm separation between the electrodes. However, the structure has to be more robust to withstand the increased pressure and this necessitates the use of a much thicker top plate, with a consequent increase in the X-ray absorption in the top plate. In addition, the ease with which the imaging electrode sheet may be removed is greatly reduced. An ionogram of the human hand is shown in Figure 4.24.

4.10 DIGITAL RADIOLOGY

We have seen that the basic X-ray system operates on an entirely analogue basis with virtually no opportunity for image processing. Clearly, if we can change the manner of recording the X-ray image to a digital basis, then we have the possibility of processing the image data and/or altering the display. The flexibility which comes from a digital approach is offset by the increased expense. Central to any digital system must be an image receptor capable of recording the image in digital form; this may be approached either by recording the whole image in one go, or sequentially. Among the image receptors described thus far, the image intensifier, the ionographic chamber and

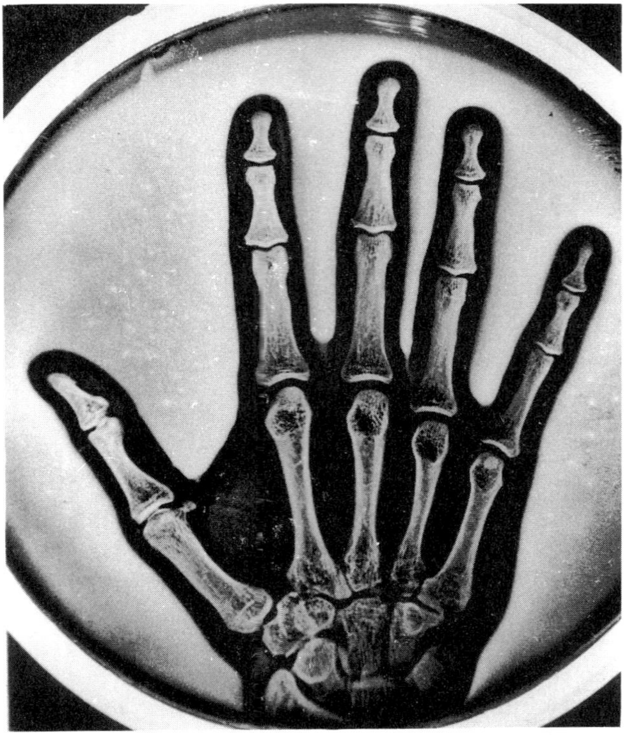

Figure 4.24 Ionogram of a hand, taken at 40 kV and 75 mA, showing the quality of the image obtainable. It is instructive to compare this figure with Figure 4.1, which ushered in X-ray radiography. The two pictures are separated by 85 years. (Reproduced from B.M. Moores, J.A. Ramsden and D.L. Asbury, *Phys. Med. Biol.* (1980) **25**: 893–902, with permission.)

the stimulated luminescence plate are ideally suited to providing a digital version of the X-ray image directly.

An alternative approach is the use of a scanning X-ray beam and receptor combination. Figure 4.25 shows, in simple outline, the basis of such a system. It is clear from the figure that a scanning system of this type must possess very good scatter rejection since it operates with two beam collimators (slits). The receptor in this instance is a narrow strip screen composed of gadolinium oxysulphide viewed in turn by 1024 (2^{10}) photodiodes responding to the visible output. Scanned projection radiographs with a digital basis are also produced by the computed

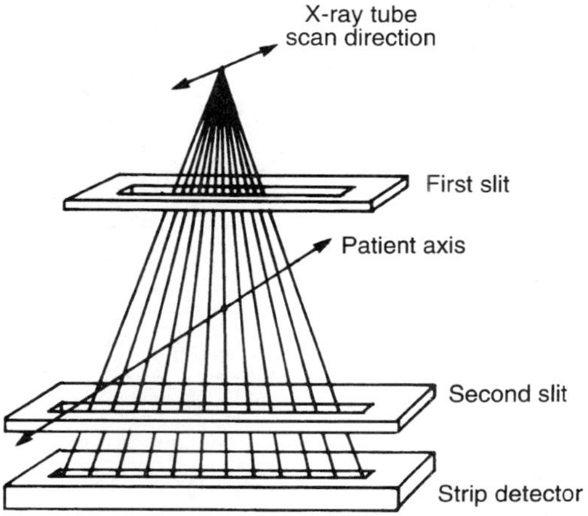

Figure 4.25 Line scanning system for digital radiography.

tomography systems which are described in Chapter 6. An obvious drawback to scanning ystems is that much of the useful X-ray tube output is wasted and long exposure times are necessary.

The basic layout of a system for digital radiology is shown in Figure 4.26. The X-ray tube and image receptor are interfaced to, and controlled by, a computer. The digitized information may then be stored or displayed on a VDU screen; alternatively a hard copy (analogue) may be produced. Although a digital radiographic system will have resolution inferior to that offered by a screen–film image receptor, this may not be clinically significant, and is more than offset by the advantages offered by the digital approach. We can readily identify four areas in which digital radiography is superior, or more flexible, namely:

1. image display
2. X-ray dose
3. image processing
4. image storage and retrieval.

Image display, in digital radiography, permits the user to set the levels of brightness on the monitor display, or density on the film record. In this way the film produced from the digitally recorded image can be properly exposed and matched to the range of pixel values in the image. Alternatively, the range of density or brightness can be enhanced artificially for a particular area of interest, or 'window', in the image.

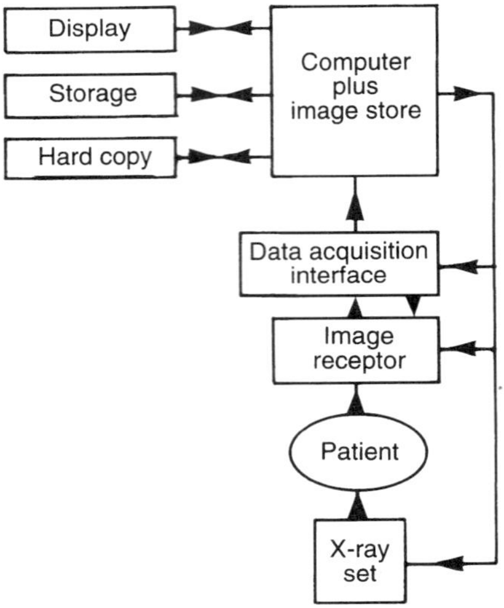

Figure 4.26 Basic elements of a digital radiology system.

An important advantage of digital radiology is the possibility of dose reduction. We have seen that in conventional radiology, the dose is determined by the sensitivity of the image receptor, and its signal-to-noise ratio, together with the film gamma. In digital radiology both of these restrictions can be relaxed. Dose reductions may be achieved by adjusting the dose to suit the noise level in the image or by using the X-ray spectrum which gives the lowest dose for a given signal-to-noise ratio; losses in contrast are then recovered by using the display techniques mentioned above.

Image processing by digital means offers the radiologist the possibility of subtraction imaging, in which complex background architecture is removed digitally to expose selected features of the radiograph. A particularly important example of subtraction imaging is that of digital subtraction angiography, which may be used to visualize blood vessels following injection of contrast media. An image of the blood vessels is obtained before the iodine-containing contrast medium is present in the imaging area. This image is then used to mask off, or subtract from, the images obtained as the contrast medium arrives; the progress of the medium through the blood vessels may then be followed. If the contrast medium has become very dilute by the time that it arrives in the region of interest, then the high contrast resolution available with digital techniques becomes very important; a contrast resolution of 1% is achievable.

An alternative use of digital image processing is to mimic the edge enhancement obtained with xeroradiography. This can be done with a digital image simply by applying high pass filtering, and it aids the perception of abnormalities. It is conceivable that, in the future, digital radiographic images will be 'interpreted' automatically, since they lend themselves well to such an objective. However, the supporting computer programs and hardware have yet to be developed and such a goal currently seems to be a long way off. Image storage and retrieval by digital systems offers further important medical advantages since digital images are readily stored on disc, and then copied or transmitted to other interested parties. The use of digital imaging systems is set to dominate X-ray practice.

4.11 X-RAY IMAGES FROM THE CLINICAL SCENE

We conclude this chapter, which has considered the best known application of physics in medical diagnosis, by assembling a sort of clinical 'portrait gallery' exemplifying the results which can be obtained. The original, and most straightforward, representation of an X-ray image is by means of a photographic film. Figure 4.27 shows an image of the human head obtained in this way. It is noteworthy for two reasons; the first is that the skull itself is relatively well defined, and the second is that the skull contents (the brain) are unresolved. The problem of resolving brain structures haunted neurological studies up to the introduction of computerized X-ray tomography (CT) in the early 1970s, when the brain could be imaged satisfactorily for the first time. Prior to the introduction of CT, the only imaging alternatives were the injection of radio-opaque materials (**cerebral angiography**) or air (**pneumoenecephalography**).

Cerebral angiography is still carried out, but nowadays in a rather different fashion, using digital fluoroscopy. The digitial image from a flat-plate video monitor is recorded on a single-sided emulsion film via a lens system (this is fluoroscopy). The digitized signal is then manipulated to perform digital subtraction, so that the arterial structure, which is carrying iodine-rich contrast material, is emphasized to the virtual exclusion of background detail. Figure 4.28 shows an image of the right internal carotid artery produced by this routine. The internal carotid artery received mention in Chapter 1 in connection with the supply of blood to the hypothalamus, as part of the body's core temperature regulating system.

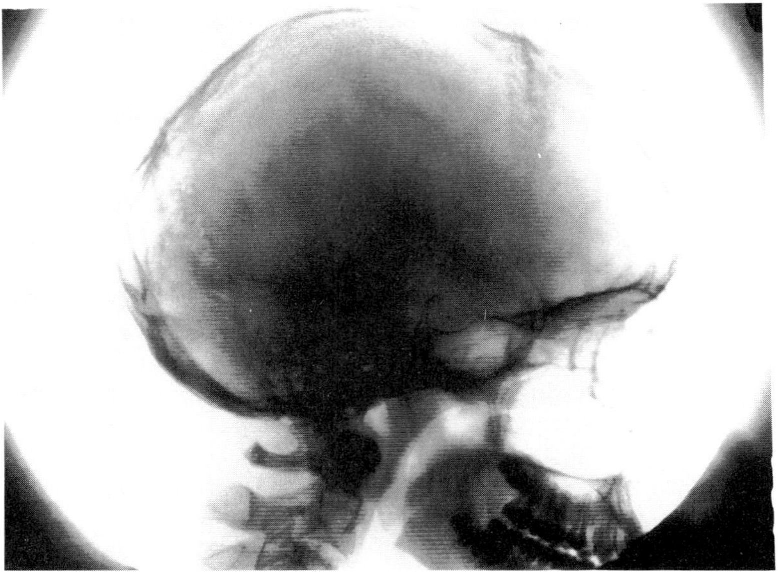

Figure 4.27 Standard X-ray image of the human skull. This image is significant in that it emphasizes the almost total lack of information provided about the contents of the skull. (Courtesy of Mr A. Ashley.)

An alternative approach to imaging, using an iodine-rich contrast medium, is intravenous digital subtraction angiography. A high volume of iodine-rich contrast medium (a **bolus**) is injected into the venous circulation, rather than directly into the artery under investigation. The bolus of contrast medium passes into the right side of the heart and is ejected into the arterial system from the **left ventricle**. During this time the contrast medium becomes diluted with blood. Despite this dilution, the blood still has increased opacity when it reaches the blood vessels of the lower limbs. Figure 4.29(a) shows a normal digitized image (with no manipulation) of the arteries just above and below the knee. Figure 4.29(b) shows the same field of view, but now the image has been subject to digital subtraction, increased system gain and increased enhancement. These digitally based, image enhancement techniques transform Figure 4.29(a), where the blood vessels are barely visible against the highly absorptive structures such as the leg bones, to Figure 4.29(b), where the blood vessels are crisply defined and all background detail has been excluded.

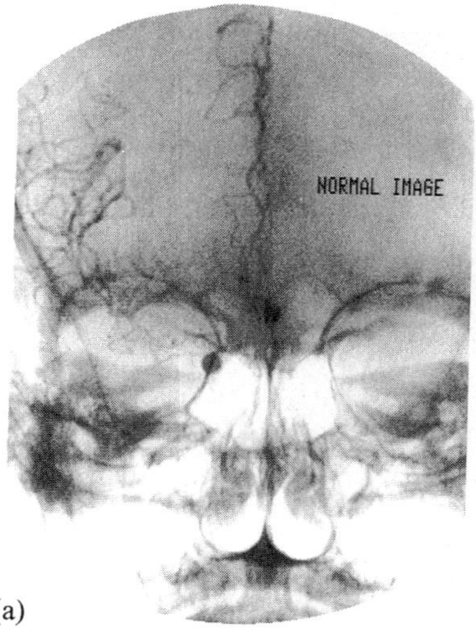

(a)

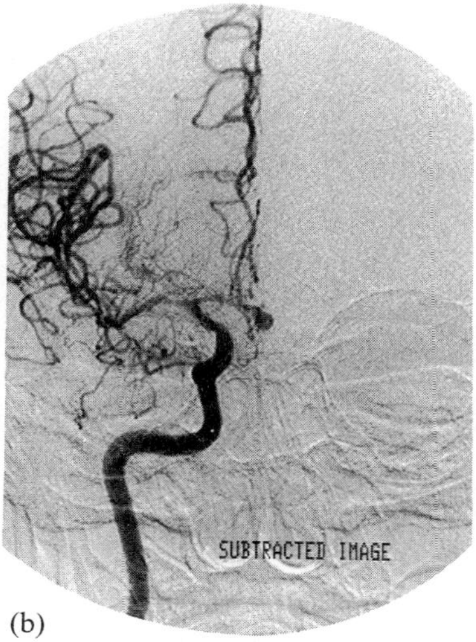

(b)

Figure 4.28 Cerebral angiogram, generated by digital fluoroscopy, following the injection of iodine-rich contrast medium via a catheter inserted into the femoral artery, and passed through the aorta to the right internal carotid artery. The image is taken from the front of the skull, showing (a) the normal digitized image, and (b) a digitally subtracted image which enhances the arterial detail. (Courtesy of Mr A. Ashley.)

A more mundane, yet straightforward, example of image enhancement is that achieved by using barium sulphate as a 'barium meal'. This approach has been used for many years, as the high opacity of barium sulphate outlines the digestive system selectively. Figure 4.30 shows the **oesophagus** 'outlined' by barium sulphate.

There are situations where the patient dose is reduced through the use of image intensifiers, the X-ray tube running continuously. Cardiac catheterization, where a catheter is passed through the blood vessels to reach the heart, is an example of such procedure. Progress of the catheter is monitored by X-ray viewing. Figure 4.31 shows an example of an image obtained in this way, during **balloon valvuloplasty** of the mitral valve. The mitral valve is identified in Figure 2.26.

PROBLEMS

1. Calculate the minimum wavelength associated with a 100 kV X-ray beam. What radiation dose would be delivered by this beam?

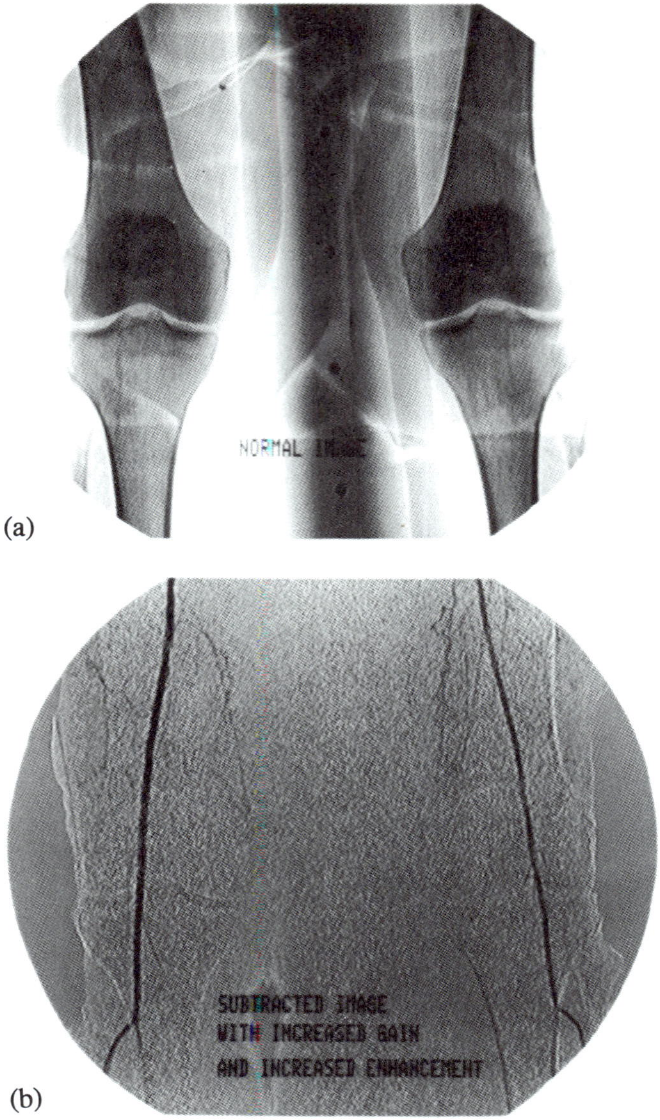

Figure 4.29 (a) Normal digitized X-ray image, with no manipulation, showing the bones in the vicinity of the knees. (b) A digitally subtracted image with system gain and digital enhancement, revealing the arteries just above and below the knee to the exclusion of all other radiographic detail. (Courtesy of Mr A. Ashley.)

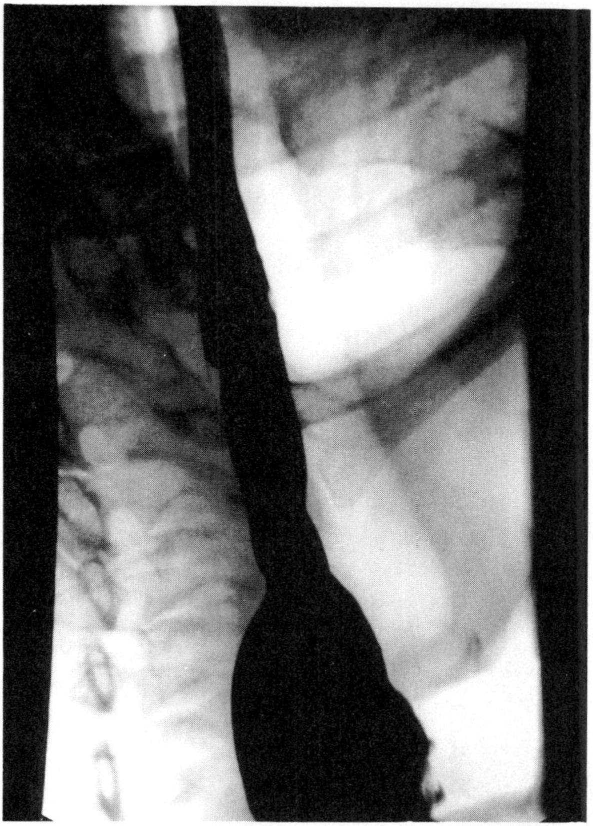

Figure 4.30 X-ray image of the oesophagus 'outlined' with barium sulphate. (Courtesy of Mr A. Ashley.)

2. Calculate the contrast which we can expect to get in an X-ray image involving muscle and blood, where the linear attenuation coefficients are $1.80\ \text{mm}^{-1}$ and $1.78\ \text{mm}^{-1}$, respectively.
3. A cubic block of muscle is imaged with X-rays of fixed energy under conditions which yield the minimum surface dose to the patient. If the volume of muscle being imaged is halved, but with no change to its geometry, calculate the percentage increase in the minimum surface dose.
4. The output beam of an X-ray source, operating at 50 kV, passes through an aluminium window 0.5 mm thick, having a mass attenuation coefficient at that energy of $3.6 \times 10^{-2}\ \text{m}^2\ \text{kg}^{-1}$. What

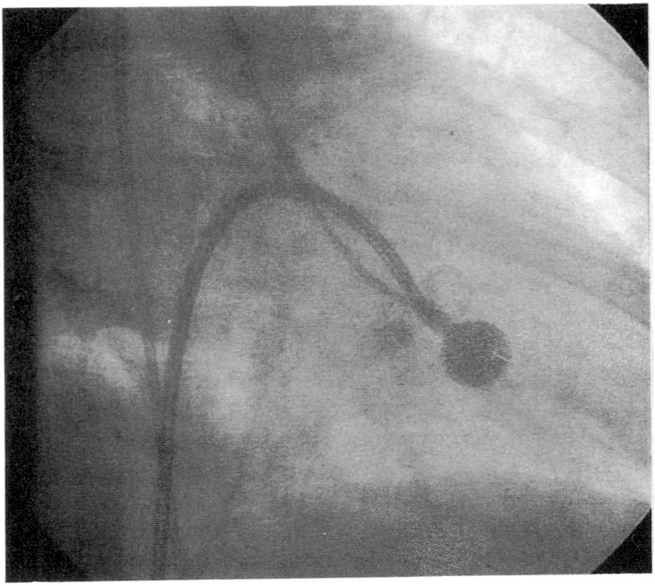

(a)

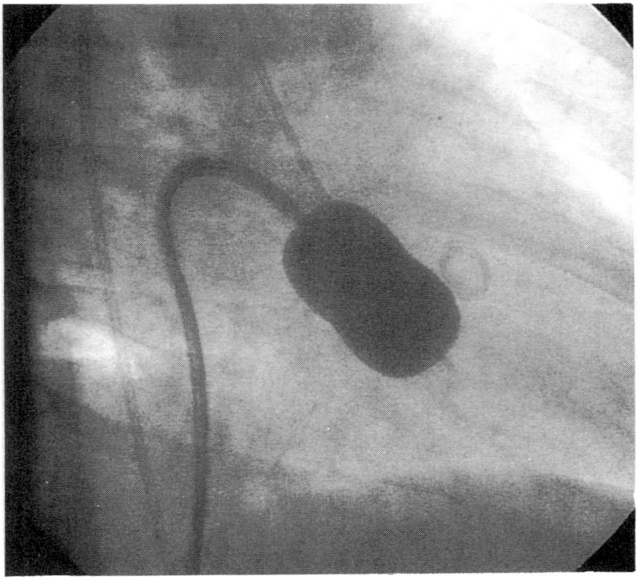

(b)

Figure 4.31 X-ray image obtained using an image intensifier showing two stages in balloon valvuloplasty of the heart mitral valve. (a) First stage of balloon inflation; (b) final stage of inflation. (Courtesy of Dr H. Singh.)

fraction of the incident intensity is transmitted? If the X-ray energy is lowered to 10 keV, the mass attenuation coefficient becomes $2.68\ \mathrm{m}^2\ \mathrm{kg}^{-1}$. What fraction of the incident intensity is now transmitted?

5

Ultrasound imaging

5.1 INTRODUCTION

Ultrasound is defined as sound waves beyond the limit of human hearing, that is, frequencies greater than around 20 kHz, although the actual cutoff varies from one person to another. In medical applications the frequency range encountered will be from about 1 MHz up to tens of MHz. The first studies of sound at frequencies beyond the range of human hearing were carried out by W.H. Wollaston in 1820, actually to determine the limits of human hearing. He introduced the first device for the production of ultrasound, a mechanical ultrasonic transducer or pipe, the Wollaston whistle. Modern medical ultrasound transducers, however, are based on the piezoelectric effect which was discovered by Jacques and Pierre Curie in 1880. It required developments in electronics (the vacuum tube), immediately prior to and during World War I, before piezoelectric transducers could be used conveniently. P. Langevin, a student of P. Curie, worked on the development of an ultrasonic system for submarine detection, although the system was not deployed until after World War I. Nevertheless, it triggered relatively rapid growth in the use of ultrasonics from 1918 onwards. Of course, a complete ultrasound system requires not only receivers (operating by the piezoelectric effect), but also suitable transmitters. The inverse piezoelectric effect, which was predicted by G.J. Lippman in 1880, provided the simple means by which this could be achieved. In effect the transmitter and receiver become one and the same. The experimental basis for Lippman's prediction was established by the Curie brothers in 1881.

Although the use of ultrasound in medical diagnosis is now second only to X-rays, this is a relatively recent phenomenon. It was 1948, some 50 years after the introduction of X-rays, before the first medical application of ultrasound was attempted by Karl Dussik, in an unsuccessful endeavour to detect brain tumours. Ultrasonic imaging, based upon the pulse-echo principle, really required the developments in electronics which accompanied the introduction of radar, just prior to and during World War II, before it could be developed successfully. Its medical use was preceded by its application in the field of non-destructive testing where, in 1940, F.A. Firestone filed a patent to use ultrasound to detect internal defects in metal structures. The first medical use of pulse-echo imaging was described in 1952 by J.J. Wild and J.M. Reid, while the first commercial equipment for ultrasonic imaging (actually developed from an industrial flaw detection system) did not appear until 1958.

Ultrasound is useful for medical imaging because it both penetrates and interacts with the body. In contrast to X-rays, which simply suffer attenuation on passage through a medium, an ultrasound beam experiences both refraction and reflection in addition. This occurs at any interfaces which the ultrasound beam may traverse; its behaviour is therefore entirely analogous to that of a light beam traversing an inhomogeneous medium. Ultrasound is, in medical applications anyway, a longitudinal pressure wave which propagates much more slowly than say X-rays or light, making transit times easily measurable, and permitting radar-like pulse-echo methods. A further important property, which stems from the small wavelength of ultrasound, and distinguishes it from sound in the audio spectrum, is that we can readily generate a narrow ultrasound beam from a compact source. These characteristics, plus the development work which has been conducted since the 1950s, have led to ultrasound being one of the most frequently used imaging techniques in medical diagnosis.

5.2* THE PHYSICS OF ULTRASOUND PROPAGATION

In medical applications of ultrasound we are always considering the propagation of longitudinal pressure waves rather than transverse or shear waves since, with the exception of bone, biological tissues act like fluids and will not support transverse waves. If we exclude bone from our considerations, the range of propagation velocities we need to consider is between approximately 1350 and 1800 m s^{-1}. It is usually convenient to assume an average speed of about 1540 m s^{-1} for ultrasound propagation in soft tissue. Figure 5.1 shows the distribution of sound velocities for a variety of body materials. The velocity of sound

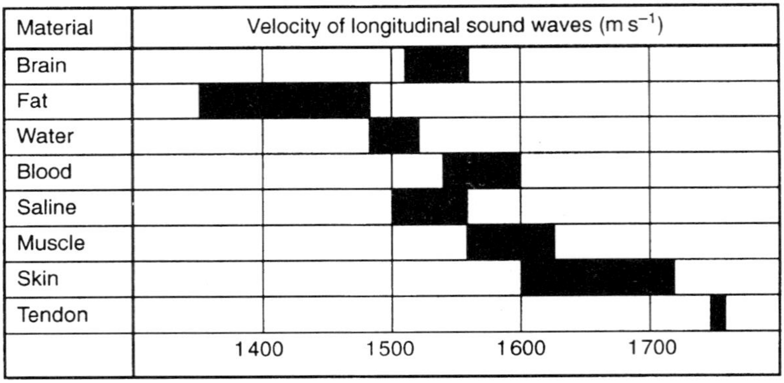

Figure 5.1 Range of values measured for the speed of longitudinal sound waves in various body media with water as a comparison.

exhibits very little dispersion so that we may consider it to be essentially independent of frequency (except in bone) and the velocity c_L is related to frequency f and wavelength λ by the usual relationship

$$c_L = f\lambda \tag{5.1}$$

If we consider the range of frequencies used in medical ultrasound, the lowest frequency will be around 1 MHz (wavelength 1.5 mm) and the highest perhaps 20 MHz (wavelength 0.075 mm). The speed of sound is temperature dependent with a positive temperature coefficient for non-fatty material and a negative temperature coefficient for fatty material.

We have already had a glimpse of the general equation of wave motion (as applied to the propagation of an action potential in an unmyelinated axon). Here we wish to apply it to the longitudinal waves generated by a piezoelectric crystal. These waves have velocity c_L, and a vibration amplitude ξ, so that we can write

$$\frac{1}{c_L^2}\frac{\partial^2 \xi}{\partial t^2} = \frac{\partial^2 \xi}{\partial x^2} \tag{5.2}$$

The general solution of this equation is

$$\xi(x, t) = \xi_1 \cos(\omega t - \beta x) + \xi_2 \cos(\omega t + \beta x) \tag{5.3}$$

where ξ_1 represents the amplitude of the forward wave (that travelling to the right) while ξ_2 is the amplitude of the reflected wave (that moving to the left); ω is the angular frequency and $\beta = 2\pi/\lambda = \omega/c_L$ where λ is the wavelength; and x is the particle displacement.

A more general solution which contains equation 5.3 as its real part is

$$\xi(x, t) = \xi_1 \exp i(\omega t - \beta x) + \xi_2 \exp i(\omega t + \beta x) \tag{5.4}$$

This equation actually contains more information than we usually need, but it will serve to develop the basic parameters of ultrasound propagation. From wave motion theory we have the result that the pressure exerted in the x direction, p_x, is given by

$$p_x = -c_L^2 \rho_0 \frac{\partial \xi}{\partial x} \tag{5.5}$$

where ρ_0 is the static density of the material in which the wave is propagating. Hence we may write that

$$p_x = c_L^2 \rho_0 i\beta[\xi_1 \exp i(\omega t - \beta x) - \xi_2 \exp i(\omega t + \beta x)] \tag{5.6}$$

Similarly the particle velocity in the x direction, u_x, is just $\partial\xi/\partial t$ so that

$$u_x = i\omega[\xi_1 \exp i(\omega t - \beta x) + \xi_2 \exp i(\omega t + \beta x)] \tag{5.7}$$

The reason for deriving these equations is that it is useful to construct an analogy between simple electrical concepts and their acoustic equivalent. In this connection we look for an analogy between the electrical impedance and the acoustic impedance. The electrical impedance is just V/I, the acoustic analogues are pressure and particle velocity, hence we may write the acoustic impedance as

$$\frac{p_x}{u_x} = \rho_0 c_L \left[\frac{\xi_1 \exp i(\omega t - \beta x) - \xi_2 \exp i(\omega t + \beta x)}{\xi_1 \exp i(\omega t - \beta x) + \xi_2 \exp i(\omega t + \beta x)}\right] \tag{5.8}$$

This expression actually defines the specific acoustic impedance Z_{sp} which is a complex quantity with both resistive (real) and reactive (imaginary) components. It is the mechanical impedance per unit area. If we have no reflected wave then equation 5.8 simplifies to

$$Z_{sp} = \rho_0 c_L = Z \tag{5.9}$$

where Z is the characteristic impedance of the medium; Z is a property of the medium only. Whereas for equation 5.8 the pressure led the displacement by 90°, in equation 5.9 the pressure and displacement are now in phase and the characteristic impedance is real and resistive: it has the units kg m^{-2} s^{-1}.

In practice, most sound fields are not uniformly distributed so it is useful to consider the power which is transmitted through unit area; this is the sound intensity I and has the units W m^{-2}. We begin with equation 5.8, which for the condition where only the forward wave is present may be written as

$$p_x = u_x Z \tag{5.10}$$

The mechanical power is equal to the time-averaged product of force and velocity. The total alternating force exerted over an area A by a sound pressure p_x is $p_x A$ and the particle velocity is u_x so that the sound power W is

$$W = \overline{u_x p_x} A \tag{5.11}$$

where the bar indicates a time average. When the pressure and particle velocity are in phase with each other their time averages can be taken separately; these are the root mean square values p_{rms} and u_{rms}. For simple harmonic motion these values are related to the peak pressure P and peak velocity U by $p_{\mathrm{rms}} = P/\sqrt{2}$ and $u_{\mathrm{rms}} = U/\sqrt{2}$. We can rewrite equation 5.11 as

$$W = p_{\mathrm{rms}} u_{\mathrm{rms}} A = \frac{PU}{2} A = \frac{U^2 Z}{2} A = \frac{P^2 A}{2Z} \tag{5.12}$$

We have defined the sound power W already so that the sound intensity I is just

$$I = W/A = \overline{up} = u_{\mathrm{rms}} p_{\mathrm{rms}} = \frac{U^2 Z}{2} = \frac{P^2}{2Z} \tag{5.13}$$

We can relate the wave intensity I to the oscillating incremental pressure p and the particle velocity u by the equation

$$I = \frac{c_{\mathrm{L}}}{2}\left(\rho_0 U^2 + \frac{P^2}{\rho_0 c_{\mathrm{L}}^2}\right) \tag{5.14}$$

where the first term in the bracket represents the kinetic energy density and the second the potential energy density, both in units of J m^{-3}. At the peak velocity U, the pressure is at ambient level and the potential energy term disappears, while at the peak pressure P the particle velocity is zero, the displacement reaches a maximum, and the kinetic energy becomes zero. We see that the pressure and velocity variations are 90° out of phase with each other.

An ultrasound beam propagating through a medium undergoes attenuation in the same fashion as a light beam passing through an optical medium, or an X-ray beam through matter. The process is again described by Lambert's law, which relates the incident intensity I_0 to the intensity I_x a distance x into the medium by the equation

$$I_x = I_0 \exp(-\mu x) \tag{5.15}$$

where μ is the linear attenuation coefficient with units m^{-1}, or more usually cm^{-1} or nepers cm^{-1}. In a similar fashion we can describe the pressure or particle velocity amplitudes so that, for example,

$$p_x = p_0 \exp(-\alpha x) \tag{5.16}$$

where α is amplitude attenuation coefficient, again with units of m^{-1}. Since we know from equation 5.12 that I is proportional to U^2 or P^2, then

$$I_x/I_0 = \left(\frac{P_x}{P_0}\right)^2 = \left(\frac{U_x}{U_0}\right)^2 \quad \text{and} \quad \mu = 2\alpha \tag{5.17}$$

It is customary to express intensity ratios and amplitude ratios in decibels (dB) so that the intensity level is expressed as $10 \log_{10}(I_x/I_0)$ dB; the amplitude attenuation is expressed as $20 \log_{10}(P_x/P_0)$ dB since I is proportional to P^2. If we wish the units of μ to be in dB m^{-1} then μ (m^{-1}) = 0.23 μ (dB m^{-1}) and similarly for amplitude attenuation, only now the multiplying factor becomes 0.115.

As with X-rays, the intensity attenuation coefficient has contributions from scattering as well as straightforward absorption, although the relative assignments to these two processes are not well known for most body tissues. In regard to amplitude attenuation however, it is found that ultrasound attenuation increases with frequency and that most body tissues show similar properties with respect to the frequency dependence of this attenuation. It may be represented by $\alpha \simeq 1$ dB cm^{-1} MHz^{-1} for a wide range of tissues.

When the intensity is reduced by absorption, the wave energy is converted into heat and the tissues suffer a consequent temperature rise. This temperature rise is the basis for ultrasound-induced hyperthermia, but in other applications is kept to a minimum. The power levels appropriate to medical ultrasound will be considered in section 5.13. The mechanisms by which the ultrasound energy is converted to heat can be divided into three classes:

1. classical viscous loss
2. molecular relaxation
3. relative motion losses.

Of these three the most important, in soft tissue, is molecular relaxation, where the incident ultrasound causes reversible changes in the structure of tissue molecules. As a rule, the more complex the molecular structure the greater the absorption.

Scattering of ultrasound within the body may be classified into three types:

1. geometrical scattering
2. stochastic or probabalistic scattering
3. Rayleigh scattering.

Geometric scattering can be described by the laws of optics so that at a large scale boundary between different media the usual law of reflection and Snell's law for refraction apply, enabling us to predict the direction of the reflected and refracted beams (Figure 5.2(a)). If the angle of incidence is θ_1, the angle of reflection θ_2, and the angle of refraction θ_3 then

$$\frac{\sin\theta_1}{c_1} = \frac{\sin\theta_2}{c_1} = \frac{\sin\theta_3}{c_2} \tag{5.18}$$

where c_1 and c_2 are the sound speeds in the two media, and $\theta_1 = \theta_2$.

If the incident sound beam strikes the interface normally then we can express the intensity ratio for the incident and reflected beam as

$$\frac{I_2}{I_1} = \left(\frac{Z_2 - Z_1}{Z_2 + Z_1}\right)^2 \tag{5.19}$$

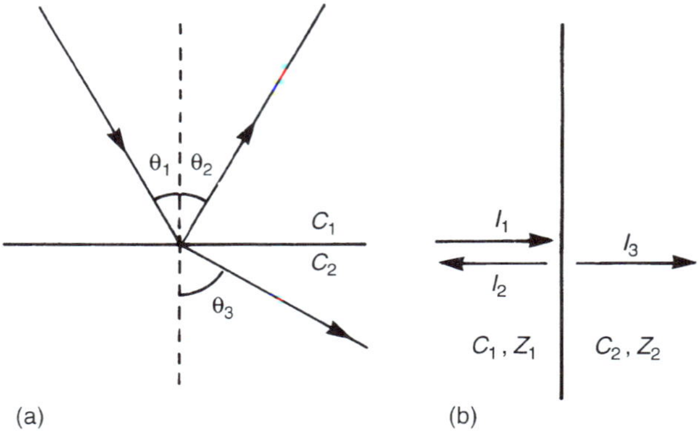

Figure 5.2 (a) Reflection and refraction at a plane boundary between two media having characteristic acoustic impedances Z_1, Z_2 and speed of sound c_1 and c_2. (b) Reflected and transmitted intensities for a sound beam incident normally upon a plane boundary between two media.

where I_2 is the reflected beam intensity and Z_1 and Z_2 are the characteristic impedances of the media on either side of the interface (Figure 5.26); this is Rayleigh's law. The corresponding expression for the transmitted beam ratio is

$$\frac{I_3}{I_1} = \frac{4Z_1 Z_2}{(Z_2 + Z_1)^2} \tag{5.20}$$

Equation 5.19 is central to the formation of images using a pulsed ultrasound beam and provides the explanation of the 'echoes' which are obtained from an ultrasound beam as it traverses the boundary between two dissimilar media. The greater the difference between the characteristic impedances of the media, on either side of the interface, the larger the fraction of reflected ultrasound, so that in the extreme case of an ultrasound beam incident normally on an interface bounded on one side by air, the reflection is essentially total. In medical terms this means that structures behind the lungs, for example, are 'invisible' to an ultrasound beam.

Stochastic scattering modifies reflections from boundaries and is highly anisotropic. It tends to predominate in internal regions of organs. Rayleigh scattering operates on much smaller scale structures, such as cells in tissues and red cells in the blood. The effect is proportional to the volume of the scatterer and varies with a f^4 frequency dependence; even so, the scattering is very weak. It is significant, however, in Doppler-shift based measurements of blood flow velocity, the subject of section 5.9.

5.3* THE GENERATION AND RECEPTION OF ULTRASOUND

Ultrasound for medical applications is generated by transducers exhibiting the piezoelectric effect. There are many materials which exhibit this effect, but the principal ones in medical usage are based on lead zirconate titanate compounds (ceramics) or polyvinylidene difluoride (a plastic). In earlier work quartz or Rochelle salt (sodium potassium tartrate) were used. A particular characteristic of these materials is a polar structure which lacks a centre of symmetry and in which the charge distribution may be unbalanced by the application of a mechanical stress or a voltage. Figure 5.3 shows a simple schematic representation of the interaction between a force or a voltage in a piezoelectric crystal. The example shown is based on quartz, but the principle applies to other piezoelectric materials as well. Application of a stress

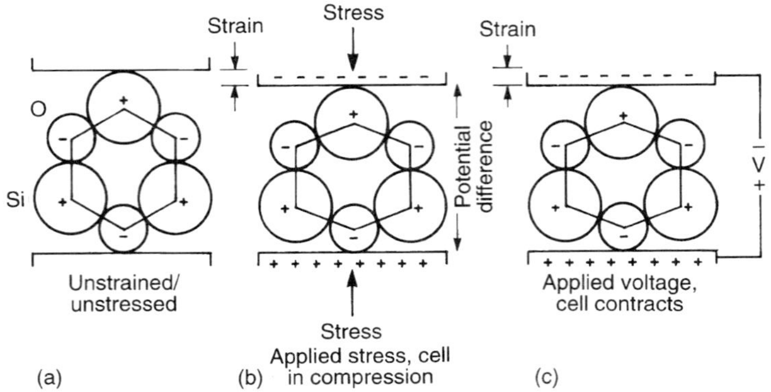

Figure 5.3 Simplified schematic diagram showing the origin for (a) a quartz type structure of (b) the piezoelectric effect and (c) the inverse piezoelectric effect.

to such a crystal distorts the charge symmetry so that the opposing faces acquire a charge and a potential difference appears across the crystal; this is the piezoelectric effect. In an equivalent fashion, application of a voltage of the same sense will yield a strain, the inverse piezoelectric effect. In fact the barium zirconate titanates are actually ferroelectric and when produced are in the form of a ceramic composed of a large number of randomly orientated polarized domains. If such a material is heated above its Curie temperature and then cooled in the presence of a strong electric field, the directions of polarization of the domains line up with the field and remain lined up after the material is cool and the field is removed. The material now possesses a permanent electric dipole and behaves as if it were a natural piezoelectric material; the above process steps are known as poling.

The output from an ultrasound transmitter is unique amongst diagnostic radiation sources in that it is coherent, with pronounced interference effects, just as one sees with laser light. Further, ultrasound receivers are necessarily amplitude sensitive, by virtue of their dependence on the piezoelectric effect, in contrast to other imaging systems where the sensor is usually intensity sensitive. This has the consequence that ultrasound receivers are able to generate coherent, interference-like fluctuations in apparent received magnitude due to signal integration over a non-planar received wave.

Most ultrasound transducers take the form of a flat plate, generally circular in section, whose front surface vibrates parallel to the surface normal, in the manner of a piston, thereby generating longitudinal

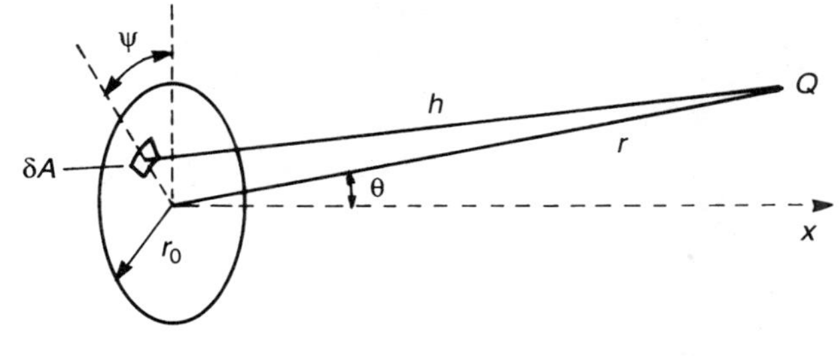

Figure 5.4 Construction used to derive the acoustic radiation pattern from a circular ceramic piezoelectric piston.

pressure waves. Discussion is simplified if we consider a circular transducer of radius r_0 set in a tight-fitting circular hole in an infinite plate. This construction prevents radiation from the rear of the piston travelling round to the front and interfering with the forward-radiated energy. Figure 5.4 shows the construction used to derive the radiation pattern from a circular piston.

The pressure P_Q at the point Q will be proportional to the radiating area and inversely dependent on the distance from the radiating surface. We can represent it in terms of the pressure specified close to the face of the radiating piston by

$$P_Q = P^* \int_A \frac{1}{h} \exp \mathrm{i}(\omega t - \beta h)\, \mathrm{d}A \tag{5.21}$$

where P^* is proportional to the pressure at the face of the piston and has the units N m^{-3} or pressure per unit length. It is not straightforward to solve equation 5.21 unless we make certain approximations. Two special cases are of particular interest, the first of which arises when $r \gg r_0$ and is known as the far-field case (Fraunhofer region). In this instance we can make further approximations, namely that $1/h \simeq 1/r$ and $h \approx r - y \sin\theta \cos\psi$.

Integration of equation 5.21 yields

$$P_Q = P^* \frac{\pi r_0^2}{r} \exp \mathrm{i}(\omega t - \beta r) \left[\frac{2J_1(\beta r_0 \sin\theta)}{\beta r_0 \sin\theta}\right] \tag{5.22}$$

where $\beta = 2\pi/\lambda$ and J_1 is a Bessel function of the first kind.

We can summarize the predictions of equation 5.22 by noting that at distances which are large compared with the piston radius, the acoustic pressure is a complicated function of θ and decreases with increasing distance; the term within the square brackets of equation 5.22 is known as the directivity function and gives the variation in pressure with angle

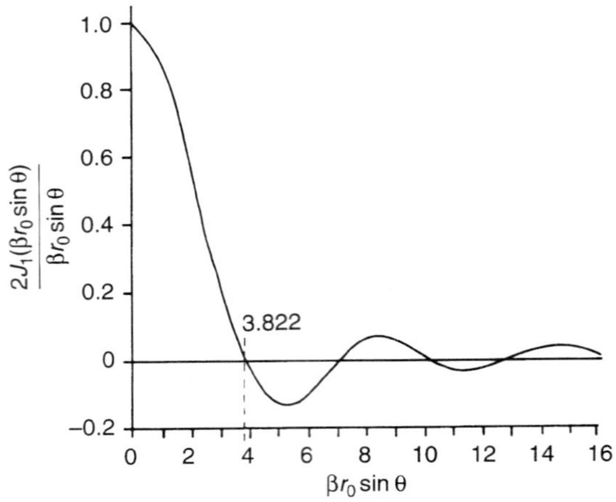

Figure 5.5 Plot of the directivity function for a circular piezoelectric piston.

θ from the axis. This function, for the circular piston, is plotted in Figure 5.5. A different, but perhaps more instructive way of plotting the pressure distribution from a flat circular piston is the polar diagram. Figure 5.6 shows the polar diagrams corresponding to the radiation from circular pistons as the ratio of piston diameter to sound wavelength is altered. An important aspect of these polar diagrams is the increasingly 'directed' nature of the pressure distribution so that as the ratio of piston diameter $2r_0$ to wavelength λ increases, the angular divergence of the main beam lobe decreases and small side lobes appear. If we imagine that this main lobe is just contained within a cone whose apex is sited at the transducer surface, then the semi-angle θ_c of the cone is given by the smallest value of $(\pi r_0/\lambda)\sin\theta$ in Figure 5.5 which brings the ordinate value to zero. This happens when

$$\frac{2\pi r_0}{\lambda}\sin\theta_c = 3.822$$

and

$$\sin\theta_c \simeq 0.61\,\frac{\lambda}{r_0} \tag{5.23}$$

Figure 5.6 shows explicitly how the 'beam character' of the ultrasound output depends sensitively on the ratio r_0/λ. It also explains why we need ultrasound pressure waves for medical diagnostic work, since in the audio range, even if we chose to work at 20 kHz, the equivalent wavelength is 75 mm and the minimum transducer diameter for an ultrasound beam would be about 150 mm – far too big to be useful.

If we restrict our considerations to the pressure on the axis of the transducer then we can solve equation 5.22 to plot the magnitude of P_Q, normalized to the peak pressure, against the ratio x/r_0. The result of this is shown in Figure 5.7 and we are able to distinguish the conditions where P_Q is varying rapidly with change in x/r_0, compared with that where P_Q decreases monotonically with increasing x/r_0. We can specify the position of maxima and minima by the equation

$$x_n = r_0^2/n\lambda - n\lambda/4 \tag{5.24}$$

where maxima occur for odd values of n and minima for even values. As n increases we move in towards the transducer surface.

The position of the last maximum, most distant from the transducer, is determined when $n = 1$, and if $r_0 \gg \lambda$ is given by

$$x \simeq r_0^2/\lambda \tag{5.25}$$

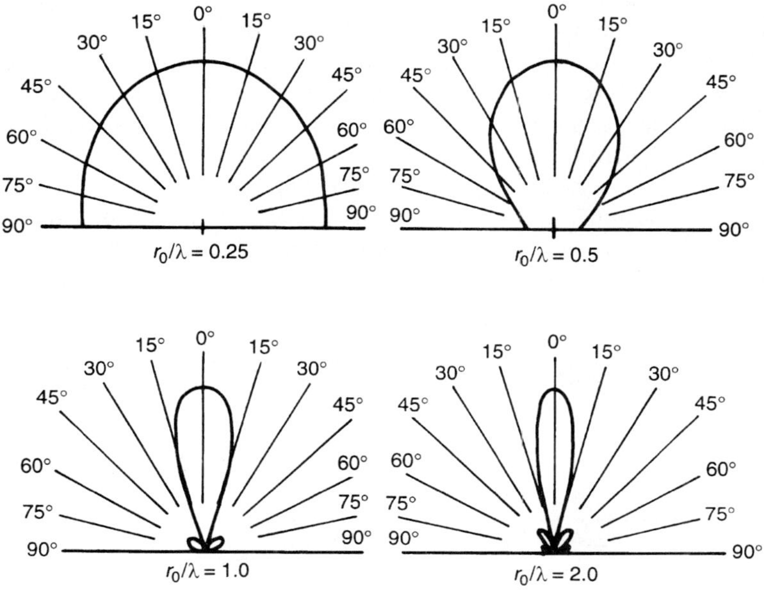

Figure 5.6 Polar diagrams for the radiation from circular piezoelectric pistons, as a function of the ratio r_0/λ.

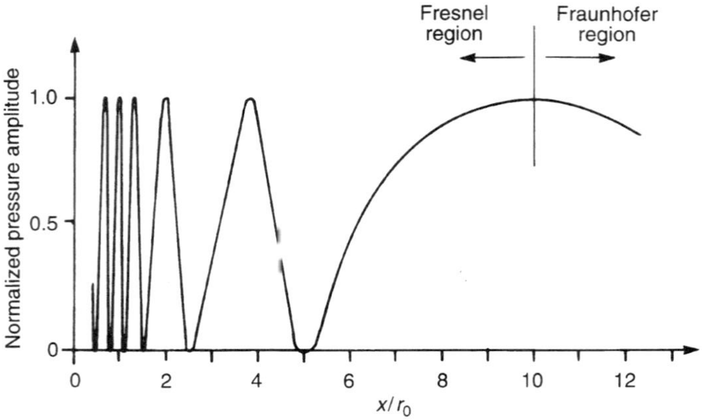

Figure 5.7 The pressure distribution, normalized to the peak pressure, along the axis of radiation for a circular piston of radius r_0.

The last axial maxima is usually regarded as the boundary between the near-field, or Fresnel region, and the far-field. Equation 5.25 sets the extent of the near-field in which most of the transducer energy is confined within a transducer radius of the central axis and the beam may be considered to be parallel. In the far-field the wave field is more uniform, but tends to a spherically divergent wave. The distribution or beam shape is summarized schematically in Figure 5.8. Of course these results are derived for a continuous wave source, but in practice a pulsed source is used. The effect of pulsing the source is to simplify the overall structure of the sound beam, if we assess it in terms of a single field parameter, such as the peak pressure amplitude. To summarize, the near-field interference structure and the far-field side lobes are

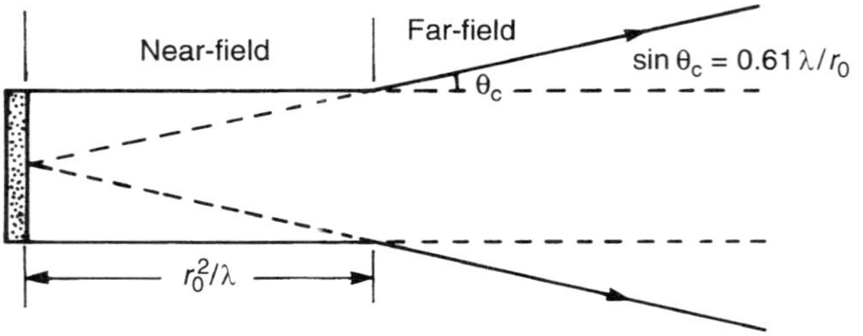

Figure 5.8 Simplified field distribution pattern for a circular piezoelectric piston distinguishing the near and far field contributions to the beam profile.

significantly smoothed; in the same way the near-field/far-field boundary becomes ill defined. All of these effects depend on the exact pulse parameters used to assess the acoustic field.

The restricted nature of the above results becomes even more apparent when one realizes that, in medical applications, the source is often focused. The manner in which focusing is achieved is the subject of section 5.5. Instead of the plane wavefront we have considered thus far, a spherical wavefront is launched, and this modifies the acoustic field distribution in a number of ways. First, the position of the last on-axis maximum of the equivalent plane-wave source moves closer to the transducer than either the geometric focus, determined by the radius of curvature R of the transducer, or that computed by equation 5.25. Accompanying this shift is an increase in the on-axis maximum pressure amplitude at the focus and a decrease in the width of the sound beam at the focal position. Past the focal point the beam diverges, but now the divergence exceeds that for a plane-wave source, specified by equation 5.23, so that we find as focusing is introduced, the depth range over which a narrow beam is maintained becomes less. The stronger the focusing, the narrower the beam width at the focus, but the shorter the range over which a narrow beam is maintained. In this connection it is common to find transducers described in terms of their 'strength of focusing' so that strong focusing corresponds to $R < 0.25x_1$, medium focusing to $0.25x_1 \leq R < 0.5x_1$, and weak focusing to $R \geq 0.5x_1$, where x_1 is the distance from the transducer face to the near-field/far-field boundary, established by putting $n = 1$ in equation 5.24.

5.4* THE ULTRASOUND TRANSDUCER

The critical component in any ultrasound imaging system is the ultrasonic transducer responsible, on the one hand, for generating the ultrasound signal, and on the other, for receiving the ultrasound echoes. It is necessary that it operates with high efficiency and generates an accurate electrical equivalent of the received ultrasound echoes.

The basic features of a conventional ultrasound transducer are shown in Figure 5.9(a). The transducer itself will usually be cut from a slab of lead zirconate titanate ceramic, known as PZT, which has been poled to give maximum mechanical response at right angles to the planar surfaces. These surfaces are silvered to provide contact electrodes. Under these conditions a voltage applied to the electrodes will produce a mechanical strain in the fashion described in section 5.3 and Figure

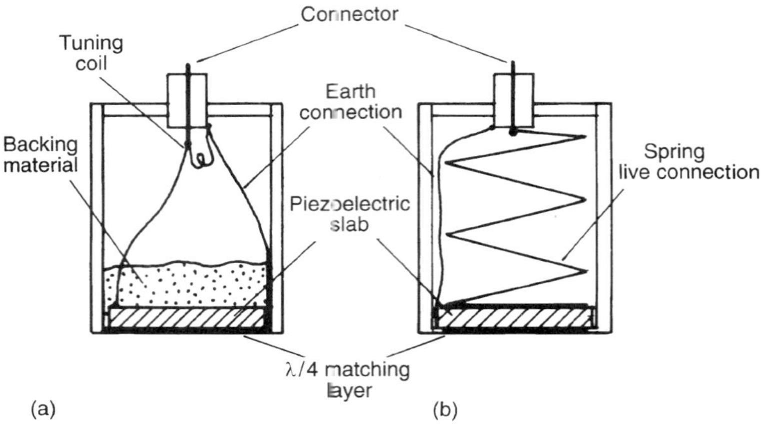

Figure 5.9 (a) Section through a typical ultrasound transducer, showing means by which damping is applied to permit pulsed operation. (b) Transducer mounting to permit maximum efficiency of operation in continuous wave mode.

5.3. Conversely, a mechanical strain will produce a voltage difference between the silvered surfaces.

An alternative material which offers some useful advantages is PVDF (polyvinylidene difluoride). This material has some exceptional properties, amongst which are its low mechanical Q and low acoustic impedance. Its output voltage is approximately 20 times greater than that of PZT and it has a frequency response from DC to GHz. At ultrasound frequencies, pressures of several MPa may be generated.

Transducers based upon lead zirconate titanate are usually operated in a so called free-ringing mode or half-wave resonance where the transducer thickness is $\lambda/2$ and λ is the wavelength of longitudinal sound waves in the piezoelectric material. As a rule of thumb, at a resonant frequency f (MHz) the transducer thickness T (mm) is numerically equal to $2/f$. Once the transducer is operating it is necessary to ensure that its vibrational energy is efficiently coupled into the medium which is being investigated. The ultrasonic transducer, of course, delivers power from both its front and rear face. In order to achieve the maximum power output $W_{\max}$ from a transducer backed by a medium of characteristic impedance Z_1 and intended to deliver its power predominantly into a fronting medium of characteristic impedance Z_2, we can use equation 5.26 which specifies the resonant power output in terms of the above impedances, the drive voltage V, the transducer capacitance C and the piezoelectric constant h' relating field developed to the applied mechanical strain.

$$W = \frac{(4h')^2 C^2 V^2 Z_2}{(Z_1 + Z_2)^2} \tag{5.26}$$

Maximum power will be delivered when $Z_1 = 0$, i.e. when the backing impedance is zero. This is most nearly achieved using air and Figure 5.9(b) shows how this is accomplished in practice. However, if the transducer is to operate in the pulsed mode, which is usually the case, then short pulses are required and the high mechanical Q factor conferred by an air backing is not suitable. We define Q_{mech} as

$$Q_{\mathrm{mech}} = f_0/(f_1 - f_2) \tag{5.27}$$

where f_0 is the resonant frequency and f_1, f_2 are the frequencies at which the radiated power has dropped to half its value at f_0. A low Q_{mech} can be obtained by making the backing impedance Z_1 essentially equal to that of the transducer itself. In practice exact equality is not possible, but a factor of two difference is usually adequate damping and Q_{mech} of 2–4 is possible. Typically, Araldite or other similar epoxy resin, containing 15% tungsten powder, is used. In fact we wish to avoid reflecting energy back into the transducer and the backing should be a good absorber. Good absorption properties may be obtained by using plasticized epoxy, embedded fine scatterers (aluminium powder), or shaped backs. The above comments apply specifically to transducers based upon PZT (lead zirconate titanate). The plastic film transducers based upon PVDF (polyvinylidene difluoride) have inherently low Q_{mech} and do not need any special backing in order to improve the bandwidth. Both electrical and mechanical Q contribute to the overall Q so that it is possible, using electrical tuning, to manipulate the electrical Q for the best compromise between sensitivity and resolution. This will usually be done by inserting an inductance across the essentially capacitive transducer (Figure 5.9(a)).

We have so far considered the problem of ensuring that the ultrasonic transducer operates with maximum effectiveness by requiring that the energy deposited in the backing material is a minimum. However, it is equally important that the energy emitted from the front of the transducer be coupled efficiently into body tissues. If we consider a transducer operating in the continuous-wave mode, then we can use a result from transmission line theory and write that the ideal matching impedance will be obtained using a matching layer $\lambda/4$ thick having a characteristic impedance Z_m given by

$$Z_m = (Z \cdot Z_{ti})^{1/2} \tag{5.28}$$

where Z is the transducer impedance, Z_{ti} the tissue impedance. Once again the matching problem is confined to transducers built from PZT, since no matching layer is required for PVDF transducers. In all cases, however, the transducer must be interfaced with the human body by means of a coupling medium which operates to some extent as a lubricant, but more importantly serves to exclude air from the transducer–body interface. Proprietary couplants are available for this purpose.

5.5 ULTRASOUND BEAM FOCUSING

It is often necessary to produce a focused ultrasound beam. This may be done in four distinct ways:

1. by means of a bowl-shaped transducer
2. by application of a lens structure to the radiating surface of a planar transducer
3. by use of a suitable reflector
4. by means of a multielement phased array.

Examples of the structures required for each of these approaches are shown in Figure 5.10.

The simplest approach to focusing is probably through a transducer, manufactured from the start, with the required curvature (Figure 5.10(a)). Since acoustic waves behave in many ways like light waves, in that they can be reflected and refracted, we would expect to be able to construct lenses and mirrors analogous to optical systems and this proves to be so.

For convergent beams of ultrasound, however, it may be necessary to use an optically divergent lens since the lens material may propagate longitudinal sound waves at a different speed from that of the tissue into which its output will be delivered. Inspection of Table 5.1 and Figure 5.1 shows that the speed of sound in Perspex or polystyrene exceeds that in all human tissues, fat, blood, etc. Hence it will be necessary to construct a lens made from either of these materials with a plano-concave section (optically divergent). If, on the other hand, silicone rubber is used, the lens will need to be plano-convex (optically convergent), since the speed of sound in this material is less than that in most body materials. It is also necessary that any material selected for

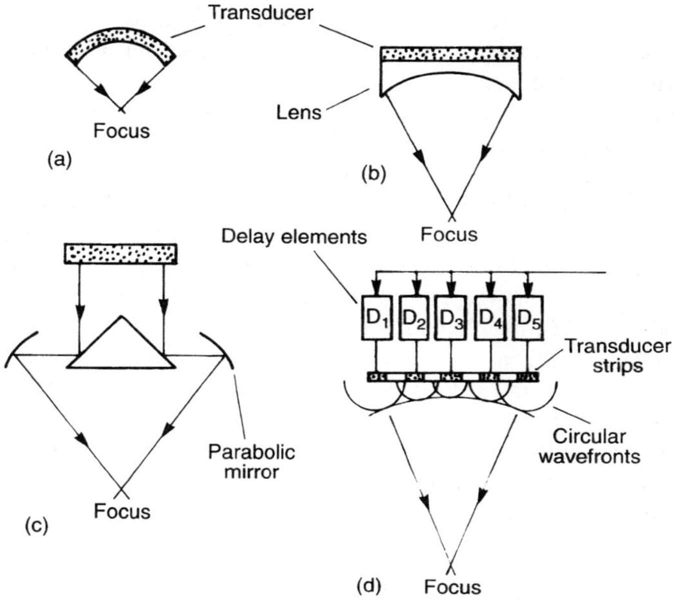

Figure 5.10 Ultrasound beam focusing using (a) a fixed-focus, curved piezo-electric bowl; (b) an attached lens structure; (c) a reflector system (due to Barone); (d) a phased array.

Table 5.1 *Physical constants for some common materials*

Material	*Density* ρ_0 (kg m^{-3})	*Longitudinal sound velocity* c_L (m s^{-1})	*Characteristic acoustic impedance Z* (kg m^{-2} s^{-1})
Air	1.29	331	425
Water	1.0×10^3	1.5×10^3	1.5×10^6
Quartz	2.65×10^3	5.9×10^3	15×10^6
PZT	7.5×10^3	4.6×10^3	34.5×10^6
PVDF	1.78×10^3	2.2×10^3	3.91×10^6
Silicone rubber	1.2×10^3	1.03×10^3	1.23×10^6
Aluminium	2.69×10^3	6.26×10^3	16.9×10^6
Perspex	1.2×10^3	2.67×10^3	3.2×10^6
Polystyrene	0.98×10^3	2.35×10^3	2.3×10^6

use as a lens should have a characteristic impedance as similar as possible to that of the tissues to which it will be coupled, and also a low absorption coefficient. Polystyrene meets these requirements better than Perspex, but both materials find use as a lens material. Figure 5.10(b) shows examples of this approach. In each case the lens must be bonded to the transducer, entirely excluding air spaces.

We are not confined to the use of lenses for ultrasound focusing, but if we choose to use mirrors we need materials whose characteristic impedance is as different as possible from that of the intervening material. If the intervening material is water say, then steel mirror envelopes will suffice. Figure 5.10(c) shows a reflector focusing system which avoids the need to put the ultrasound source in front of a reflecting mirror and instead uses a central conical reflector coupled to a parabolic section mirror. The incident plane wave is converted into a cylindrical wave by the 90° cone, and finally into a spherical wave converging on a focus. This arrangement, due to Barone, does not lend itself to use in medical ultrasound applications.

5.6 MULTIELEMENT TRANSDUCERS

A more important approach to focusing, is that based upon multi-element transducers (Figure 5.10(d)). Here the transducer is sectioned off into narrow strips, or annular rings, each strip or ring being fed separately, either directly from the master oscillator or through an adjustable delay. Reference to Figure 5.11(a) shows that when no delays are used lateral scanning may be achieved if the signal is switched from one array element to the next, or from a group of elements to an overlapping group.

When delay elements are used we have what is termed a phased array. Now we are provided with a potentially very powerful method of creating ultrasound images, which provides for:

1. beam steering without transducer motion;
2. dynamic focusing, either along the transducer normal or either side of the normal;
3. digital computer control.

Beam steering is shown schematically in Figure 5.11(b), and beam focusing in Figure 5.11(c). Each element is assumed to behave like a 'Huygens' source so that by suitable choice of phase delay we can obtain a planar wavefront which is directed to either side of the transducer normal, or even a curved (focused) wavefront. Under computer control, with properly designed digital delays, it is possible to achieve essentially instantaneous changes in direction or focusing.

As may be imagined there are problems which stem from this approach, the chief amongst which are grating lobes, since the far-field, or Fraunhofer, diffraction pattern is actually obtained from the Fourier transform of the source aperture. In consequence, we can think of the main beam and associated grating lobes as lying under a sinc-function envelope. As the main beam is steered away from the transducer normal it follows the sinc-function profile and diminishes in magnitude,

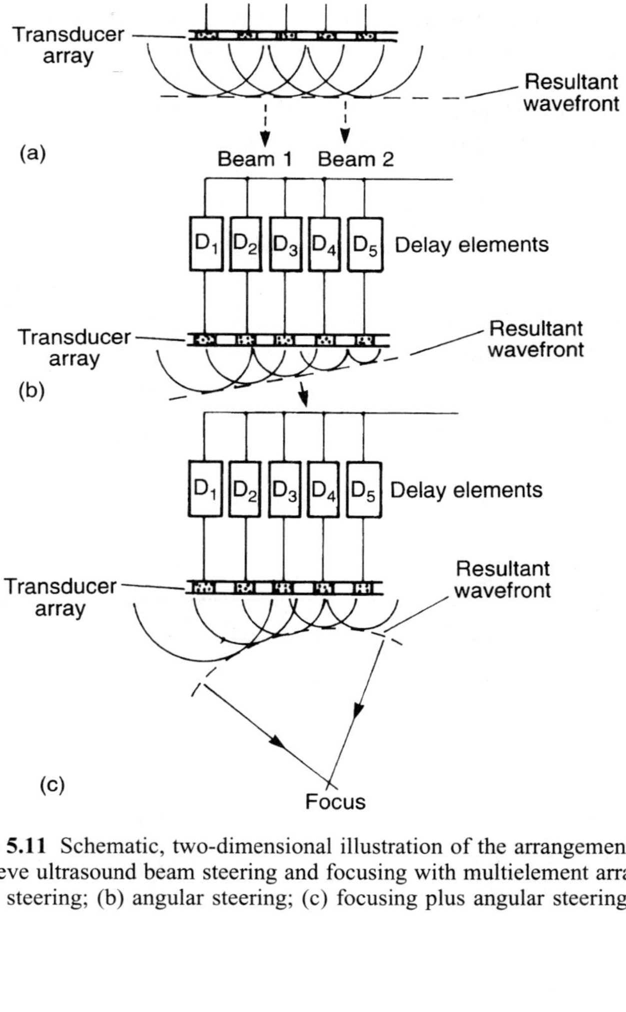

Figure 5.11 Schematic, two-dimensional illustration of the arrangements used to achieve ultrasound beam steering and focusing with multielement arrays. (a) Lateral steering; (b) angular steering; (c) focusing plus angular steering.

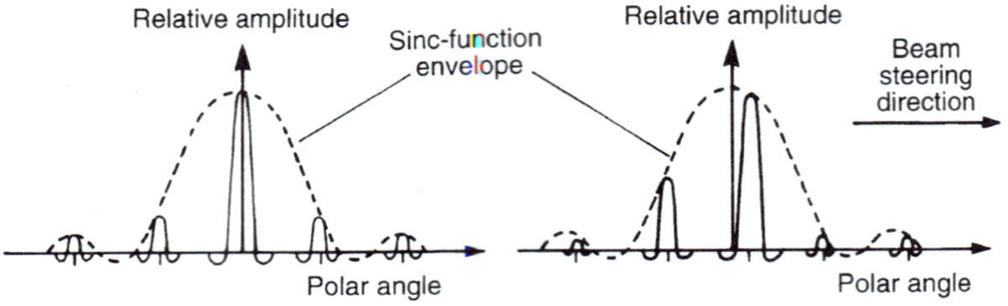

Figure 5.12 (a) Sinc-function envelope which embraces the main and grating lobes from a multielement array. (b) Distribution of magnitudes between main lobe and grating lobes as beam steering is executed.

while the associated grating lobes on the surface normal side of the main beam grow in magnitude, again following the sinc-function envelope. Similarly, the grating lobes on the other side of the main beam diminish in magnitude. This effect is shown schematically in Figure 5.12. In practice the main beam lobe and the first grating lobe become equal in magnitude at a steering angle $\theta_{\max}$, the maximum useable steering angle, where

$$\sin \theta_{\max} \simeq \lambda/2r_0 \tag{5.29}$$

The physical arrangements of some typical phased array structures are shown in Figure 5.13. In each case, the gap between the elements is as small as possible and the width of each element is $\leq\lambda/2$ so that each element can behave as a more or less omnidirectional source. This condition corresponds to the first and second distributions shown in

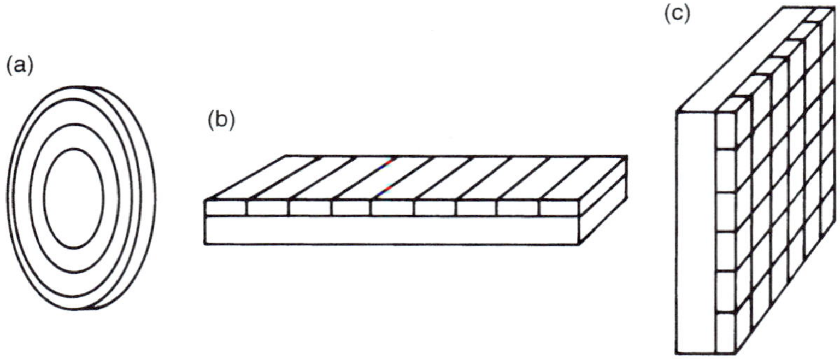

Figure 5.13 Transducer array configurations. (a) Annular array which, in principle, produces a point focus; (b) linear array, which can produce a line focus; (c) two-dimensional array capable of lateral motion in two dimensions, linear or point focusing with or without steering.

Figure 5.6. Figure 5.13 shows only a few of the phased array structures which have been described.

5.7 PULSE-ECHO IMAGING

Ultrasonic imaging of body structures may be thought of as 'radar for people', in the sense that the principles which are applied to construct the image are exactly analogous to those used in radar, the only differences being the nature of the wave disturbance (a longitudinal pressure wave), the velocity of the wave (very much less than in radar) and the wavelength of the wave (less than used in radar). Even phased arrays are based upon radar techniques!

The basic, simplified, structure of a pulse-echo imaging system is shown in Figure 5.14(a). It comprises a transmitter which will operate

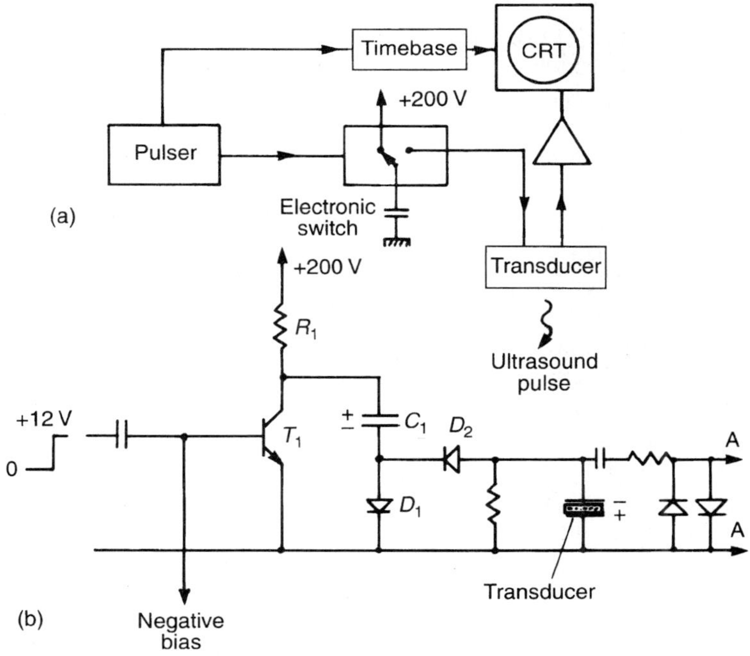

Figure 5.14 (a) Simplified block diagram showing the essential elements of a pulse-echo system operating in the A-scan mode. The input pulse sets the pulse repetition frequency and triggers the electronic switch which discharges the storage capacitor C_1 across the transducer. (b) Basic circuit for implementing the block diagram above. The echo amplifier is connected to the points marked A–A.

in a pulsed mode allowing the piezoelectric crystal to oscillate for 2–4 cycles at its natural frequency (determined by its thickness). This is most simply achieved by discharging a capacitor across the essentially capacitive transducer using an electronic switch (transistor) triggered by a master oscillator which sets the pulse repetition frequency f_{pr}. Following the generation of a pulse there will be a stream of echoes from the interfaces which the sound pulse has traversed. Clearly, all of the imaging information is contained in one clock cycle. This has the effect of limiting the depth to which one can image, so that the pulse repetition frequency, or clock cycle, is linked to the imaging depth D_{max} by

$$f_{pr} = \frac{c_L}{2D_{max}} \tag{5.30}$$

Here D_{max} is in mm and f_{pr} is in kHz, if c_L is in m s^{-1}. A pulse repetition rate of 1 kHz is often convenient.

A simple circuit which illustrates the principles involved is shown in Figure 5.14(b). The switch here is transistor T_1 which is supplied from a 200 V supply. In between trigger pulses from the master oscillator (operating at 1 kHz), the transistor is turned off, i.e. does not conduct. Capacitor C_1 charges through R_1 and D_1 to the full 200 V. It is prevented from discharging across the transducer during this time by diode D_2, which is reverse biased. When a positive pulse from the master oscillator is applied to the base of T_1, it saturates and its collector potential becomes essentially zero so that the full 200 V is switched across the transducer, since D_2 is now conducting and D_1 is reverse biased. This momentary voltage step causes the transducer to 'ring' at its natural frequency for a few cycles; this is the transmitter pulse. Inspection of the circuit of Figure 5.14(b) shows that the transmitter pulse is also applied to the echo detector amplifier which is connected across the transducer, since the transducer has a dual function, that of both transmitter (inverse piezoelectric effect) and receiver (piezoelectric effect). Careful design is required to protect the echo amplifier. Apart from the selection of a suitable pulse repetition frequency we have also to give some thought to the actual operating frequency, since this will determine the resolution and penetration. Frequencies in the range 3–5 MHz prove convenient to image the liver and other abdominal organs, or the heart. Structures lying less deep may be imaged using frequencies in the range 4–10 MHz, whilst the eye, being essentially directly accessible, may be imaged using frequencies in the range 7–15 MHz.

The arrangement of Figure 5.14(b) constitutes what is known as an A-scan (A-mode), in which the information available is essentially the distance to the reflecting interface (time for echo) and the degree of acoustic mismatch (echo strength), and in itself does not form the most descriptive mode of display. By far the most commonly used mode of display is the B-scan (B-mode) and this is usually what is being referred to if ultrasound imaging is mentioned. Figure 5.15 illustrates the essential characteristics of this approach to ultrasound imaging. A general feature is that a position signal is now provided for the transducer and this appears as a displacement along the x axis of a CRT. The A-scan distance (time) information is now applied to a timebase operating along the y axis, while the amplitude (peak height) modulates the brightness (z axis). A two-dimensional image is built up by moving the transducer in the x–y plane. This is a dramatic achievement because the image constitutes a cross-section of the human body in the chosen x–y plane and is therefore the first example of a tomogram or 'slice' image (from the Greek *tomos* meaning slice and *gram* meaning message). Clearly the plane x–y can be chosen arbitrarily so that we can build up a 3-D 'view' of a body structure by combining tomograms. This was the first example of a physics-based technique providing a cross-section image. The X-ray equivalent was not available till much later.

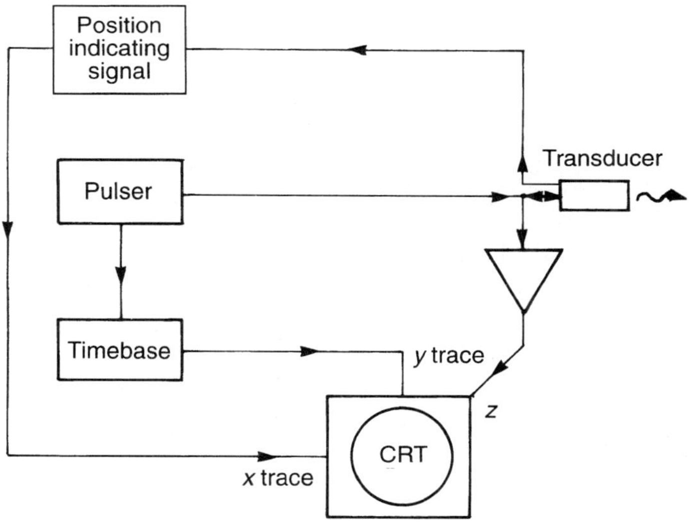

Figure 5.15 Essential elements (simplified) in the experimental arrangements required for B-scan.

Where it is necessary to image a moving structure a different approach is required known as the M-scan (M-mode or T-M mode). In this mode the A-scan signal, for echoes arriving from along a constant line of interest through the body, is applied to the *y* axis of a CRT display where the signal amplitude is used to modulate the brightness *z*-axis. A slow timebase is applied to the *x* axis, generating a time–motion image (T-M), Figure 5.16. A typical example of the application of such an imaging approach would be in examination of the pattern of movement of heart valves.

5.8 SIMPLE SCANS AND COMPOUND SCANS

Several scanning methods exist and these are crudely classified as simple scans and compound scans. Two types of simple scan exist and these are depicted in Figure 5.17(a), (b). The simple linear scan (Figure 17(a)) requires a large 'window' of access to the body, but gives the best images, with a wider field of view near the skin. The simple sector scan has the advantage that it is easier to move the beam rapidly if the scanning is mechanical, and a much narrower access window may be used. Examples of this kind of imaging are imaging the heart through **intercostal** spaces. Compound scanning (Figure 5.17(c)) gives a more complete display of boundaries and improves the effective lateral

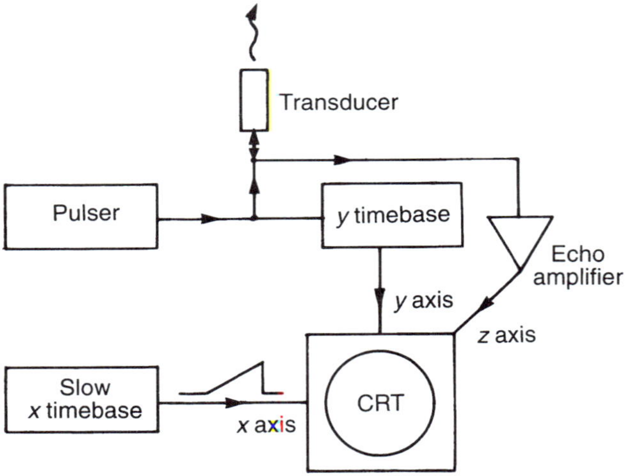

Figure 5.16 Simplified block diagram showing essential elements required for ultrasound imaging of moving structures, the so-called M-scan or T-M scan.

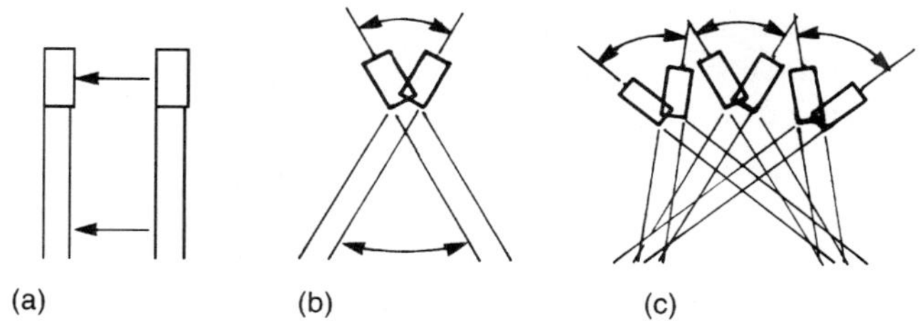

Figure 5.17 Some of the scanning methods used in pulse-echo imaging. (a) Simple linear scan; (b) simple sector scan; (c) compound scan formed by averaging a number of sector scans.

resolution by averaging the noise or speckle in the image. Unfortunately, it requires a large 'window' of access to the object and is complex; fast scans are difficult. Moving objects cannot be imaged this way, since the noise averaging process depends on the object remaining fixed.

5.9 DOPPLER-SHIFT BASED ULTRASOUND

An important application of ultrasound is that centred upon the measurement of blood flow velocities using Doppler-shift based measurements. These measurements are possible since blood is not a simple, well-behaved Newtonian fluid like water, but a suspension of individual scatterers. The red blood cells, or **erythrocytes**, may be modelled as biconcave disks approximately 8 μm in diameter, with a maximum thickness of 2 μm; there will usually be about 5×10^9 red cells per ml of whole blood. The ratio of total red cell volume to the total blood volume is about 0.45 and is known as the **haematocrit**. For a typical ultrasonic Doppler frequency of 3 MHz, the wavelength in blood is about 0.5 mm, so that an individual red blood cell is two orders of magnitude smaller than the wavelength of the incident ultrasound. Consequently, it may be considered to be a Rayleigh scatterer, for which the backscattered power increases as the fourth power of the frequency (section 5.2). Ultrasonic blood flow detection is only possible because the flowing blood interacts with the beam of ultrasound rather like a random array of discrete targets swept along by the moving medium in which they are suspended. If we consider a single scatterer moving with velocity v and direction θ with respect to the direction of the transmitted, continuous sound wave described by the equation

$$\xi_{\mathrm{t}} = \xi_0 \cos \omega_0 t \tag{5.31}$$

where ξ_t is the transmitted amplitude, ξ_0 the incident wave amplitude, and ω_0 the angular frequency, then for a separation distance x, the received wave ξ_r will be described by

$$\xi_r = \xi_0 \cos\left[\omega_0\left(t - \frac{2x}{c_L}\right)\right] \tag{5.32}$$

where the factor 2 appears because the round trip distance is $2x$. In addition x is changing at a rate $v \cos\theta$ and at any time t equals $tv \cos\theta$. Hence equation 5.32 becomes

$$\xi_r = \xi_0 \cos\left[\omega_0\left(t \pm 2\frac{tv\cos\theta}{c_L}\right)\right] \tag{5.33}$$

$$= \xi_0 \cos\left[\omega_0 t\left(1 \pm \frac{2v\cos\theta}{c_L}\right)\right] \tag{5.34}$$

If we contrast equation 5.34 with equation 5.31 we note that we can write it in the form $\xi_r = \xi_0 \cos\omega t$ where

$$\omega = \omega_0\left(1 \pm 2\frac{v\cos\theta}{c_L}\right) \tag{5.35}$$

or

$$f = f_0\left(1 \pm 2\frac{v\cos\theta}{c_L}\right) \tag{5.36}$$

Therefore the Doppler-shift $f - f_0 = \Delta f$ is given by

$$\Delta f = \pm 2 f_0 v \cos\theta / c_L \tag{5.37}$$

Figure 5.18 shows the geometry of the Doppler-shift measurement.

It should be noted that the Doppler-shift is double that which would be obtained if the 'source' were moving and the 'observer' stationary, the conventional situation. Here the apparent 'source' reflects the sound waves rather than emits them. For incident ultrasound frequencies in

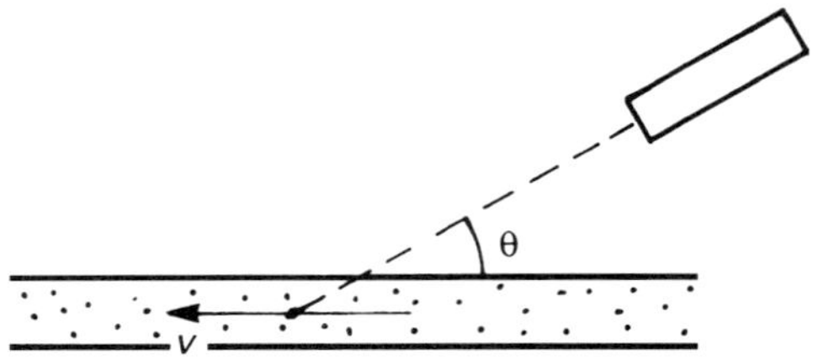

Figure 5.18 Geometry of the Doppler-shift measurement.

the range 1–10 MHz and $\upsilon \cos\theta$ in the range 0–1 m s^{-1}, Δf is 0–13 kHz, which is within the audio spectrum. We can therefore, by pure chance, contemplate 'listening' to blood flow and this is, in fact, a common method for the subjective interpretation of Doppler-shift signals, applying in virtually all practical physiological situations.

There are two particular problems with the application of Doppler-shift measurements, however: first is the selection of f_0 and the second is the determination of $\cos\theta$. Since Rayleigh scattering is the chief mechanism at work, we have an f^4 dependence which helps with the choice of f_0, coupled with which we choose to optimize the signal-to-noise ratio. The backscattering efficiency of blood increases with frequency while the useful depth of penetration decreases with frequency. This results in an optimum frequency for a given depth (Figure 5.19). The more difficult problem is determination of $\cos\theta$, which is not usually easy except for one or two situations where the ultrasound beam can interact with a blood vessel tangentially, so that $\theta = 0$ and $\cos\theta = 1$. Thus it is possible to measure blood flow in the **ascending aorta** in this way.

There are some subsidiary problems associated with Doppler-shift measurements. The first is that the quantity measured is actually a

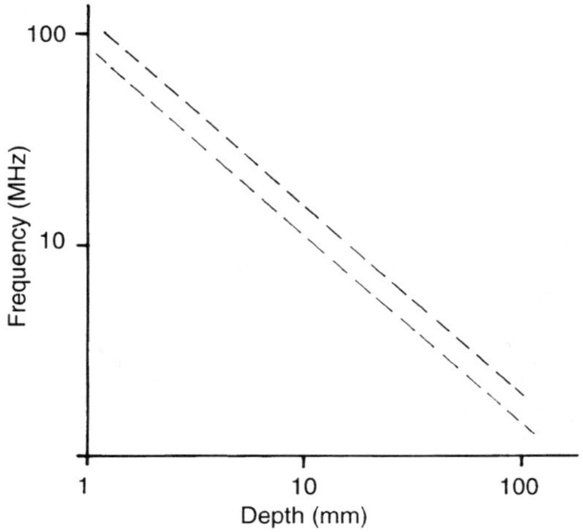

Figure 5.19 Range of frequencies (as a function of depth in muscle tissue) which will maximize the signal-to-noise ratio for the detection of ultrasound scattered from blood. (Adapted from Baker, D., Forster, F.K. and Daigle, R.E., 1978, Ultrasound, its application in medicine and biology (ed. F.J. Fry), Elsevier, New York.)

velocity rather than a flow rate; it is the latter which is usually required. Second, there are many moving scatterers each producing a different Δf, so that what is recorded is a Doppler-shift spectrum which varies with time during a cardiac cycle. In fact there is useful information to be obtained from these spectra, in particular, the shape of the spectrum and its variation with time and/or position can be a useful guide to the performance of an artery. Third, the derivation of equation 5.37 neglects the finite width of the ultrasound beam and the fact that scatterers may move across the beam. As a result, instead of the line spectrum which we might expect, we actually obtain a broadened spectrum. Finally, we have assumed a continuous-wave ultrasound source; this is often not the case.

5.10 MEASURING THE DOPPLER-SHIFT

Doppler-shift measurements may be based upon continuous-wave or pulsed sources. The geometry of these two different sources are shown in Figure 5.20. Conceptually, the simplest approach is that based upon a continuous-wave source and this is shown in Figure 5.21. A continuous wave is transmitted from one element of the transducer whilst a second element receives the backscattered signal. The difference frequency is extracted by mixing the main signal with the scattered signal, followed by low-pass filtering and amplification. The low-pass filter is required because the mixer produces the sum and difference frequencies and only the difference frequency is required. Unfortunately,

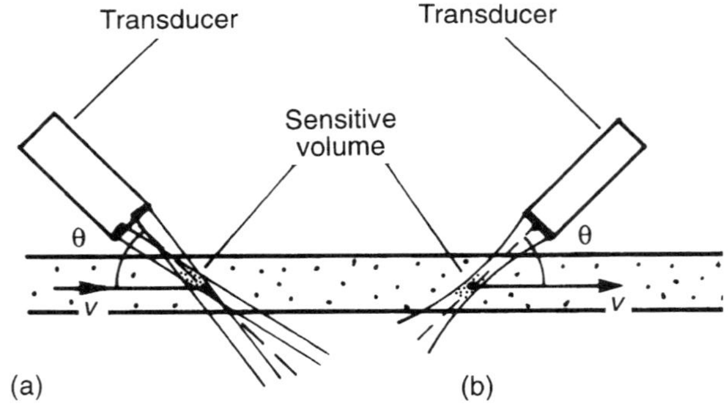

Figure 5.20 Two types of Doppler system, indicating the sensitive volume from which the scattered signal is obtained. (a) The continuous wave system using two overlapping beams to define the sensitive volume; (b) pulsed-wave system in which the sensitive volume is determined from beam width and pulse length.

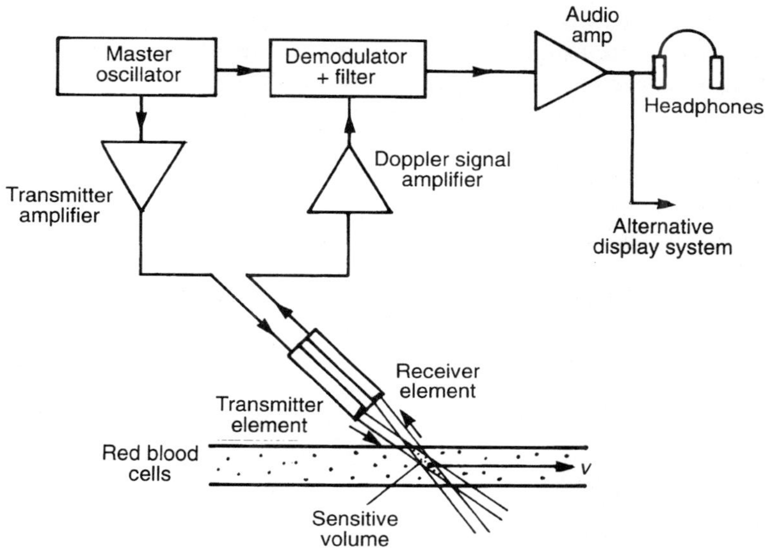

Figure 5.21 Block diagram showing the essential components of a continuous-wave ultrasound, Doppler-shift, blood flow velocity system.

this simple approach does not distinguish between flow towards and flow away from the transducer. This disadvantage may be removed by mixing the scattered signal with a reference frequency so chosen that flow forward is represented by frequencies above the reference value and flow backwards by frequencies below the reference value. Of course, this method produces an output frequency even for $v = 0$ and is not therefore suitable for audio presentation. It is however possible to discriminate electronically (by phase quadrature) between flow towards or flow away from the transducer, as a separate measurement, and combine this information with the Doppler-shifted frequency. This is an approach originally seen in conventional radar systems. Forward flow signals are presented to one ear, and reverse flow signals to the other, through headphones. Real-time spectrum analysis is now possible so that the Doppler signal may be presented as a display of frequency (ordinate) versus time (abscissa) with the amplitude indicated as a 'grey' level. This form of representation is known as a 'sonogram'. A display of this sort can also be used to generate derived characteristics such as the mean frequency or peak frequency. These may in turn be displayed so that their time variation is picked out.

A major drawback associated with the Doppler systems described above is that they convey no spatial information except in so far as they have enhanced sensitivity for a particular depth. This drawback may be overcome by adopting a pulsed-wave rather than a continuous-wave approach, thus combining the Doppler system with the pulse-echo system to produce a signal from which both distance and velocity data may be obtained. A single transducer is used for both transmission and reception. Figure 5.22 shows a block diagram of the circuit elements required to implement a pulsed Doppler system. The signal from the master oscillator is gated under the control of a clock which also determines the pulse repetition frequency. The drive signal, a few cycles long, is applied to a single element transducer which also acts as the receiver. The echo signal is amplified and passed to another gate which, in effect, permits one to sample the echo signal at a time which

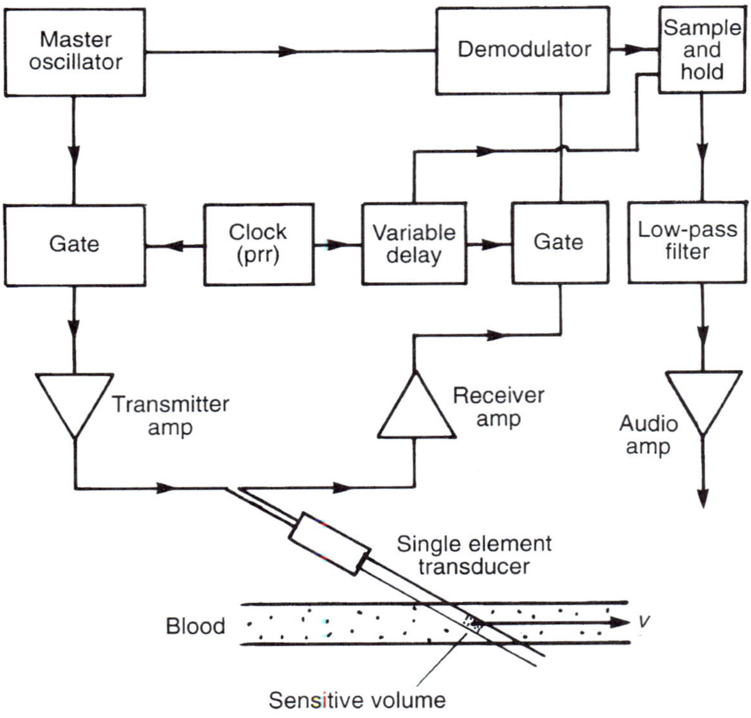

Figure 5.22 Block diagram showing the basic elements of a pulsed Doppler system.

is variable with respect to the clock signal which triggered the transmitted pulse. The output from the gate is thus a portion of the echo signal which relates to a particular depth in the body structure. This depth-selected signal is then demodulated and filtered to give the Doppler-shift frequency as before.

Unfortunately, there are performance limitations on all pulsed Doppler systems, and these may be understood if we combine equation 5.30 for the pulse repetition rate and equation 5.37 for the Doppler-shift frequency, with the additional requirement that $\Delta f \leq f_{pr}/2$ to avoid aliasing of the detected audiofrequencies (we need the Nyquist frequency) thus

$$f_{pr}/2 = \frac{c_L}{4D_{max}} = \frac{2f_0(\upsilon \cos\theta)_{max}}{c_L} \tag{5.38}$$

and

$$(\upsilon \cos\theta)_{max} = c_L^2/8f_0 D_{max} \tag{5.39}$$

If we assume that $\theta = 0$ and $c_L = 1500$ m s^{-1}, then we may express υ_{max} as

$$\upsilon_{max} = 2.8 \times 10^5/f_0 D_{max} \tag{5.40}$$

where f_0 is in MHz, D_{max} in mm and υ_{max} in mm s^{-1}. Equation 5.40 shows that υ_{max} decreases rapidly as either f_0 or D_{max} increase and this proves a severe limitation on the applicability of pulsed Doppler methods in the study of deep-lying blood vessels, if range/velocity ambiguities are to be avoided.

An inherently much more significant quantity than the flow velocity is the flow rate, but in order to determine this we need to make assumptions about the blood flow velocity profile together with estimates or measurements of the blood vessel inner diameter. This latter quantity can be obtained by means of pulse-echo ultrasound imaging, but there remains the problem of providing a reliable estimate of cos θ in the Doppler-shift equation. A possible way around some of these problems has been suggested by P.N.T. Wells and co-workers. They proposed an attenuation-compensated, volume flowmeter which operates using two concentric ultrasound beams generated by an annular array similar to that depicted in Figure 5.13(a). One of the beams from this array is sufficiently wide to embrace the whole of the blood vessel in question. The backscattered power is then proportional to the projected blood vessel area, while the mean Doppler-shift frequency gives the resolved velocity component.

We may write that wide-beam backscattered power (WBP) is proportional to $A/\cos\theta$. Therefore WBP $= KA/\cos\theta$, where A is the cross-section area and θ the measuring angle; K is a constant of proportionality. Mean Doppler frequency (MDF) $= \bar{v}\cos\theta$.

Hence the product of these two quantities gives

$$\text{WBP} \times \text{MDF} = K\frac{\text{A}}{\cos\theta}(\bar{v}\cos\theta) = KA\bar{v}$$

which is just some unknown constant times the flow rate $A\bar{v}$. If we measure the backscattered power from the second ultrasonic beam, which is much smaller in diameter and intercepts only a fraction of the overall blood vessel area, then the signal received, the narrow-beam back-scattered power (NBP), will be scaled by some geometric factor which is constant, and can be determined by calibration. Let us call this factor x. The actual flow rate is now given by

$$\frac{\text{WBP} \times \text{MDF}}{\text{NBP}} = \frac{KA}{\cos\theta}\cdot\frac{\bar{v}\cos\theta}{xK} \qquad (5.41)$$

$$= \frac{A\bar{v}}{x} \qquad (5.42)$$

This approach eliminates the dependence upon θ and estimates of blood vessel diameter (area), since x is determined in the calibration process; it may be considered an absolute method.

5.11 REAL-TIME DOPPLER COLOUR FLOW IMAGING

We have seen in the preceding sections how ultrasound may be used in a radar-like pulse/echo mode or, alternatively, in a Doppler-shift mode. However, the problems associated with blood flow measurement and display invite a combination of these two approaches; this was first demonstrated in 1978.

A curved or phased array transducer (section 5.6) is used to provide a sector scan spanning the blood vessel structure of interest, while the scan line of a pulsed Doppler system is swept slowly through the imaged plane to distinguish the flow signals in the blood vessel (Figure 5.23). These latter are distinguished by false colour. Colours not only possess luminosity (brightness or shade), but also hue and saturation. Hue is determined by the wavelength of the light, while saturation indicates the degree to which a colour departs from white and approaches a pure spectral colour. When a hue is desaturated, or diluted, with white light it is classified as a tint. The eye can discriminate between hues in the middle part of the visible spectrum (390–740 nm) with wavelength differences of 1–2 nm, so that about 130 steps of hue difference can be perceived across the visible spectrum. About 20 tints can be identified as each hue is diluted with white. Colour presentation thus offers a rather finely graded visible scale.

Different instruments, of course, adopt different schemes of colour coding. In general, however, increasing flow velocity towards the ultrasonic probe is colour-coded red–orange–yellow–white, whilst increasing reverse-flow velocity is displayed in dark blue through to light blue and white. Actually this is the reverse of the colour assignment one might expect, since with light, observation of the Doppler-shift results in blue for approaching motion and red for receding motion. Nevertheless, colour scales are attractive since they can be interpreted without reference to a coding key.

5.12 ULTRASOUND POWER MEASUREMENT

No discussion of the application of ultrasound in medical diagnosis would be complete without a description of the power levels which

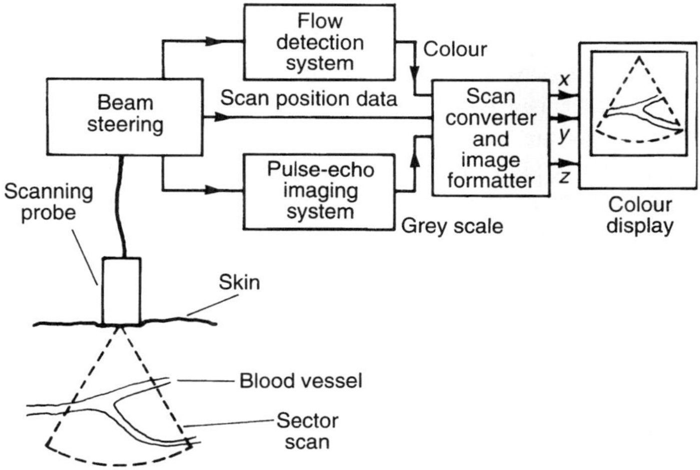

Figure 5.23 Block diagram of a real-time, two-dimensional, colour flow imaging system. A phased array or curved array transducer produces a sector scan. Distance information is extracted by the pulse-echo system, and velocities are determined from Doppler-shift. These two signals, together with a scan position signal, are combined to yield the overall display in which the blood flow is colour coded in the blood vessel image.

have been found to be safe in medical use and the means by which these power levels are measured. The effects of ultrasound on bone and body tissue are very dependent on the power levels, with potentially catastrophic effects at the highest power levels.

It is generally accepted that in diagnostic ultrasound, power densities below 100 mW cm^{-2} are non-destructive. In the range 1–20 W cm^{-2} there are destructive effects which depend primarily on the type of biological tissue being irradiated as well as the duration of irradiation. Above 20 W cm^{-2} ultrasound produces damaging and destructive effects within the biological system. The typical power levels used in medical applications are set out in Table 5.2. It will be seen from this table that the sound levels used therapeutically and surgically are in a range where we should expect to see changes in cell behaviour which are clinically significant. However, since we are primarily concerned with the application of ultrasound in medical diagnosis, we need not concern ourselves with these power levels over much, except perhaps to take note of some of the destructive mechanisms at work. Within the panorama of processes at work during absorption of a compressional sound wave there are three which become particularly significant at higher power levels. These are absorption processes by which acoustic energy is converted into heat; non-linear effects which cause the generation of high frequency components and shock-waves; and cavitational losses due to energy dissipation associated with small bubbles in the system. This last mechanism is particularly destructive of fluid-filled tissues. Heating effects are very damaging to bony structures, which have, anyway, a much higher absorption coefficient than soft tissue.

The measurement of acoustic power is usually carried out by one of three basic approaches, namely:

Table 5.2 *Summary of power density ranges for medical procedures*

	Power density (W cm^{-2})			
Procedure	0	0.1	1.0	2.0
Diagnostic (perivascular Doppler)	—	—		
Other	—			
Therapeutic		—	—	
Surgical			—	
	← Safe →		← Destructive →	

1. radiation force measurements
2. calorimetric measurements
3. optical diffraction measurements.

Of these three approaches probably the first two are the most common, and will be described here. Radiation force measurements depend on the fact that the time-averaged total power $\langle W \rangle$ radiated by the transducer is assumed to be related to the radiation pressure P by the equation

$$\langle W \rangle = c_L P/k \tag{5.43}$$

where c_L is the sound velocity in the medium (usually water) and k is a constant equal to unity for total absorption of the ultrasound, or two for total reflection. Partial absorption/reflection yields k values between 1 and 2; power levels from a fraction of a milliwatt to tens of watts may be determined.

Figure 5.24(a) shows one embodiment of the principles employed in radiation pressure measurements, and described by M.A. Perkins in 1989. A hollow conical target is suspended on jewelled bearings in a water-filled container. The equilibrium position is detected electro-optically using photodiodes, and a restoring force is applied electromagnetically using a magnet and a balance coil. The equilibrium position is maintained during irradiation by applying a feedback signal to the coil surrounding the magnet on the suspended cone (Figure 5.24(b)). The feedback signal may be calibrated in terms of the radiation pressure exerted on the conical target. This target is surrounded by a cylindrical, total absorber to prevent the ultrasound radiation reflected from the suspended cone from re-impinging on the cone and resulting in an erroneously high reading. This type of approach to ultrasound power measurements is capable of yielding a sensitivity of 1 mW.

An alternative approach to ultrasound power measurement is that based upon the temperature rise associated with absorption of the ultrasound in a suitable medium; the calorimetric approach. Calorimetric methods, like radiation-force methods, have the particular attraction that they can, in principle, be carried out using direct reference to fundamental physical quantities, that is, they may be absolute methods. In the case of calorimetric methods, calibration may be effected by electrical means where the electrical power delivered is determined very precisely.

In calorimetric methods the ultrasound power is absorbed in a convenient absorbing medium, such as castor oil, and the consequent

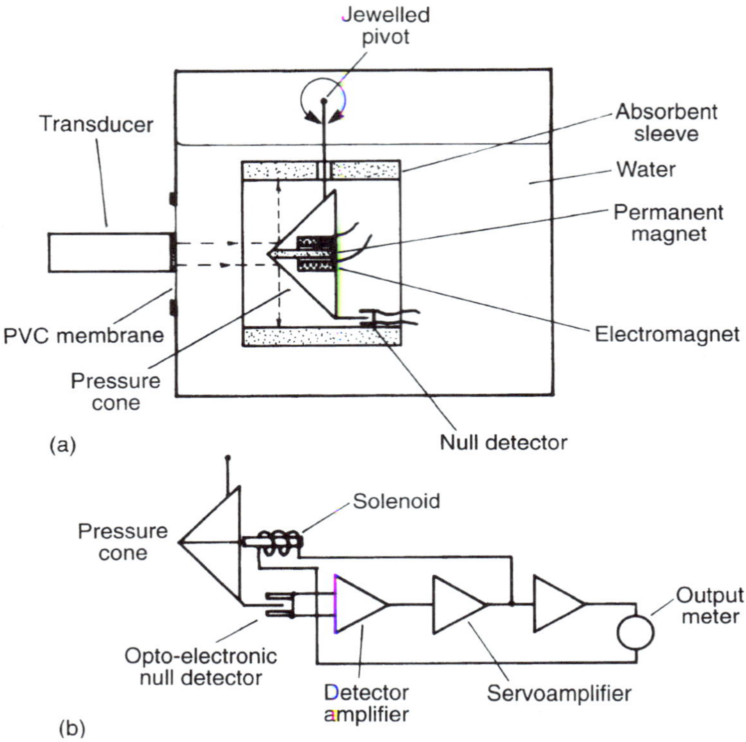

Figure 5.24 (a) Simplified cross-section of a radiation-force balance for the measurement of ultrasound power. The deflection of the pivoted cone is restored by a servo-system operated by the null detector. The reflected ultrasound is absorbed by the surrounding sleeve. (After M.A. Perkins, *Phys. Med. Biol.* **34** (1989) 1645.) (b) Circuit arrangement for radiation-force balance.

temperature rise measured. Again a sensitivity of 1 mW in the range up to 10 W can be achieved. The actual calorimeter can be based on constant-flow, differential measurement, or an absolute measurement. Figure 5.25 shows, schematically, a section through an absolute calorimeter in which the ultrasound power is absorbed in a cup (castor oil-filled), attached to a stem so designed that the temperature difference established across the stem of the calorimeter is numerically equal to the power (in watts) dissipated in the bowl. The lower extremity of the calorimeter is maintained at constant temperature (essentially room temperature) using a Peltier device (heat pump) driven by an electronic control circuit. A differential, copper–constantan thermocouple, is used

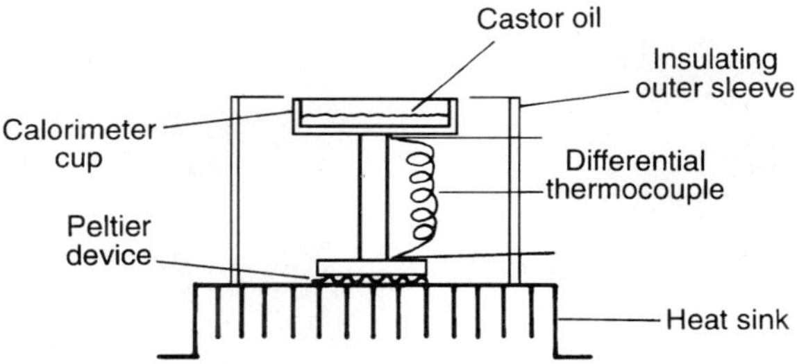

Figure 5.25 Absolute calorimeter for ultrasound power measurement. The lower surface of the calorimeter is maintained at a constant temperature (electronically) while the temperature difference arising between the top and bottom surfaces is measured to determine the ultrasound power absorbed in the castor oil layer. (After T.A. Delchar and R.J. Melvin, *Meas. Sci. Technol.* **5** (1994) 1533.)

to measure the temperature difference resulting from the input of ultrasound power. An electrical (resistive) heater wound around the receiver bowl serves to permit initial calibration.

The temperature difference $\Delta\Theta$ across a length l of conductor having cross-section area A and thermal conductivity K, subject to a power input W, is given by

$$W = K\frac{A}{l}\Delta\Theta \qquad (5.44)$$

For a given material the ratio l/A may be set numerically equal to K, so that W is numerically equal to $\Delta\Theta$. The resulting calorimeter is absolute, and extremely linear (0.1%)

5.13 CLINICAL APPLICATIONS OF ULTRASOUND

In this section we examine ultrasound images which typify some of the principal imaging approaches. The ultrasound B-scan shown in Figure 5.26 is for a human embryo, a fairly typical and recognizable subject.

Time–motion T-M imaging or M-scan imaging of mitral valve motion in a normal heart is shown in Figure 5.27 and the synchronization with the EEG signal can readily be seen. The mitral valve controls blood flow into the left ventricle of the heart. A measurement which is usually made in conjunction with the M-scan is that of the blood flow velocity. Figure 5.28 shows the pulsed Doppler signal obtained for line

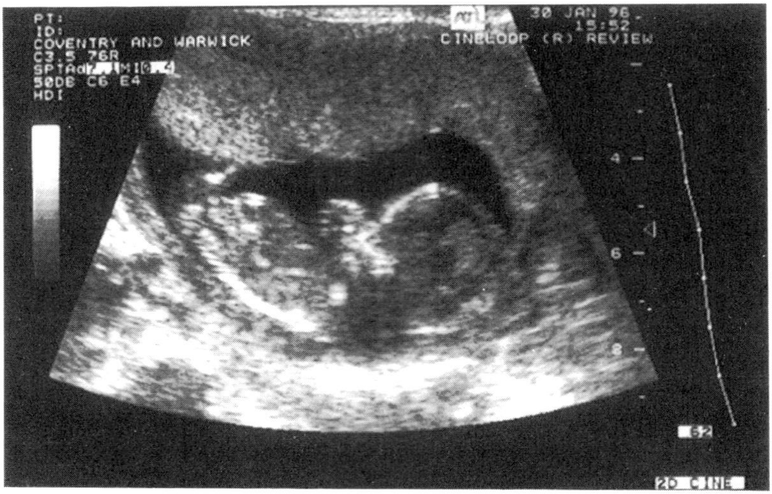

Figure 5.26 Ultrasound B-scan image of human embryo recorded with an incident ultrasound frequency of 3.5 MHz. (Courtesy of Mr A. Ashley.)

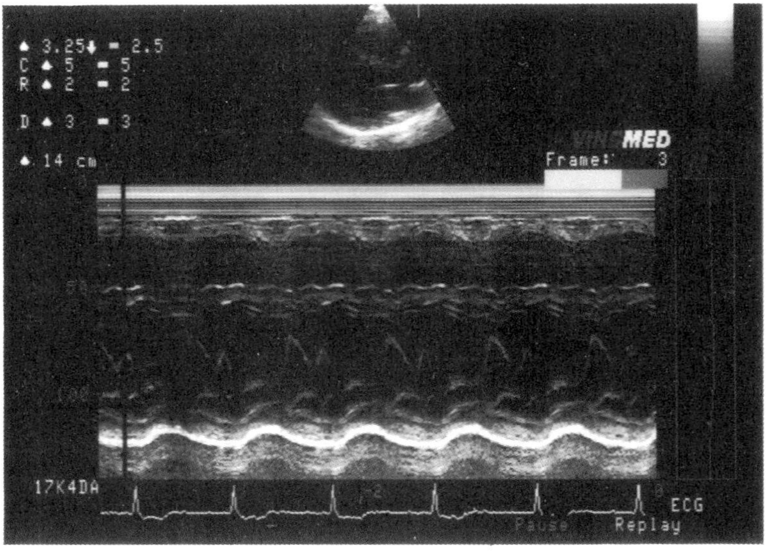

Figure 5.27 Ultrasound M-scan imaging of the mitral valve in a normal heart. Incident ultrasound frequency 2.5 MHz. The uppermost part of the image is the chest wall, the mitral valve signal is arrowed. The EEG signal is displayed along the bottom of the image. (Courtesy of Dr H. Singh.)

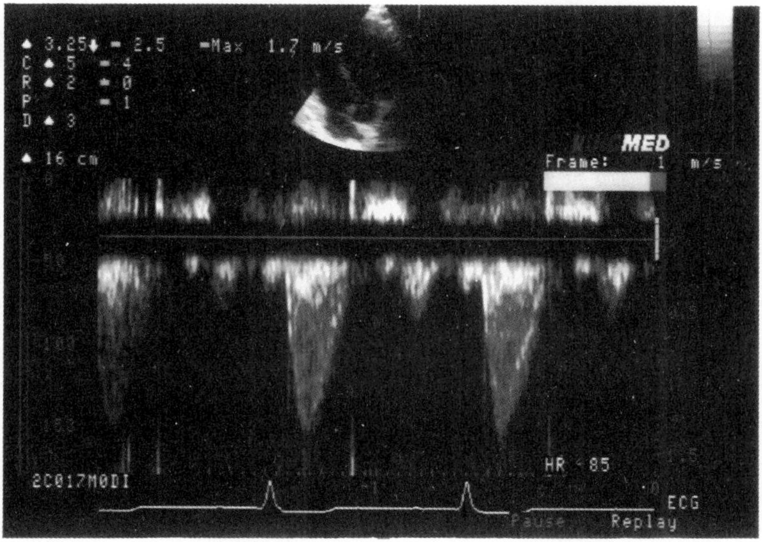

Figure 5.28 Ultrasonic measurement, using pulsed Doppler, of the blood flow velocity (in m s^{-1}) through the mitral valve. Flow downwards represents flow away from the transducer. (Courtesy of Dr H. Singh.)

of sight through the mitral valve and reflects the blood flow velocity (in m s^{-1}) through the valve. The image is again that from a normal heart.

PROBLEMS

1. Calculate the fraction of incident sound intensity transmitted through an aluminium–water interface for an ultrasound beam, incident normally.
2. A 1 MHz PZT transducer requires a matching layer to minimize acoustic mismatch at body transducer interfaces. If body tissue has a density of 1.01×10^3 kg m^{-3} (a) calculate the characteristic impedance of the required matching layer. (b) The matching layer is constructed from material with a density of 2.4×10^{-3} kg m^{-3}, calculate the thickness of the matching layer. You should assume that the velocity of longitudinal sound waves in body tissue is 1.54×10^3 m s^{-1}.
3. A PZT ultrasound transducer is mounted in a tube of 3 mm internal diameter, and is to be used, with warm saline solution as a coupling medium, in order to assess the 'echo' from the ear drum in patients

suffering from 'glue ear'. If the maximum range of the transducer is 10 mm, calculate (a) the required operating frequency and (b) the thickness of the transducer element. Assume the properties of saline to be identical to those of water.

4. In problem three above, what fraction of the incident ultrasound intensity will be reflected if the ear is normal, that is the middle ear is air-filled?
5. A body structure lying at a depth of 300 mm is to be imaged in the B-scan mode. What is the maximum pulse repetition rate which may be used to provide an unambiguous image? Assume sound speed is 1.54×10^3 m s^{-1} in tissue.

6

X-ray transmission computed tomography

6.1 INTRODUCTION

The preceding chapter has described the first application of physics in medical diagnosis which was able to yield an image of a section or slice through the body in a chosen plane. This slice image, or tomogram, was obtained using ultrasound beams in the radar-like pulse-echo mode, a technique whose introduction into medical diagnosis occurred nearly six decades after the first use of X-rays. Despite the scope of the ultrasonic beam approach, there were, and still are, some parts of the body which cannot be imaged satisfactorily by this means, perhaps the most important example being the brain. For the brain the use of ultrasound in the pulse-echo mode poses particular problems, because an ultrasound pulse entering the head experiences strong reflection at the scalp–skull interface, and the resulting echo bounces back and forth within the skull. Furthermore, throughout most of the cranial skull, the irregular water-filled spaces such as the **ventricles** constitute pronounced refracting, reflecting and dispersing structures so that, in consequence, the path of an ultrasound beam is grossly disrupted. For this reason one cannot usefully scan the human head/brain and produce an interpretable, clear image.

We have seen that conventional X-ray radiography also has drawbacks, the most important of which are:

1. the absence of depth information – the X-ray image is simply a shadowgraph;
2. the lack of contrast presented by many body structures, principally soft tissue.

This latter drawback means that X-ray radiography will not readily provide images of brain structures or defects. Although an X-ray radiograph of the head/brain provides a very sharp and clear image of the skull, it provides no structural information on its contents unless special contrast procedures are adopted. Two such procedures have been widely used in the past, **cerebral angiography** and **pneumoencephalography**. In the first procedure a contrast medium was introduced, usually via the carotid artery in the neck; in the second the **cerebrospinal fluid** was wholly or partly replaced by air, outlining the ventricles. Both of these procedures were, and are, hazardous and traumatic for the patient. The second drawback listed above means that conventional X-ray radiography will not readily provide images of brain structures or defects.

The solution to the problems of imaging the brain was described by G.N. Hounsfield and J. Ambrose (a neuroradiologist) in 1973, in a paper entitled 'Computerised Transverse Axial Scanning Tomography'. The commercial development of X-ray computed tomography which followed has been described as the most important development in diagnostic radiology since the introduction of the planar radiograph. Hounsfield shared the 1979 Nobel Prize for Physiology and Medicine with the American A.M. Cormack, who had also been working on this problem in the 1960s. The advance embodied in the work of Hounsfield and Cormack solved, in one go, not only the problem of imaging low contrast areas of the body, but also the provision of depth information, and the imaging of the brain. The important theoretical concept underlying computerized tomography (CT) is usually attributed to the Austrian mathematician J. Radon, who, whilst working on gravitational theory, published the seminal paper '*Uber die Bestimmung von Funktionen durch ihre lntegralwerte langs gewisser Mannigfaltigkeiten*', or in translation, 'On the determination of functions from their integrals along certain manifolds'.

This paper was published in 1917, and is the first example of a fundamental, theoretical contribution to the application of physics in medical diagnosis. The theoretical ideas found in this early work have been applied more widely in the newer imaging systems based on other modalities, such as magnetic resonance imaging (MRI), or emission computed tomography (ECT); these topics are dealt with in later chapters. The basic ideas underlying computerized tomography will be discussed, largely from the point of view of the problems raised in X-ray imaging. It is worth noting, in passing, that the first 'reconstructed' images, that is images reconstructed from projections, were actually obtained independently of Radon's work by R.N. Bracewell, who used them in the field of radioastronomy in 1956, and by De

Rosier and Klug, who applied the reconstruction idea in 1968, invoking a 'projection theorem' (originally due to W.L. Bragg) in the solution of structures from transmission electron microscopy images. Ironically, the first computerized tomography system was probably built in Russia in 1958 by B.I. Korenblyum and co-workers and received no recognition in the West.

6.2 TRADITIONAL OR CLASSICAL TOMOGRAPHY

Prior to the introduction of computerized tomography, sectional imaging had in fact been accomplished using either linear tomography or axial transverse tomography. In the former technique, a section of a patient is obtained by placing a photographic film on one side of the patient, parallel to the image plane of interest, with the X-ray source on the other. By moving the X-ray source parallel to the image plane, and the film in the opposite but parallel direction, we can ensure that points in the plane of interest project clearly onto the photographic film. The relative speeds of the film and X-ray source, together with the relative spacings of the film and source on either side of the patient, determine the position of the image plane (the fulcrum of the moving system). Thus if the film and X-ray source are positioned equidistant from the patient and move with equal, but opposite velocities, the image plane will be situated midway between the film and source. Points in planes either above or below the plane chosen for imaging will be blurred; Figure 6.1(a) shows the geometry of this arrangement.

An alternative arrangement, and one more closely related to computerized tomography, is transaxial tomography or axial transverse tomography. This is depicted in Figure 6.1(b). The patient sits in a special rotating chair, in an upright position; the X-ray film lies flat on a rotating horizontal platform adjacent to the patient. The X-ray beam is directed obliquely through the patient and onto the film. The X-ray source remains stationary during an exposure, but both patient and film are rotated about their axes in the same direction and at the same velocity. Only those points actually on the focal plane indicated in Figure 6.1(b) remain clear throughout rotation. Points above and below this plane are blurred. The thickness of the section which is imaged is determined by the angle between the X-ray tube and film. As this angle is made more oblique the imaged section becomes thinner, points in the image plane of the patient map onto the film so that a point in the image plane, on the axis of rotation of the patient, maps onto a point on the film corresponding to the axis of rotation of the film. In clinical

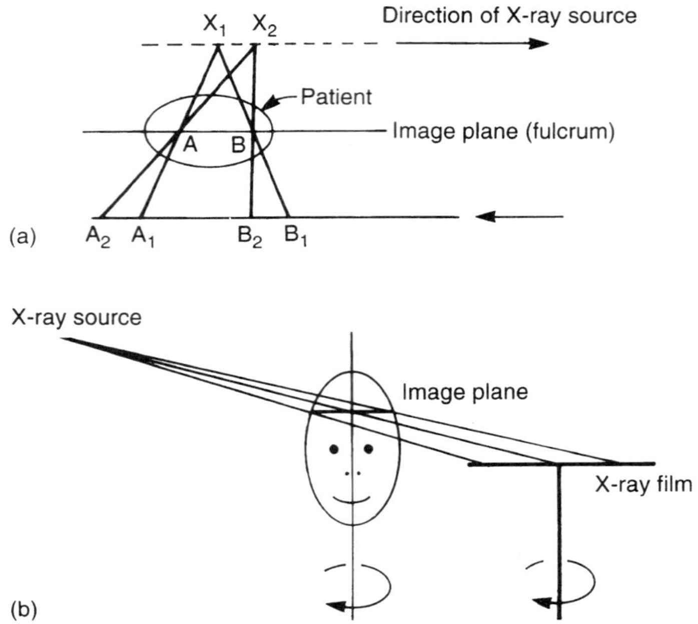

Figure 6.1 (a) Basis for linear tomography in which the X-ray source and X-ray film are moved in parallel, but opposite, directions on either side of the patient. Only details in the image plane are recorded without blurring. Thus A projects as A_1 and then A_2 a short time later. (b) Axial transverse tomography in which a fixed X-ray source is directed obliquely through the patient and the details of the image plane recorded on the X-ray film. The patient and the X-ray film are rotated at the same speed. Neither technique provides a sharp image.

practice it is usual to keep the patient stationary and horizontal, while rotating the X-ray source and detector about the patient, and maintaining the relative geometry described above.

6.3 X-RAY COMPUTED TOMOGRAPHY: THE CONCEPT

The traditional method of X-ray imaging, which produces a shadowgraph, is still widely used and is often quite adequate for diagnostic purposes. Difficulties only arise when it is necessary to image those parts of the body where the structural elements differ very little in their absorptivity to X-rays. Although there are some ways around this difficulty, principally by using radio-opaque material injected into the tissue, or by using digital radiography, as described in Chapter 4, there remain parts of the body where the X-ray imaging capability is quite inadequate: the brain is the most obvious example. If we are to build up an image of a slice or section through the body then we must alter the experimental approach quite radically in so far as we must confine the X-rays to the layer, slice or section, whose structure is to be determined. In other words we must use a 'pencil' of X-rays to scan the chosen layer from a number of different directions, but always within the plane of the layer. For each direction the attenuation of the X-rays is measured along a set of evenly spaced 'ray paths'. The measurement for each individual ray path gives a 'ray integral' or 'ray sum'. If we execute stepwise, parallel, translations of the X-ray beam and detector, as shown in Figure 6.2(a), then we obtain the 'projection' of the structure of the plane in the given direction. By rotating the X-ray source–detector combination through successive angular steps, still within the same plane, a large number of projections may be obtained. An image of the brain obtained in this way is shown in Figure 6.2(b).

Although this was the approach adopted in the early work of Hounsfield, and is conceptually straightforward, it is not capable of yielding an image in less than 4–5 min. This is due to the fact that something like 160 lateral steps of X-ray source and detector are made in each angular position and there are 180 angular positions (Figure 6.3). The total number of readings in this combination is 28 800, and these have then to be processed to retrieve the slice image. The rate-limiting step in this process is obtaining the data, so all improvements have to come from changes to the mechanical arrangements for data recording. For the moment we will stick with stepwise lateral motion plus angular rotation while developing the theoretical concepts. In practice the 'slice' is usually approximately 2 mm thick and the resulting image is as if the slice had been removed from the body and imaged with X-rays normal to the slice plane. The resulting image displays the human anatomy with a spatial resolution of about 1 mm and a level of discrimination in the linear attenuation coefficient μ corresponding to attenuation coefficient changes of as little as 0.005. To match this, a conventional X-ray would have to render visible on film intensity differences of approximately 0.02%; this performance is two orders of magnitude better than can actually be achieved. In fact, all of the image information is actually contained in the transmitted photon flux recorded by the conventional X-ray film. However, the limitations of film as a detector on the one hand, and the human eye as a discriminator on the other, together with the problems of overlapping image detail, prevent the film (radiograph) from yielding all of the recorded information. X-ray computerized tomography (X-ray CT), on the other hand, can resolve soft tissue detail, and structures which overlap may be distinguished separately. Hounsfield's ideas first appeared commercially in 1972 as the EMI Mk1 CT 1000 head scanner, and revolutionized diagnosis of conditions within the skull. An example of the sort of image which may now be obtained as a section through the skull is shown in Figure 6.2(b).

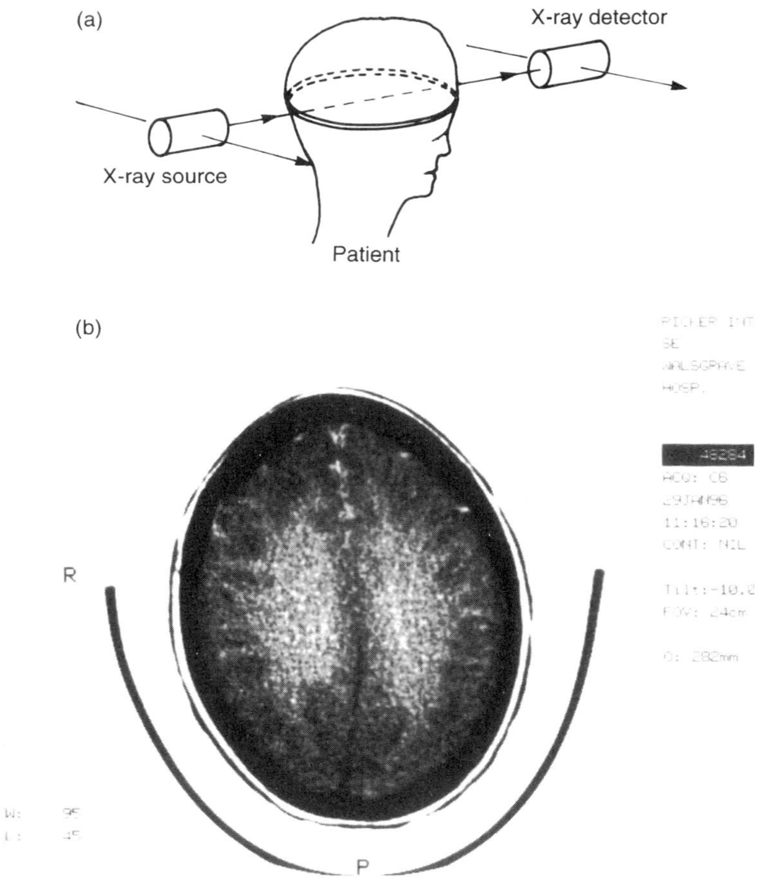

Figure 6.2 (a) Principle of the scanning arrangement used for computerized X-ray tomography. (b) Image obtained for a section through the skull using computerized tomography (actually at the C6 level indexed in Figure 6.20).

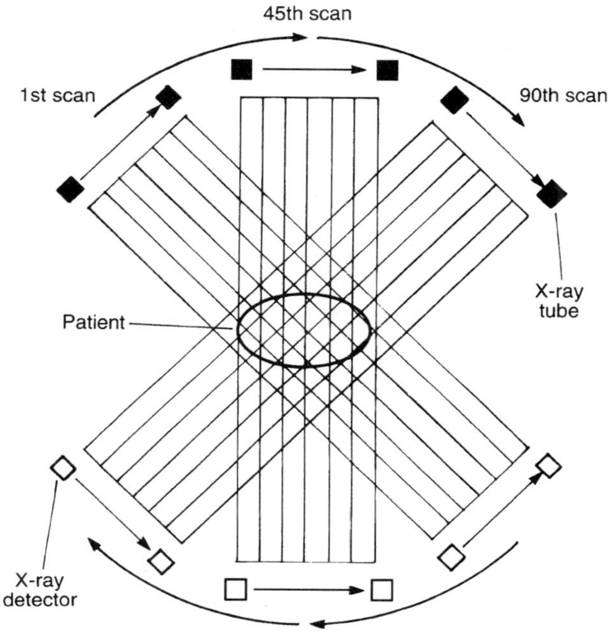

Figure 6.3 Scanning basis of a first-generation computerized tomography X-ray system (basis for the EMI Mk I CT 1000 head scanner, 1972).

6.4 X-RAY COMPUTED TOMOGRAPHY SCANNING SYSTEMS

The skull represents a special imaging situation in that not only had there never been a really satisfactory method of imaging its contents, but it could be immobilized quite readily, so that the long scanning time of the early instruments (approximately 4.5 min) was not a serious drawback. Clearly though, any extension of the technique to other parts of the body must require a more rapid scan, since generally it is not possible to maintain complete immobility, particularly of those parts which are subject to natural internal motion, such as breathing, etc. However, there are some further somewhat hidden advantages which attach to the early, or first-generation, scanner. These arise from the manner in which the data is obtained. A finely collimated X-ray source defines a pencil beam of X-rays which is then measured by a well-collimated detector. This source–detector combination measures parallel projections one at a time as it is stepped laterally across the patient.

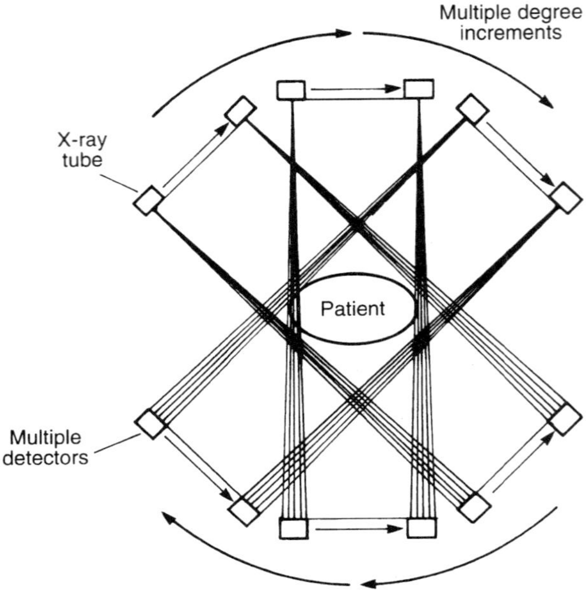

Figure 6.4 Second-generation fan beam incremental scanning system. The fan beam subtends an angle of approximately 10° and rotation takes place in 10° intervals. Fewer lateral steps are required.

Since there is only one detector, there is no problem of having to balance multiple detectors as is the case in later, faster systems, and the scatter rejection is necessarily higher than in all subsequent systems, because of the two-dimensional collimation at both source and detector.

The first reduction in scanning time came with the introduction of a single X-ray source illuminating a bank of detectors and defining an angle of approximately 10° (Figure 6.4). Now, although the lateral, stepwise motion is maintained, many more projections are measured simultaneously at each step; the number of projections is equal to the number of detectors. Fewer lateral steps are required and the angular change is increased to approximately 10° (the fan angle) with the consequence that an image may be produced in a greatly shortened time, typically about 20 s. A timespan of this magnitude is sufficiently short for patients to hold their breath, and images can then be obtained from the chest and abdomen. This arrangement constitutes a second-generation scanner.

An obvious improvement to the second-generation scanner is to widen the included angle of the fan beam to embrace even more detectors and simultaneously the whole of the patient. With this change the gantry carrying the source and detectors need only rotate, without stopping, since no lateral motion is required. Consequently, the data may be gathered in a much shorter time, typically around 4–5 s. Although this means that it is relatively easy for a patient to maintain immobility during a scan, the large rank of detectors requires very careful balancing if image artefacts are to be avoided. Usually, xenon detectors are selected owing to their stable operating characteristics. The X-ray source is pulsed in this arrangement; since there will typically be several hundred detector tubes, each pulse provides several hundred measurements. This arrangement constitutes what may be thought of as a third-generation scanner, and is shown schematically in Figure 6.5.

In the fourth-generation scanner (Figure 6.6) a further structural simplification is achieved by having only the X-ray source rotating around the patient, but now within a stationary ring of detectors, some 1000 in number. Although the scan time does not change significantly from that achieved with third-generation scanners, there is an advantage in that the X-ray fan beam passes outside the extremities of the patient and this means that every X-ray detector samples the unattenuated beam at some time. This permits 'real-time' calibration and the problem of image artefacts disappears.

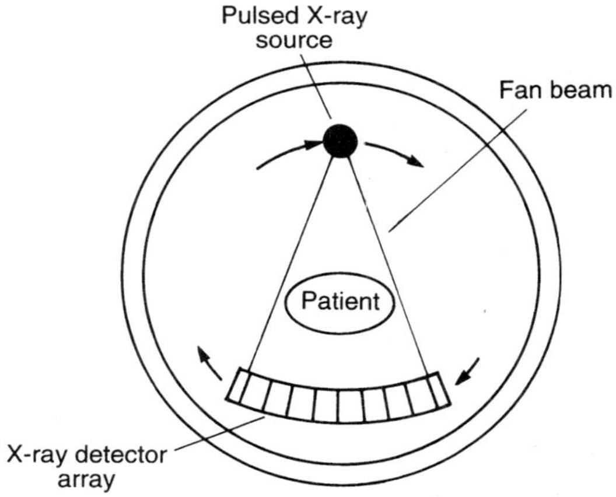

Figure 6.5 Third-generation fan beam 'spinning' scanner. The X-ray source is now pulsed and rotates without stopping.

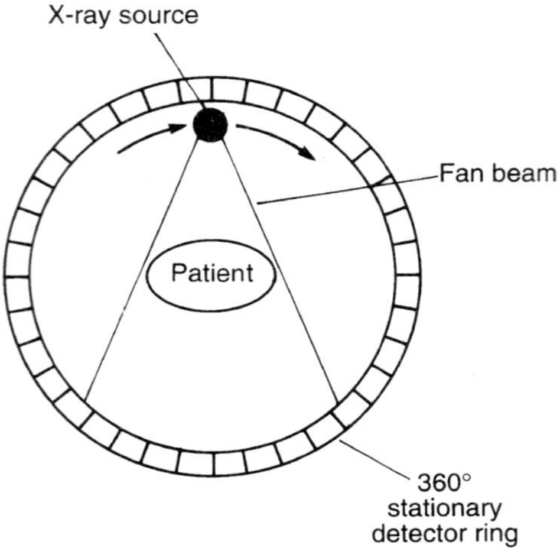

Figure 6.6 Fourth-generation fan beam 'spinning' scanner. Unpulsed source rotates without stopping within a ring of stationary detectors.

The scanners described thus far permit imaging of most body structures (with breath retention). However, the heart and organs, which are generously endowed with blood supply, have a pulsatile motion which cannot be captured for an X-ray image unless the data acquisition time can be reduced to around 100 ms or less. This is achieved in the fifth-generation scanner developed by D.P. Boyd *et al.* in 1979, known as the Imatron. The fifth-generation scanner has no moving parts in the strict mechanical sense. Instead an entirely unconventional type of X-ray tube is employed in which the position of the electron beam focus, on the X-ray tube anode, is under the control of external deflection coils and the X-ray tube anode subtends an angle of approximately 210° around the patient. The anode is now a sector of a circle; Figure 6.7 shows the physical arrangement employed in the Imatron C-100. The particular advantages of this arrangement are:

1. the high speed with which the X-ray source (point of focus of the electron beam) can be moved around the patient, under electronic control;
2. the ability to choose adjacent 'slices' of the patient simply by splitting the anode into separate parallel arcs and moving the electron focus (electronically).

Because the only moving 'part' is an electron beam, the data acquisition time is reduced to a matter of some 50 ms. By this means any part of the human body can be imaged. Another unconventional feature of the Imatron is the geometric relationship between the target ring and the detector rings. In a conventional X-ray CT system the X-ray source and the detectors are strictly coplanar with the imaging plane. The fan beam geometry of the Imatron, results in a minimum angular sampling

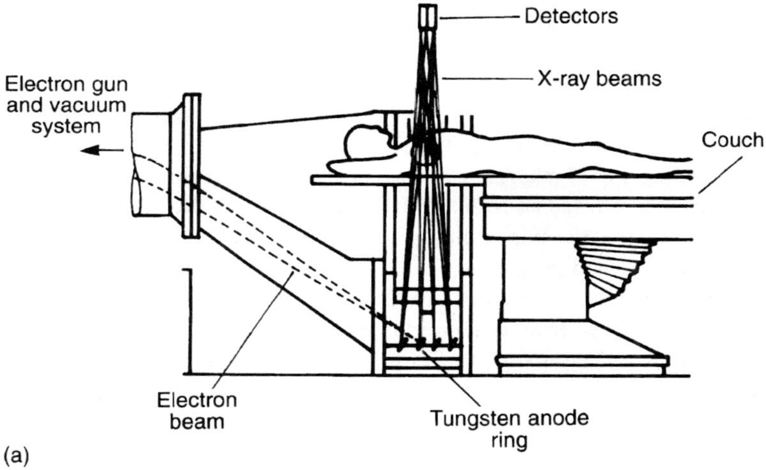

(a)

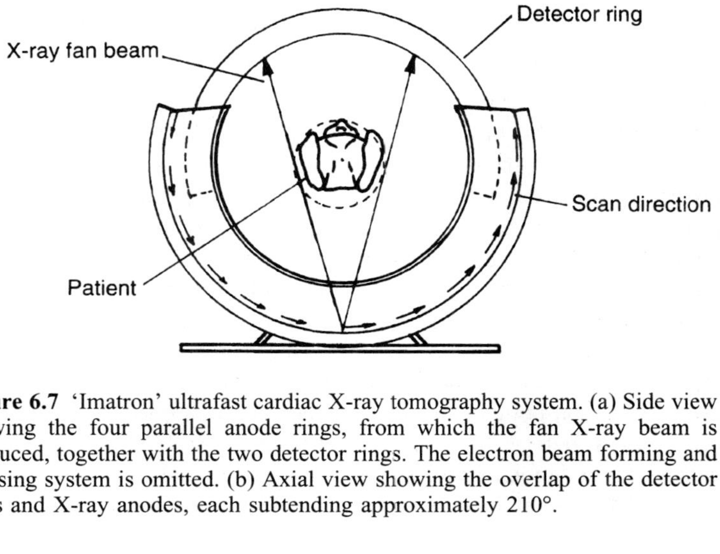

(b)

Figure 6.7 'Imatron' ultrafast cardiac X-ray tomography system. (a) Side view showing the four parallel anode rings, from which the fan X-ray beam is produced, together with the two detector rings. The electron beam forming and focusing system is omitted. (b) Axial view showing the overlap of the detector rings and X-ray anodes, each subtending approximately 210°.

requirement which exceeds 180°; the stationary detector ring cannot be exactly coplanar with the target ring, but must be offset axially (Figure 6.7(a)). Collimation of the X-ray beam to the detectors is accomplished by a pair of collimators, which have the effect of creating an unconventional three-dimensional fan beam, or cone beam, cross-connecting the target plane and the detector plane. Image reconstruction is performed using a standard two-dimensional fan beam filtered backprojection algorithm, which does not specifically take account of the special geometry used.

The Imatron Cine-CT has four separate target rings and two contiguous detector rings. A single target scan (50 ms) produces a pair of adjacent tomographic slices. Eight tomographic slices can be obtained by scanning the four target rings sequentially, in combination with the two detector rings. The result is that eight separate, though not exactly parallel, levels are imaged almost simultaneously (<230 ms), to cover an 8 cm thick image volume.

The preceding descriptions give no impression of the computerized tomography scanner as it is presented to the patient. A typical scanner is characterized by a smooth framework surrounding a large hole into which is inserted the patient's head or body. The smooth framework disguises the complex gantry, X-ray source and bank of X-ray detectors. The properties of the different scanning systems are summarized in Table 6.1.

6.5 THE IMAGING PROBLEM

Before we consider the various routes by which a sectional image may be constructed, starting from transmission measurements through the patient, it is useful to examine some of the background details of the imaging problem. In order to create an image of an object we have to convey information from the object to the image. Information can, in principle, disperse to all image locations from each object location. To be useful, however, an imaging modality should ensure that the principal contribution to a given image point will come from a single object point, a one-to-one transfer. In practice, other neighbouring points will contribute smaller amounts of information, the contribution decreasing rapidly away from the main contributing location; we can think of this as the neighbourhood principle. This idea is usually described quantitatively by a point-source response function (PSRF), which will be narrow for a 'good' imaging system in which the neighbourhood is confined to a small region. Thus, in order to construct an image we

Table 6.1 *Summary of computerized tomography scanners*

Type	*Mode*	*Comments*	*Scan time*
First-generation Pencil beam 'incremental' scanner	Pencil beam, 160 lateral steps followed by 1° rotation for each set of readings.	No detector calibration problem, source and detector move together on gantry.	4–5 min
Second-generation Fan beam 'incremental' scanner	Fan beam subtending 10° angle, reduced number of lateral steps followed by 10° rotation for each set of readings.	As above but now multiple detectors. Data collected for several sets of parallel rays, total number of rays increased.	~20 s
Third-generation Fan beam 'spinning' scanner	Wide angle fan beam encompasses patient, no lateral steps, source now pulsed and rotates without stopping.	Detector balance and calibration problem; several hundred detectors required. Several hundred readings obtained for each pulse.	4–5 s
Fourth-generation Fan beam 'spinning' scanner	Wide angle fan beam (non-pulsed) source rotates without stopping within ring of stationary detectors.	Real time calibration of detectors possible.	4–5 s
Fifth-generation 'Imatron'	Special type of scanning X-ray source, no mechanical moving parts.	Essentially simultaneous multiple slice images possible.	~50 ms

need a means by which we can establish a physical link between object and image. This link, in medical imaging, may take a variety of forms, for example γ-ray photons in the gamma camera, infrared photons in thermal imaging, or attenuation of X-ray photons in an X-ray image.

The function which relates the object plane to the image plane is known as the space variant point-source response function (SVPSRF).

In some cases this function may actually be space invariant, that is the same for all locations. Actually an image is the convolution of the object distribution with the point-source response function (PSRF), which carries the information from object plane to image plane, so we might expect that the process of deconvolution will be helpful in removing any signal degradation. Since convolution of two functions is identical to the product of the appropriate Fourier transforms, we note that in Fourier space the object can be split up into its component spatial frequencies. A sharp image contains spatial frequencies which are higher than those in a more blurred image. The application of these ideas involving convolution and/or Fourier transforms will be outlined in the following sections for the particular example of imaging with X-ray beams.

It is our aim to produce an image of a slice through the patient. In order to do this we will divide this slice up into arbitrary square cross-section blocks or volume elements of dimension equal to the slice thickness, usually known as *voxels* for short. The objective is to assign values of μ, the linear attenuation coefficient, to each voxel, and then display these values as an array. Within the array the value of μ will be represented as a level of greyness in each square, where each square is known as a picture element or *pixel* for short. Before these data are displayed on a video screen, it is usual to rescale them in terms of a 'CT number' (CT = computerized tomography) which is defined as

$$\text{CT number} = \left(\frac{\mu_{\text{tissue}} - \mu_{\text{water}}}{\mu_{\text{water}}}\right) 1000 \qquad (6.1)$$

Equation 6.1 shows that the CT number of any particular tissue is the fractional difference of its linear attenuation coefficient relative to water, measured in tenths of a per cent. The CT numbers of most soft tissues are very similar and close to zero (these are the tissues which cannot be imaged by conventional X-ray), but they may be distinguished easily by computerized tomography. In addition, at the point of display, small differences in CT number can be amplified optically by increasing the contrast of the display, that is the range of brightness assigned to a particular CT range can be altered at the whim of the operator. By this means even quite small differences in CT number can be arranged to drive the image from black to white, as is shown in Figure 6.8. This facility is important since, in effect, it permits performance equivalent to that which could only be obtained in conventional radiography by using a film with a gamma roughly 100 times larger than is actually available.

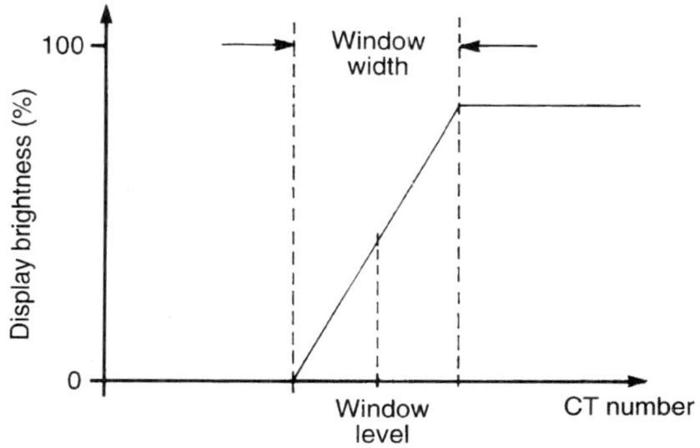

Figure 6.8 The 'windowing' facility allows the display brightness to be adjusted to any range of CT values, by adjusting the window 'level' and 'width'.

When an X-ray beam passes through the body, its attenuation at any point depends on the material at that point and on the energy distribution of the beam. In X-ray computerized tomography (CT) the spectrum or energy distribution of the beam is polychromatic (since we are using the type of X-ray source described in Chapter 4) and it changes (hardens) as the beam passes through the patient. An important consequence of this change is that the attenuation of a beam at any point will depend on the direction from which the beam has come. Clearly, for a monochromatic beam this would not be the case and we could assign an unique value for the attenuation to every point within the patient. The reconstruction of the distribution of attenuation coefficients would then be a well-defined problem. We could say that the CT number assigned to a voxel is a property of the tissue occupying that voxel and does not depend on the position of the voxel in the slice.

Suppose that we have a monochromatic X-ray source with a known photon energy. For a fixed position of the source and detector pair we can count the number of photons which get from the source to the detector and use this as a calibration measurement. Let this number be a calibration C_{M}. If the experiment is repeated with a patient in the beam then the number of photons reaching the detector will be reduced and form an actual measurement A_{M}. We define the monochromatic ray sum m for this beam by

$$m = -\log_e(A_{\mathrm{M}}/C_{\mathrm{M}}) \tag{6.2}$$

and we refer to the set of m values for all source and detector pair positions as the monochromatic projection data. We shall show that relative linear attenuation inside the slice, at a given energy, can be estimated accurately from the monochromatic projection data.

Actual X-ray sources are polychromatic and we define the polychromatic ray sum for this type of source in exactly the same way as we did for the monochromatic source, the calibration measurement being C_P and the actual measurement A_P. The polychromatic ray sum p is then given by

$$p = -\log_e(A_P/C_P) \tag{6.3}$$

In practice we will obtain the set of p values for all source and detector pair positions (the polychromatic projection data). However, what we require are the values of m. This raises the question: Does p determine m uniquely? In general the answer to this question is no. Alternatively, we can enquire if, given p, can we approximate m well enough to lead to clinically useful CT numbers? The answer appears to be yes and the manner in which this is accomplished is set out in the following sections. In order to achieve a reconstruction, certain simplifying assumptions are made:

1. slices are infinitely thin;
2. for any particular source and detector pair position, all X-ray photons travel in the same straight line, which lies in the infinitely thin slice.

The first assumption, in effect, removes the distinction between voxels and pixels so that the final image will represent the relative linear attenuation, pixel by pixel, as a level of greyness. In fact we can write, with reference to Figure 6.9, that

$$m \simeq \int_{\text{detector}}^{\text{source}} \mu(x, y)\, \mathrm{d}y \tag{6.4}$$

In fact because $\mu(x, y)$ is zero for points x, y outside the reconstruction region we can arbitrarily extend the limits of the integral in equation 6.4, without invalidating the equality. Our problem reduces to the calculation of $\mu(x, y)$ from estimates of its integrals along a number of lines, namely the monochromatic projection data. The solution to this problem was provided, in principle, by Radon's theorem and formula of 1917, which determines a picture from *all* of its line integrals. We are able to provide only a *finite* set of projections, or line integrals and these are really estimates rather than accurate values. The estimates are affected by factors such as the width of the X-ray beam, hardening of

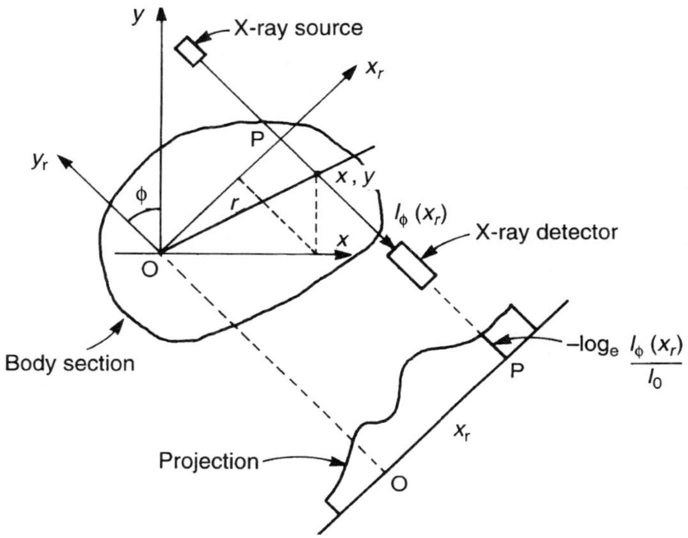

Figure 6.9 Projections in computerized tomography.

the beam, scattering, photon statistics and detector inaccuracies (failure to count all arriving photons).

The finite width of the beam can by itself lead to considerable inaccuracies in estimating $\mu(x, y)$ through the partial volume effect, depicted in Figure 6.10. Suppose we have a monochromatic, X-ray

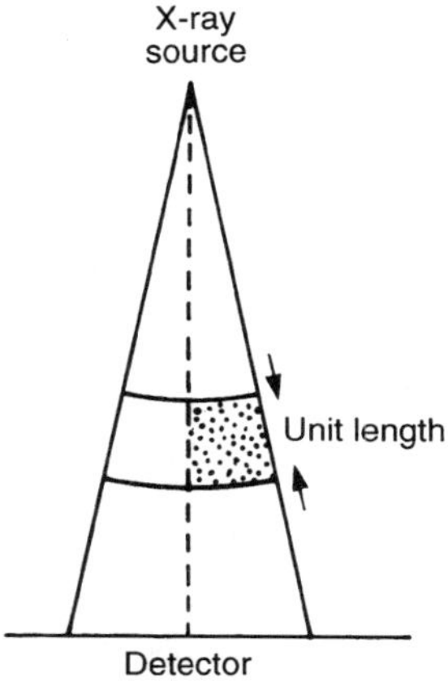

Figure 6.10 Illustration of the partial volume effect.

point source illuminating a line segment detector and that the segment is divided in half by the dashed line. The material within the sector is assumed, for argument's sake, to have an attenuation coefficient of zero everywhere, except for the element of unit length, shown dotted, where the attenuation coefficient is two. If equal numbers of photons enter each half section, then in the left-hand half all are transmitted without attenuation, so that the ratio I/I_0 is unity, while in the right-hand half, during passage through unit length, the intensity is reduced and the ratio I/I_0 is $\exp(-2) = 0.135$. The value for m, calculated using equation 6.2, becomes 0.567, although the true average value is actually unity; the rather large error is introduced by the beam being only partially blocked by attenuating material.

6.6* RECONSTRUCTION BY SIMPLE BACKPROJECTION

One way in which we can attempt to reconstruct an image from projections at different angles is to project each one backwards, as shown schematically in Figure 6.11. This can be done mathematically or by an analogue method. Since the actual density distribution is unknown, one is, in effect, assuming that the experimentally measured absorption along a ray path is due to a uniform density distribution along its length. The backprojection image thus obtained is not a good facsimile of the original object. This can be seen from Figure 6.11, where the simple, circular patch is reconstructed as a star-shaped image. We have smeared one-dimensional projection data uniformly

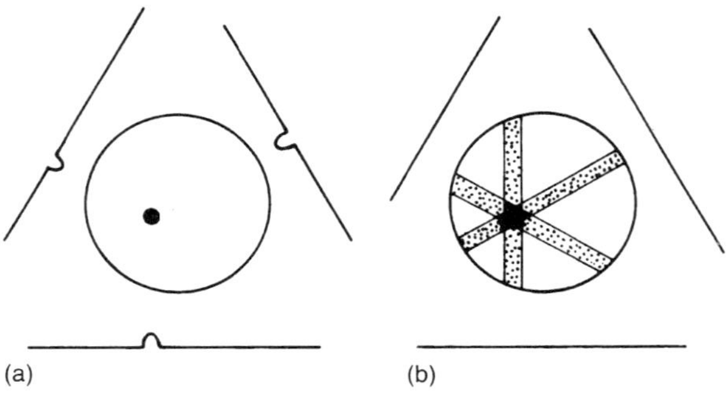

Figure 6.11 (a) Projections obtained from a single high density circular point embedded in a circular section. (b) Image reconstruction by backprojection of the three projections in (a). In each backprojection the actual density distribution is unknown, but assumed constant along the projection path. In effect the three one-dimensional projections are smeared out to form a two-dimensional image although the projections are incorrectly reconstructed as a 'star'.

back into two-dimensional space. The star shape can be described as the *point spread function* of the method. Although with a very large number of projections the star effect can be minimized, there still tends to be an unacceptable level of general blurring of detail in the image. The increasing density of lines toward the centre represents the point spread function which has a $1/r$ dependence (Figure 6.12). The density at any point in the image reconstructed in this way is the sum of all the ray projections which pass through the point, so that the method is also referred to as the *summation method* or the *method of linear superposition*. We can express the reconstruction mathematically as

$$\mu'(x, y) = \sum_{j=1}^{n} p(r_j, \phi j)\, \Delta\phi \tag{6.5}$$

or

$$\mu'(x, y) = \sum_{j=1}^{n} p[(x \cos \phi_j + y \sin \phi_j), \phi_j]\, \Delta\phi \tag{6.6}$$

where $r_j = x \cos \phi_j + y \sin \phi_j$ and n is the number of summations, each separated by an angular interval $\Delta\phi$ so that $\Delta\phi = \pi/n$. It should be noted that, in using this expression to calculate the value of $\mu(x, y)$ at some point, say x_1, y_1, the value of only one ray integral through x_1, y_1

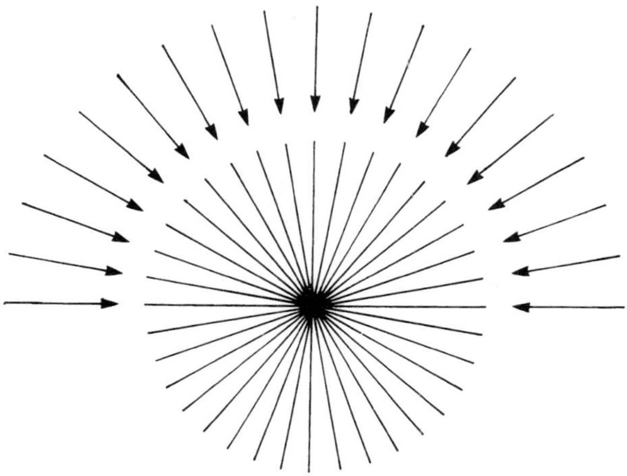

Figure 6.12 Backprojection of a large number of delta functions which results in a point spread function (PSF) which approaches a $1/r$ functional form, as indicated by the increasing line density towards the centre of the pattern.

is used with the consequence that the summation is only for the number of projections n.

Backprojection alone cannot provide a generally satisfactory method of reconstruction in computerized tomography since there is likely to be blurring of sharp features as referred to above. There is however a more basic reason why simple backprojection will be unsatisfactory and this lies in the fact that the backprojection and attenuation coefficient distribution we are seeking to reproduce do not have identical dimensions; the former is dimensionless while the latter has dimensions L^{-1}, the actual value depending on the measuring units used. However, the backprojection is independent of the measuring units used. The lack of correct dimensionality can be made good by appropriate normalization, as will be shown in section 6.9.

6.7* ITERATIVE METHODS OF RECONSTRUCTION

One of the ways of overcoming the shortcomings we have seen in the backprojection method would be to find a way of removing the totally unjustifiable assumption that there is a uniform density, or constant value of μ, along each ray path. A conceptually simple approach to this problem, and one which was popular in the early days of CT (it was used by Hounsfield), is that of iteration. Although this method is no longer much used for X-ray CT, it does still find application in emission computed tomography, the subject of Chapter 7. This approach is based on solving the equation

$$-\log_e(I/I_0) = \int_{\text{detector}}^{\text{source}} \mu[x, y]\,\mathrm{d}y = p_\phi[x_r] = p_\phi(r; \phi) \qquad (6.7)$$

by the mathematical procedure of iteration. It is a method of successive approximations in which corrections are applied sequentially to an arbitrary starting solution so as to bring the new solution into progressively closer agreement with the experimental data. We note first that the integration represented in equation 6.7 is really a summation because the measurements are made at discrete spatial (lateral) positions and the pencil beam of X-rays is of finite cross-section. The layer being imaged is comprised of pixels and instead of rays one is working with strips; Figure 6.13 illustrates the concept.

For the jth ray in Figure 6.13 the ray integral can be expressed as the ray sum,

$$p_j = w_{1j}\mu_1 + w_{2j}\mu_2 + \cdots \qquad (6.8)$$

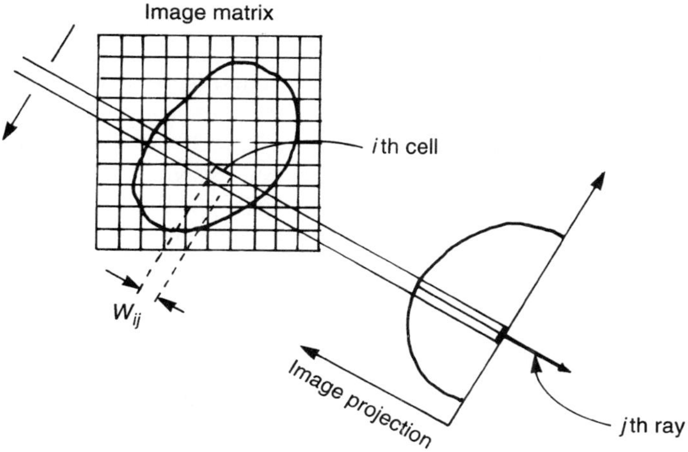

Figure 6.13 The array of pixels superimposed on the section to be imaged. The contribution of the *i*th cell to the *j*th ray is referred to as w_{ij} and is shown in bold: the measured ray sum is the height of the image projection in the *j*th ray strip.

or

$$p_j = \sum_i w_{ij} \mu_i$$

Here w_{ij} are weighting factors that allow for the fact that each cell does not necessarily contribute equally to the absorption along a given strip. The contribution is that part highlighted in Figure 6.13 as a bold section on the ray path. In fact w_{ij} is the average pathlength traversed by the *j*th beam in the *i*th cell or pixel. The values of these coefficients need only be calculated once and then stored for future use. In practice equation 6.8 represents approximately 10^5 simultaneous equations and the solution method is to adjust the values of μ_i iteratively until the computed projections p_c, closely resemble the experimental projections p. These final values of μ_i are taken to be the solution and are assigned to the appropriate pixels to provide the image.

This approach to a solution does not provide an unique solution, and one reason is that there are usually far more equations than there are unknowns; a multiplicity of solutions may exist. The iterative algorithms differ in the manner by which the corrections are calculated and applied, so that corrections may be applied additively or multiplicatively; they may be applied immediately after calculation, or stored and used at the end of each round of iteration. A later example will serve to illustrate the general principles involved.

At this stage it is important to pose the question: Do we need all of these views or, alternatively, what is the minimum number of views required to reconstruct the object? If the object comprises only one point then it is clear that two views or projections at different angles will suffice. However, as the object complexity increases this is not the case, as Figure 6.14 demonstrates. Here we have eight points arranged variously on a square grid. Views or projections taken in the horizontal and vertical directions yield identical results and the three 'objects' appear to be identical. It is only the addition of a further diagonal view, in each case, which enables us to distinguish between them. We might expect, therefore, some sort of relationship between object complexity and the required number of projections. In general it can be shown that:

1. any object is completely described by the continuous set of its projections;
2. there is a straightforward interpretation of the information that is contained in each projection.

If we consider the situation where an object of circular cross-section is sampled using parallel-beam data collection (the original EMI approach described in Figure 6.3), then the number of samples per projection is N_s where

$$N_s = 2a/d \tag{6.9}$$

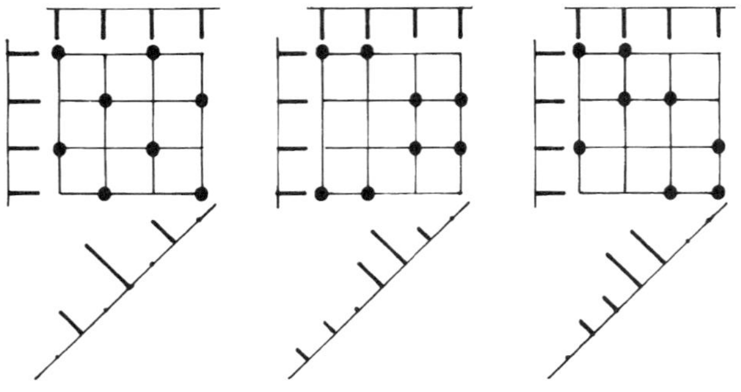

Figure 6.14 Three different objects comprising eight points which yield the same horizontal and vertical ray sums. An additional diagonal projection is required in each case to distinguish between them; these are displayed beneath each array.

$2a$ is the diameter of the circular object and d the step spacing. If we now consider the number of angular views N_ϕ, then we can write

$$N_\phi = \pi/\Delta\phi \tag{6.10}$$

since we need only move our detector through 180°. Any further rotation simply duplicates views already stored. It is tempting to propose further that any angular rotation $\Delta\phi$ should be adjusted to be equivalent to a linear change at the object perimeter also equal to d. We may now combine equations 6.9 and 6.10 to yield the total number of data points N

$$N = N_s N_\phi = 2\pi \frac{a^2}{d^2} \tag{6.11}$$

Figure 6.15 shows four three-point projections, taken from an imaginary nine-point object (circled), in which the weighting factors w_{ij} are assumed to be unity. Proceeding clockwise, the first correction is made to the horizontal ray sums which are divided by the number of pixels (three) and the resulting value assigned as a first guess. This gives a perfectly correct prediction for the horizontal ray-sums (ringed) as it must, but the vertical ray sums are incorrect. The same procedure is then applied to the vertical ray sums, that is the error is distributed equally, either additively or not, as may be appropriate; this corrects the vertical ray sums. This process is repeated for the diagonal ray sums to complete the first iteration. Simple inspection of the result of the first iteration reveals that it closely approximates, but does not equal, the original distribution of values for the linear attenuation coefficients. The rms deviation associated with each step in the iterative process reduces from 1.4, at the first step, down to 0.5 at the fourth step. This is an example of immediate, additive correction.

6.8* ANALYTICAL METHODS OF RECONSTRUCTION

Analytical methods of reconstruction are based on the Fourier coefficients of the image being related to the Fourier coefficients of the projections. For our purposes it is easier, and more transparent for the reader, to consider the first-generation X-ray CT scanning system based on parallel beams and stepwise translation plus angular rotation as our model system. The later, more advanced systems simply add unnecessary geometric complexity to our considerations, although it is the same data-set which is ultimately gathered.

The data needed to reconstruct the image are transmission measurements through the patient. We assume that we have a very narrow

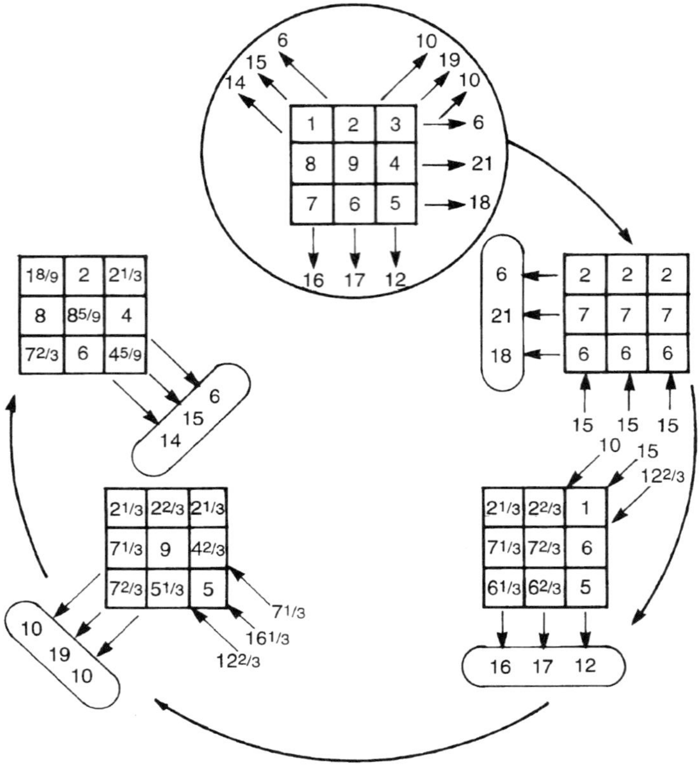

Figure 6.15 An example of iterative reconstruction taking two diagonal ray sums in addition to the horizontal and vertical ray sums. The difference between the actual ray sums and required ray sums are distributed equally between pixels.

pencil beam of monochromatic X-rays which pass through the patient and reach the detector without scattering. We define two coplanar reference frames, the x–y frame, which is fixed with respect to the object (patient) and the rotating x_r–y_r frame in which the y_r direction is always the direction of the incident X-ray beam. Both systems have a common origin located at the centre of rotation of the scanning gantry. The angle of rotation of the frame x_r–y_r with respect to the frame x–y is just ϕ. If we are considering Cartesian co-ordinates we shall use square brackets thus [ ], while polar co-ordinates will be indicated ( ).

Once again we start with equation 6.7 which we set equal to $p_\phi(x_r)$ where the quantity $p_\phi(x_r)$ is called the Radon transform or projection of

$\mu(r)$ or $u[x, y]$. Our task is to reconstruct slices of objects from their projections by inverting equation 6.7. Actually the transmission data are sampled both in x_r and ϕ, but we shall assume that $p_\phi(x_r)$ is a continuous, one-dimensional, function of x_r for a given projection angle ϕ, or a continuous two-dimensional function of ϕ and x_r. It is helpful in this endeavour to see how data are mapped from the x–y space of real objects, to the x_r–ϕ space of the projection data. Projection space may be represented either in polar co-ordinates or Cartesian co-ordinates. In the polar representation, which is often called the Radon-space representation, each point represents the value of a line integral through the object, the direction of the line integral being at right angles to the x_r axis (Figure 6.16). If we consider a particular point P within the object, then there may be as many as 180 paths of integration through it (assuming the scanning configuration of Figure 6.3) so that the point P, in object space, is mapped into Radon space by determining the locus of all the integral paths through the point P. This locus is a circle of diameter r with the points $x_r = y_r = 0$ at one end of its diameter and the point P at the other. Figure 6.17 displays these relationships.

The projection data-set $p_\phi(x_r)$ has three properties which we should note:

1. it is bounded
2. the source–detector combination may be interchanged without affecting the value of a transmission measurement: all possible projection data are contained in projections gathered over a continuous semicircle of projection angles so that

$$p_\phi(x_r) = p_{\phi+\pi}(-x_r) \tag{6.12}$$

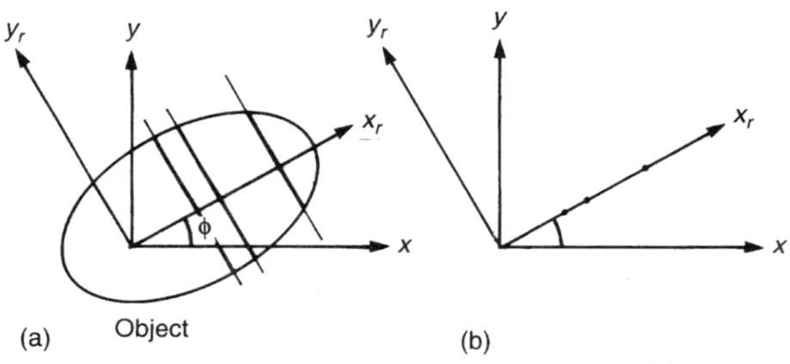

Figure 6.16 Mapping between object space (a) and Radon space (b) where the line integral corresponding to the path l transforms to a point in Radon space.

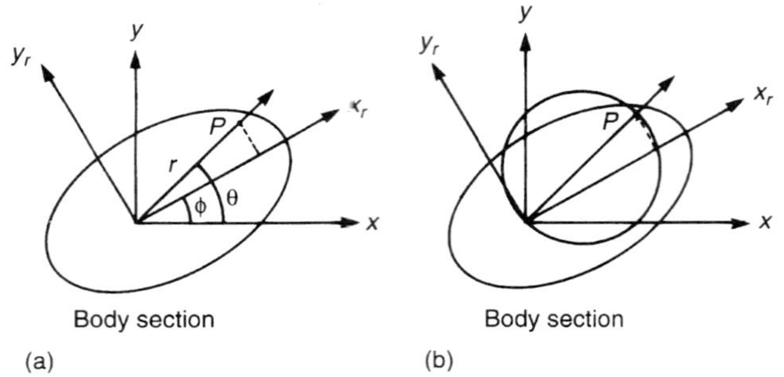

Figure 6.17 (a) The geometry of the scanning system viewing a general point P. (b) The locus of all projections through the point P (r, θ) in object space is a circle of diameter r, with its diameter at an angle θ to the x axis: ϕ takes on values between 0 and π according to the equation $x_r = r\cos(\theta - \phi)$.

3. The integral over the projection data

$$\int_{-\infty}^{+\infty} p_\phi(x_r)\,\mathrm{d}x_r = \text{const.} \tag{6.13}$$

since

$$\int_{-\infty}^{+\infty} p_\phi(x_r)\,\mathrm{d}x_r = \int_{-\infty}^{+\infty}\int_{-\infty}^{+\infty} \mu(r)\,\mathrm{d}x_r\,\mathrm{d}y_r \tag{6.14}$$

$$= \int_{-\infty}^{+\infty}\int_{-\infty}^{+\infty} \mu[x_r, y_r]\,\mathrm{d}x_r\,\mathrm{d}y_r \tag{6.15}$$

where we have substituted for $p_\phi(x_r)$ and the right-hand side is the two-dimensional integral of the object over all space. This integral clearly must be independent of the orientation of the frame in which integration is carried out.

Once again we distinguish, symbolically, between Cartesian and polar representations so that, the density function μ will be represented in Cartesian co-ordinates as $\mu[x, y]$, or $\mu_\phi[x_r, y_r]$ for a rotating frame. In polar co-ordinates the representation will be $\mu_\phi(x_r)$. Similarly for the projection p we will write $p_\phi[x_r, y_r]$ in Cartesian co-ordinates and $p_\phi(x_r)$ in polar co-ordinates. The Fourier transform of a quantity will be indicated as the upper case form of the appropriate symbol.

The general density function $\mu[x, y]$ may be expressed as a two-dimensional Fourier integral

$$\mu[x, y] = \int_{-\infty}^{+\infty}\int_{-\infty}^{+\infty} M[u, \upsilon] \exp 2\pi\mathrm{i}(ux + \upsilon y)\,\mathrm{d}u\,\mathrm{d}\upsilon \tag{6.16}$$

This describes $\mu[x, y]$ as a superposition of sinusoidal density 'waves' crossing in various directions in the x–y plane. The real and imaginary parts of the exponential represent the cosine and sine terms respectively, where u and v are the spatial frequency terms conjugate to x and y (u and v have the dimension L^{-1}).

The Fourier coefficients $M[u, v]$, of equation 6.16 (where M is just capital μ) are given by the 'inverse' transform, thus

$$M[u, v] = \int_{-\infty}^{+\infty} \int_{-\infty}^{+\infty} \mu[x, y] \exp(-2\pi i(ux + vy)) \, dx \, dy \quad (6.17)$$

The imaging problem is: How do we get from the projection function $p_\phi(x_r)$ to $M[u, v]$? It turns out that we can establish a relation between $p_\phi(x_r)$ and $M[u, v]$ using a theorem known as the *central slice theorem*.

The monochromatic raysum or projection of equation 6.7 may be expressed as its Fourier transform $P_\phi(u_r)$, thus

$$P_\phi(u_r) = \int_{-\infty}^{+\infty} p_\phi(x_r) \exp(-2\pi i u_r x_r) \, dx_r \quad (6.18)$$

where the subscripts indicate that we are now working quite generally with a rotating frame of reference.

We can substitute for $p_\phi(x_r)$ using equation 6.7, simultaneously exchanging the limits of integration for $\pm\infty$. This latter change is permitted since the value of μ between the source and the object, or the object and detector is zero. Equation 6.18 can now be written as

$$P_\phi(u_r) = \int_{-\infty}^{+\infty} \int_{-\infty}^{+\infty} \mu_r[x_r, y_r] \exp(-2\pi i u_r x_r) \, dx_r \, dy_r \quad (6.19)$$

or

$$P_\phi(u_r) = \int_{-\infty}^{+\infty} \int_{-\infty}^{+\infty} \{\mu_r[x_r, y_r] \exp(-2\pi i(u_r x_r + v_r y_r)) \, dx_r \, dy_r\}_{v_r=0} \quad (6.20)$$

In equation 6.20 we have inserted the Fourier space variable v_r conjugate to y_r, but set it equal to zero, that is the transform is to be evaluated along the u_r axis in all cases. The right-hand side of equation 6.20 is now recognizable as the two-dimensional Fourier transform of the density function, that is $M[u_r, v_r]_{v_r=0}$ evaluated along the line $v_r = 0$.

To summarize, the one-dimensional Fourier transform of a projection is equal to a section, or slice, of the two-dimensional Fourier transform of the object itself. The section is that along the u_r axis, which is conjugate to the x_r axis, for which the projection is specified. This result is known as the *central slice theorem*. The link we sought is now established as

$$P_\phi(u_r) = M[u_r, v_r]_{v_r=0} \tag{6.21}$$

This is an important result because it means that if projections are obtained experimentally in different directions, then a reconstruction of the two-dimensional density function $\mu[x, y]$ can be achieved directly, by Fourier synthesis of the Fourier components of all the projections. This theorem tells us that our projections contain sufficient information to allow the general object to be reconstructed, provided that we have an infinite number, or continuum, of projection data. In practice, of course, we do not have a continuum of projection data. The finite size of the X-ray beam effectively band-limits the data and a satisfactory reconstruction can be obtained using a finite, rather than an infinite, number of projection angles and projection positions.

Figure 6.18 illustrates the application of the central slice theorem. Consider some general, bounded object (patient); this object can always be synthesized from a linear superposition of its two-dimensional spatial frequency components. If we contemplate a single cosine spatial frequency component, then only when the projection direction is parallel to the wave crests does the projection differ from zero. For that particular direction the full cosine distribution is projected onto the x_r axis. The Fourier transform of this single component will comprise a pair of delta functions, so that for a cosinusoid of wavelength $2\pi/u'_r$ and amplitude A, the delta functions will appear equispaced about $u_r = 0$ at $\pm u'_r$. The original object is, of course, a superposition of many component waves of various phases, periods and directions, but only those waves which are parallel to the first one will have their transforms located on the u_r axis, and these are the only waves which will change the projection $p_\phi(x_r)$.

6.9* FILTERED BACKPROJECTION

We have seen previously (in section 6.6) that simple backprojection produces a blurred image of the viewed object. It resembles the object except that each point in the object has been replaced by a cusp-like patch. The irradiance of this patch falls off inversely with the distance from its geometric centre. This patch, the 'point spread function' (PSF), degrades the image quality to such a degree that it becomes useless for medical imaging purposes. However, because the point spread function is everywhere the same, we can contemplate convolving it with a suitable correcting function and projecting the filtered function. It turns out that the required algorithm is structurally simple, indeed filtered backprojection is the most commonly used process for image reconstruction. We can write the two-dimensional inverse polar Fourier transform yielding $\mu[x, y]$ as

$$\mu[x, y] = \mu(r, \theta)$$
$$= \int_0^{\pi} \int_{-\infty}^{+\infty} M(w, \phi) \exp[2\pi \mathrm{i} w r \cos(\theta - \phi)] |w| \, \mathrm{d}\phi \, \mathrm{d}w \quad (6.22)$$

where reference to Figure 6.17(a) will make clear the co-ordinate system used to describe the general object point P, and the polar co-ordinates in the integral are $|w| \, \mathrm{d}\phi \, \mathrm{d}w$. In this instance w is conjugate

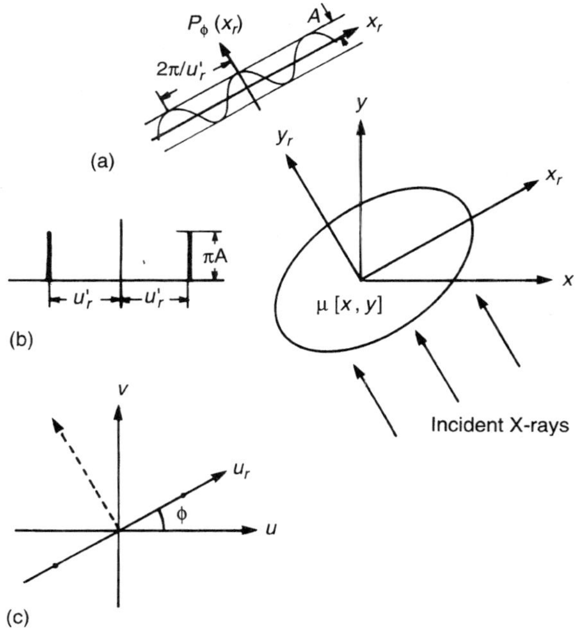

Figure 6.18 The object density distribution $\mu[x, y]$ can be built up from its Fourier components, one of which, a cosinusoid, is shown here. The Fourier transform of this component is a pair of delta functions shown in (b) and as points on the u_r axis in (c).

to r, i.e. the spatial frequency along the r co-ordinate. We note also that

$$w = (u^2 + v^2)^{1/2} \tag{6.23}$$

and

$$\phi = \tan^{-1}(v/u) \tag{6.24}$$

Equation 6.22 can be separated into two parts. We can therefore write that

$$\mu[x, y] = \int_0^{\pi} p_\phi^*(x_r) \, \mathrm{d}\phi|_{x_r = r\cos(\theta - \phi)} \tag{6.25}$$

where

$$p_\phi^*(x_r) = \int_{-\infty}^{+\infty} M(w, \phi)|w| \exp(2\pi \mathrm{i} w x_r) \, \mathrm{d}w \tag{6.26}$$

Equation 6.26 defines a new quantity $p_\phi^*(x_r)$, which is called the filtered projection. The reason for this title will become clear later. Examination of equation 6.26 shows that it is the one-dimensional Fourier transform of the product of $M(w, \phi)$ and $|w|$. This being so, we should be able to write it as the convolution of individual Fourier transforms of $M(w, \phi)$ and $|w|$. The transform of $M(w, \phi)$ can be obtained readily using equation 6.21 (the central slice theorem) so that

$$M(w, \phi) = P_\phi(u_r) \tag{6.27}$$

and

$$p_\phi(x_r) = \int_{-\infty}^{+\infty} P_\phi(u_r) \exp(2\pi \mathrm{i} u_r x_r) \, \mathrm{d}u_r \tag{6.28}$$

The transform of $|w|$ requires more care since it is not a sufficiently well-behaved function for its transform to exist. However, since the experimental system band-limits the measurements, we can truncate the value of $|w|$ also. The approximate nature of this function is shown in Figure 6.19 As a final step we can write that

$$p_\phi^*(x_r) = p_\phi(x_r) * h(x_r) \tag{6.29}$$

where $h(x_r)$ is the transform of $|w|$. Equation 6.29 represents the one-dimensional convolution of our projection data with a filter function $h(x_r)$. The filtered backprojections are then backprojected into $[x, y]$ and the individual backprojections summed to create the image; Figure 6.19

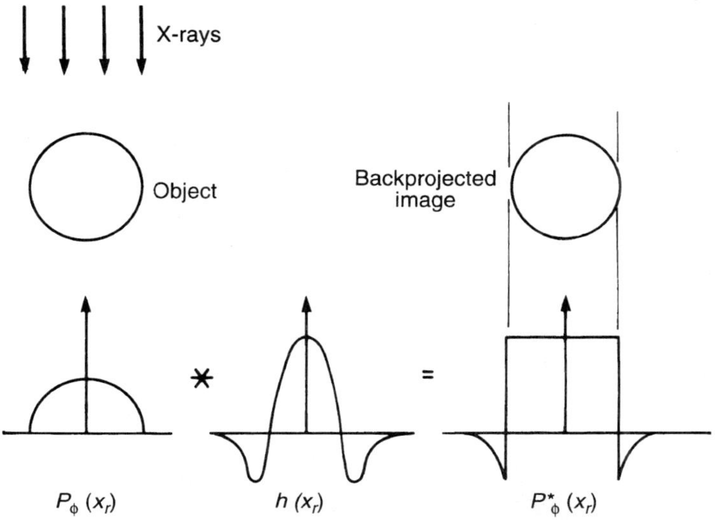

Figure 6.19 The projection $p_\phi(x_r)$ of a simple, circular object is convolved with a filter function $h(x_r)$, where $h(x_r)$ is the transform of $|w|$, to yield a filtered backprojection $p^*_\phi(x_r)$. The filtered backprojections are then backprojected and summed to create a reproduction of the original object. The filter function and backprojected functions are indicative only – in particular the negative tails are larger than they would be in reality.

represents this process graphically. The process may be summed up as 'convolution and backprojection' or 'filtered backprojection'. To summarize, the projection data are convolved (filtered) with a suitable processing function $h(x_r)$ before backprojection. The filter function has negative side lobes on either side of a positive core, so that in summing the filtered backprojections, positive and negative contributions cancel outside the central core. With the tails eliminated in this way the backprojected image will resemble the original object quite closely. The data from each view can be convolved and backprojected independently of the other views, to produce an image as good as that obtainable by any other reconstruction technique.

6.10 IMAGING LIMITS

In computerized tomography there is a trade-off between the signal-to-noise ratio S/N, the patient dose D, the spatial resolution ε and the slice thickness h, given by

$$D = k\left(\frac{S}{N}\right)^2 \frac{1}{\varepsilon^3 h} \tag{6.30}$$

where k is a constant of proportionality.

Equation 6.30 has its roots in the Poisson statistics of the photon flux. The value of the constant of proportionality is relatively independent of the processing algorithm adopted for reconstruction. Furthermore, all modern commercial systems attain the imaging performance implied by equation 6.30, this means that the only route to improved values of signal-to-noise ratio, or resolution, will be by increasing the patient dose. Since this is not generally acceptable we cannot expect

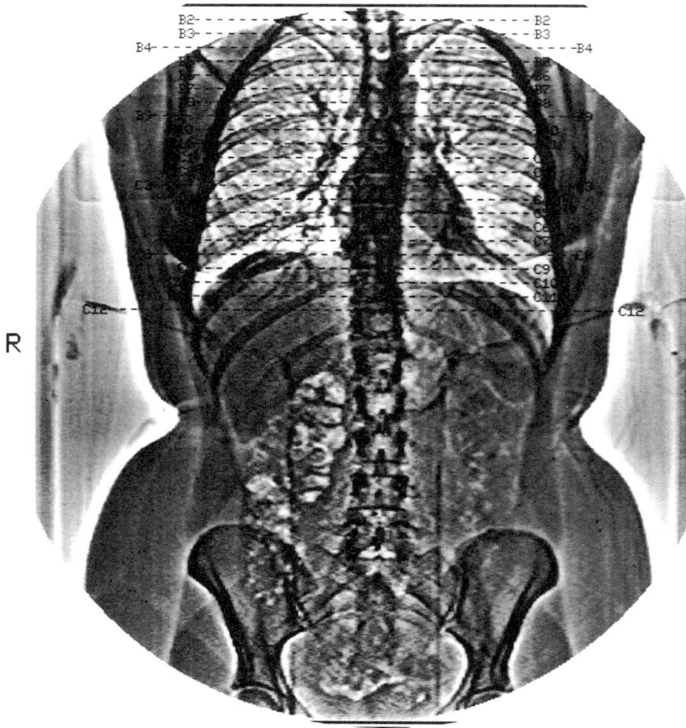

Figure 6.20 Anterior view of the human torso obtained using CT. Superimposed on this view is a labelled grid which identifies sections through the body. The increased contrast obtained using CT can be seen by comparison with Figure 4.30, which also views part of the torso.

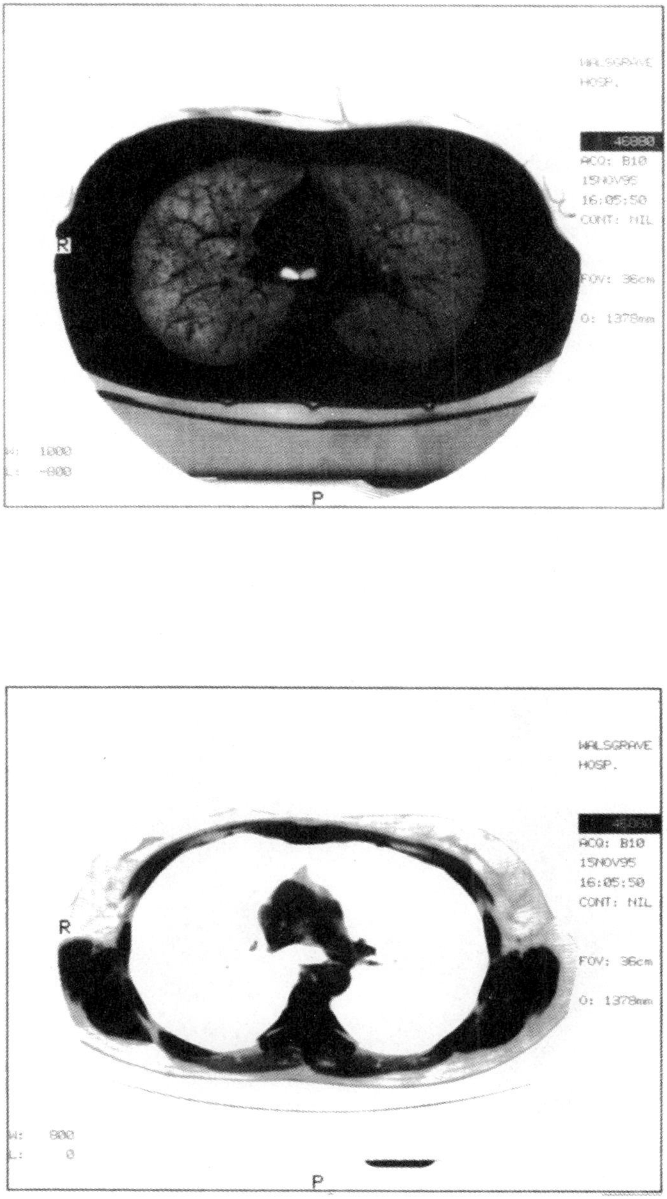

Figure 6.21 Two sectional views through the human torso, both corresponding to the B10 level in Figure 6.20, but imaged under slightly different conditions.

improvements in resolution past the present values of around 1–2 mm.

6.11 CLINICAL EXAMPLES

We have already seen, in Figure 6.2(b), an example of an X-ray tomogram, in this case a section through the skull. We conclude this chapter with further examples of tomograms, but now obtained from other parts of the body as well. Figure 6.20 shows the image obtained for the human torso overlaid with an identifying grid by which particular sections through the chest can be identified. Clearly discernible, without any artificial contrast enhancement techniques, are the major organs of the body. We are at liberty to image any of the sections identified and Figure 6.21 shows two views corresponding to the B10 level, imaged under slightly different conditions.

A similar approach can be applied elsewhere in the body and Figure 6.22 shows the skull sagittal image overlaid with a numbered grid, while individual tomograms corresponding to the levels C3 and B4 are shown in Figure 6.23. The flexibility and the level of detail provided by

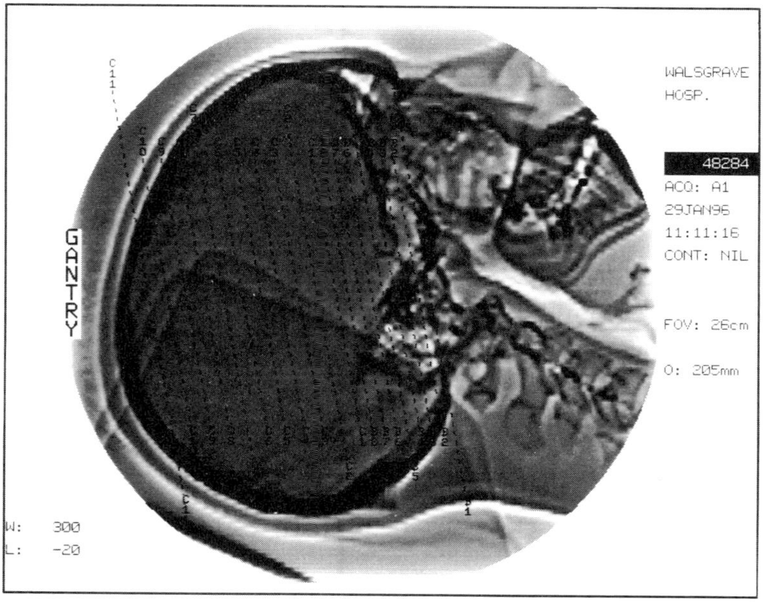

Figure 6.22 Sagittal view of the human skull with individual levels identified by means of a grid overlay.

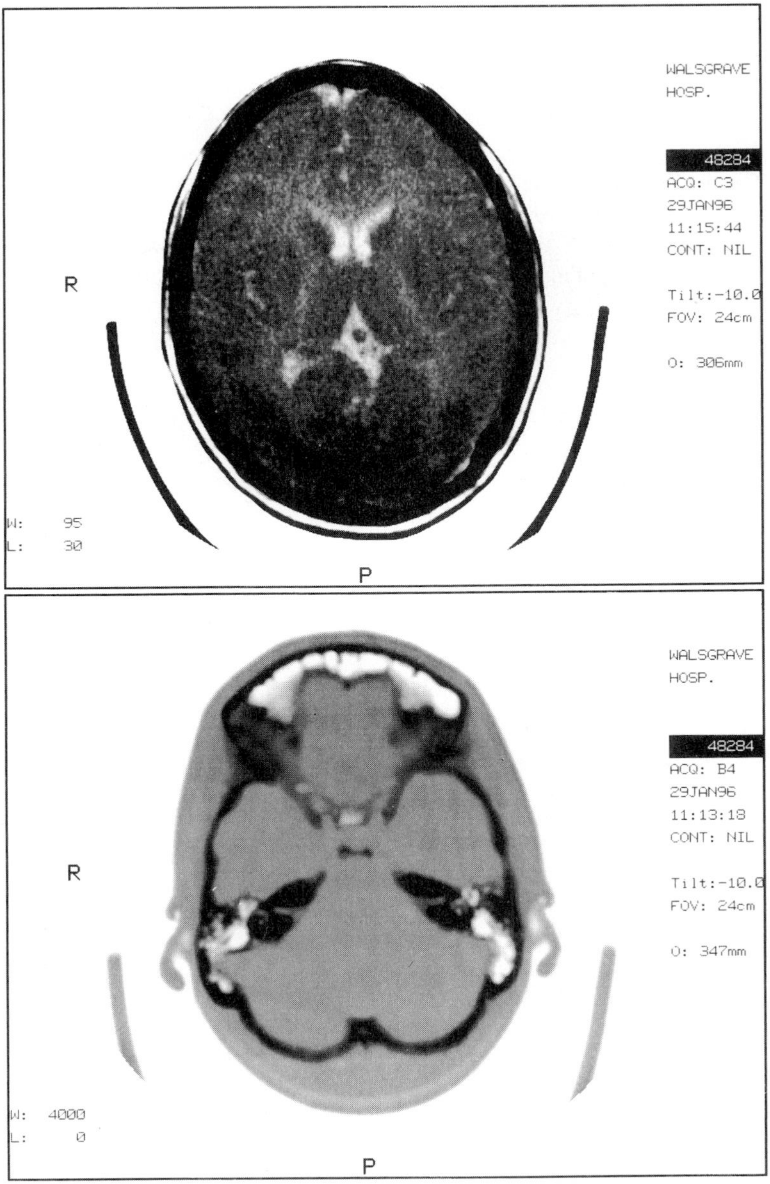

Figure 6.23 Two X-ray tomograms taken at different levels in the brain corresponding (a) to C3 and (b) to B4.

these images is very striking, particularly when contrasted with the sagittal view of the skull obtained by conventional X-ray radiography and shown in Figure 4.27. These images provide full justification for the assertion that X-ray CT is the most important development in diagnostic radiology since the introduction of the planar radiograph. The techniques applied to image reconstruction in X-ray CT are of perfect generality, consequently we may apply them in imaging systems which do not use X-rays. Chapters 7 and 8 consider their applications to emission computed tomography and magnetic resonance imaging, respectively.

7

Imaging with radioisotopes

7.1 INTRODUCTION

In the preceding chapters we have examined the role of physics in medical diagnosis, first through the innate or endogenous thermal, electrical and magnetic behaviour of the body, and second from the point of view of applying external agencies such as longitudinal ultrasound pressure waves or X-radiation. In this chapter we examine the scope and basis for a new view of the body, this time a view of its physiological functions, but now arising from the ingestion of radioisotopes which emit γ-radiation from within the body. The introduction in the late 1930s of radioisotopes such as ^{131}I provided the starting point for this kind of imaging, and it has subsequently blossomed with the use of a wide range of radioisotopes.

Radioisotope imaging involves a number of techniques which produce pictures of the distribution of radionuclide-labelled agents within the body. These agents, which are known generically as radiopharmaceuticals, provide information on the physiological and metabolic function of individual organs in the body, in contrast to other imaging modalities which provide anatomical detail. This difference arises because the distribution of radiopharmaceuticals is determined primarily by such factors as metabolic processes, blood flow and/or means of administration. A further important difference flows from the fact that, unlike X-ray imaging where both the emission and detection position of the X-ray photon are known precisely, in radioisotope imaging it is solely the X-ray or γ-ray detection position which is known. An image can only be constructed by providing some form of X-ray or γ-ray collimation capable of defining the origin and direction of the emerging photons.

The use of radioisotopes has an important temporal dimension too, since the spatial distribution of radiopharmaceuticals is a function of time, and can be studied on a scale which may range from milliseconds to minutes. From this sort of measurement we can assess the basic function of the particular organ or organs being studied. The technique is then known as dynamic planar scintigraphy.

Finally, we ask the question, can we determine the distribution of radioisotopes in three-dimensions from the two-dimensional image; in other words, can we produce a tomographic image? Here the technique of emission computed tomography (ECT), which has parallels with X-ray computed tomography, provides the answer. While X-ray computed tomography depends on the determination of the attenuation coefficient for different body tissues (Chapter 6), this is actually an undesirable effect where ECT is concerned, and has to be corrected for while determining the distribution of radioactivity. The quality of the image obtainable from ECT is not straightforwardly comparable with that obtained from X-ray tomography, owing to the lower count rates encountered with radioisotopes. However, this is not a serious problem, since it is not inherently an anatomical imaging technique. The family of techniques based upon radioisotope imaging are summarized in Figure 7.1.

The radioactive decay of radioisotopes can lead to the production of α, β^-, β^+, γ and X-radiation, according to the radionuclide in use. Within this group of particles the range of α- and β-particles in body tissue is far too small for them to emerge from the body and provide any sort of image by themselves. X-rays and γ-rays, however, have no difficulty in penetrating body tissue, although at energies below 30 keV both X-rays and γ-rays are greatly attenuated by photoelectric absorption, while at energies above 30 keV Compton scattering dominates (Figure 4.8). In general, therefore, we will be concerned with the detection of photons with energies greater than about 50 keV.

As with most applications of physics in medical diagnosis, the introduction of radioisotope imaging has been indebted to, and depended on, advances in associated technology. In this case two special requirements have had to be met: first, the ability to produce radionuclides, and second, the means with which to detect them. The former requirement was met, beginning in the 1930s, with the development by E.O. Lawrence of the cyclotron (1931), and in 1938 the most widely used radioisotope, technetium (^{99}Tc), was introduced. From the mid 1940s a range of radioisotopes became available, initially from the USA, but later from the UK.

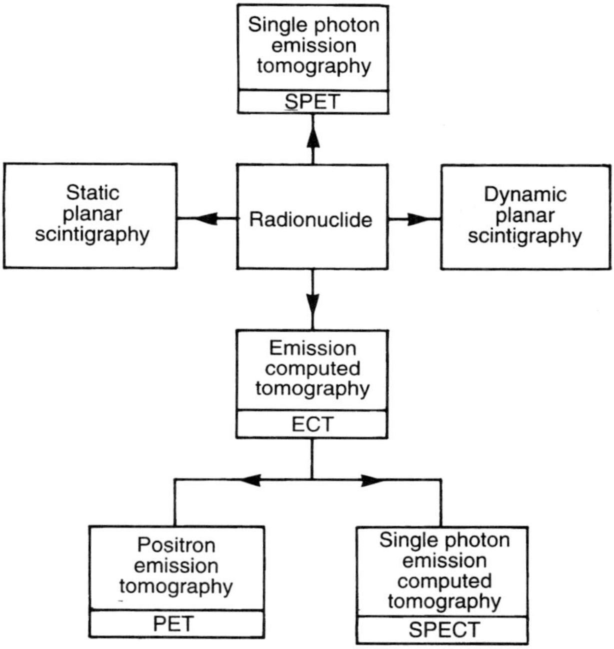

Figure 7.1 Summary of the family of radioisotope imaging techniques.

Prior to 1896 it was thought that every atom of an element persisited in its form without change, but the work of H. Becquerel and the Curies changed all that and led to the discovery of radioactivity and the elements, polonium, radium, and later, actinium. As a consequence, γ-ray detectors have been available for a long time, although not usually in a form suitable for imaging. In this context we can consider a number of γ-ray detectors ranging from the primitive but sensitive gold leaf electroscope, through the Wilson cloud chamber (1895) and Crooke's spinthariscope (1903) to the Geiger counter (1929). These devices were instrumental in establishing the basis of radioactivity at the beginning of the twentieth century. Thus we find that λ in equation 7.1 is inversely proportional to the rate of fall of the gold leaf in the electroscope of Figure 7.2(a), while the supersaturation of an air/water vapour mixture on adiabatic expansion via a piston provides copious condensation on any suitable condensation nuclei (α-particle or gas atom ionized by a γ-ray photon) traversing the Wilson cloud chamber

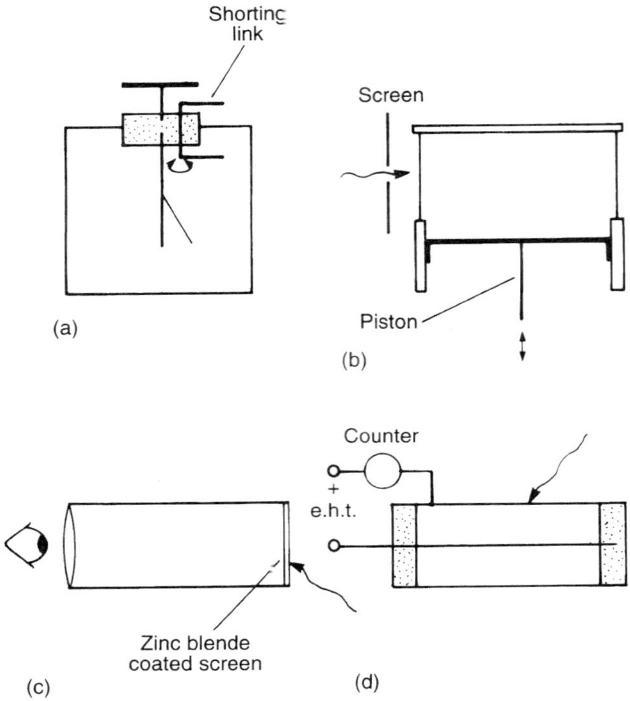

Figure 7.2 Early forms of γ-ray detector. (a) The gold leaf electroscope; (b) the Wilson cloud chamber; (c) Crooke's spinthariscope; (d) the Geiger counter. These detectors have been supplanted in modern medical use by the scintillation detector, the semiconductor detector and the multiwire gas detector.

in Figure 7.2(b). The particle tracks are thus rendered visible. Crooke's spinthariscope (Figure 7.2(c)) allows the radioactive decay products bombarding the zinc blende screen to be viewed as flashes of light through the lens at the other end of the tube. In a way this device is an early form of scintillation counter (counting done by eye). The Geiger counter of Figure 7.2(d) detects α-particles or other ionizing particles by the discharge triggered by the said particle in the tube. A current pulse is measured (counted) in the external circuit for each incident particle. All of these detectors have been supplanted by modern solid-state detectors. Before moving on to discuss the complexities of a modern radioisotope imaging system it will be useful to review the essentials of radioactive behaviour, and the formation of isotopes.

Once again we find the 'imprint' of Sir William Crookes who, writing in *Nature* in 1886, demonstrated almost prophetic insight into the nature of matter when he said:

> I conceive that when we say the atomic weight of, for instance, calcium is 40, we really express the fact that, while the majority of calcium atoms have an actual weight of 40, there are not a few which are represented by 39 or 41, a lesser number by 38 or 42 and so on.

It is astonishing to realize that this statement was not based upon any physical or chemical evidence; isotopes had not then been discovered, but Crookes foreshadowed them. We have seen, in Chapter 4, that Crookes was also involved in the pioneering work which led to the discovery of X-rays (by Roentgen). To that achievement must be added the development of what may be regarded as an early form of scintillation counter (the spinthariscope), and the isolation, in 1861, of the element thallium. The discovery of thallium resulted in Crookes being made a Fellow of the Royal Society in 1863. Thallium isotopes have subsequently found widespread use in radionuclide imaging and in the doping of scintillation counters.

7.2 RADIOACTIVITY AND RADIOACTIVE DECAY

If we have n nuclei, then their radioactive decay is essentially random in character and kinetically a first-order process, i.e. the rate of decay is proportional to the number of decaying atoms present. We can describe this process by the equation

$$\frac{\mathrm{d}n}{\mathrm{d}t} = -\lambda n \qquad (7.1)$$

where n is the number of radioactive species present, λ is the constant of proportionality, a characteristic of the species. We can integrate equation 7.1 to yield

$$n = n_0 \exp(-\lambda t) \qquad (7.2)$$

where n_0 is the number of nuclei at time $t = 0$. We define a time, the half life $t_{1/2}$, which is the time required for half the nuclei to decay, so that

$$t_{1/2} = (\log_e 2)/\lambda \qquad (7.3)$$

Each radioactive species is characterized by its half-life, as it must be, since λ is determined by the nature of the radioactive species; $t_{1/2}$ may vary from 10^{-7} s to 10^{14} years.

Often, the decay of a particular species gives a stable product in one step, but examples of chain decay are also known. If a given radioisotope decays to a second radioactive material, the activity of the first will decrease while that of the second increases. Eventually, a state of apparent equilibrium will be attained – a state in which the decrease in activity of the first will be balanced by the increase in activity of the second. In a chain of n members we can write an expression for the equilibrium situation

$$n_1 \lambda_1 = n_2 \lambda_2 = n_3 \lambda_3 = \cdots n_n \lambda_n \tag{7.4}$$

where the subscripts refer to the various species involved.

Equations 7.1 through 7.4 enable us to calculate the radioactivity of any mass of nuclei, at any time subsequent to a measurement at time zero, our reference time, but they do not explain why the material is radioactive in the first place. A characteristic property of all radioactive elements is an unstable nucleus so that we need to examine the reasons for this. Most nuclei are stable and therefore attractive forces must exist between neutrons and protons. At the same time, of course, there will be Coulombic repulsive forces between the protons. If only the attractive forces between neutrons and protons are considered, it would seem probable that any energy exchange between these particles must be a maximum when equal numbers of protons and neutrons are present. We predict, therefore, that stable neutron:proton ratios ($N:P$) should approach unity. On the other hand, as the number of protons increases, their mutual repulsion will lead to a tendency for a reduction in the number of protons. This latter effect should increase as the number of protons grows. The actual neutron:proton ratio will represent a balance between attractive and repulsive forces so that for nuclei with only a few protons (low Z), the first tendency will predominate, i.e. $N = P$. As the number of protons increases, the second effect will become more important and the ratio $N:P$ will increase slowly (it reaches $1:1.6$ for the naturally occurring nuclei). This effect is demonstrated graphically in Figure 7.3, where the number of neutrons in stable, naturally occurring nuclei, is plotted against the number of protons. An average curve drawn through these points represents the neutron–proton numbers for greatest potential nuclear stability. At low values of Z it coincides with the condition $N = P$; at high values it deviates from this condition. We can summarize the situation by saying that for a given value of nuclear charge (Z), only a limited number of isotopic species should possess marked stability. Any excess, or deficiency, in number of neutrons should then impart instability and the greater the departure

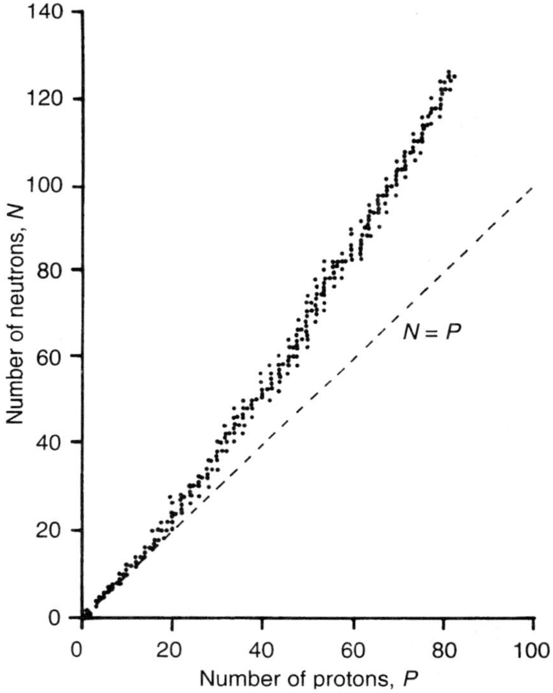

Figure 7.3 Number of neutrons plotted against number of protons for naturally occurring stable elements.

from the average curve, the greater the probable instability of that nucleus. Note that the atomic mass number A equals $N + Z$ where N is the number of neutrons and Z is the number of protons, identically P here.

Nuclear instabilities can be correlated, therefore, with the tendencies of nuclei to readjust unfavourable neutron : proton ratios to more favourable values. Nuclei with excessively high neutron : proton ratios (too many neutrons) make such adjustments by increasing their nuclear charges or decreasing their mass number A. This can be achieved in two distinct ways, namely, beta emission (β^-) or neutron emission (unlikely). Nuclei with low neutron : proton ratios (too many protons) make adjustments by decreasing their nuclear charges. There are four ways in which this may be done: positron emission (β^+), orbital electron capture (EC), proton emission (unlikely) and α-particle emission. The different ways that adjustments can be made are summarized as follows.

1. Beta emission (β^-) is just the emission of electrons, which may be regarded as having been created by the transformation of neutrons into protons. The emission of a β-particle (electron) increases the nuclear charge by one unit ($Z \rightarrow Z+1$), but has no effect upon the mass number. This process is common and characterizes nuclei lying above the general stability curve in Figure 7.3. Energy conservation in β^- decay is maintained by neutrino emission.
2. Neutron emission is difficult to detect and relatively unimportant. Neutron emission decreases the mass number by one unit ($A \rightarrow A-1$), without affecting nuclear charge.
3. Positron emission (β^+) can be regarded as resulting from the transformation of a proton into a neutron. Emission of a positron decreases the nuclear charge by one unit, ($Z \rightarrow Z-1$), but it has no effect on the mass number A. Again energy conservation is maintained by neutrino emission. This type of transformation occurs only among the synthetic radioisotopes. When the positron comes to rest it combines with an electron to yield two antiparallel γ-ray photons, each of energy 511 keV (the electron and positron are annihilated).
4. Electron capture provides an alternative to positron emission, where a nucleus captures an orbital electron, thereby converting a proton to a neutron with the same net result as is seen for positron emission. The electron captured is normally one of those lying closest to the nucleus, i.e. a K-electron, so the process is termed K-electron capture. To fill the vacancy so created an electron from a higher level drops into the K-shell and characteristic X-radiation is emitted.

   In general, whether a nucleus will attain stability through the processes of β emission or positron emission or K-electron capture, can be predicted by the stable isobar rule of Mattauch. This empirical rule states that stable isobar pairs (different atomic number but same mass number) do not exist where nuclear charges differ by only one unit. A nucleus, therefore, decays to a stable isobaric nucleus. Thus unstable $^{24}_{11}Na$ decays to stable $^{24}_{12}Mg$ by β emission, on the other hand unstable $^{13}_{7}N$ decays to stable $^{13}_{6}C$ by positron emission, while unstable $^{49}_{23}V$ decays to stable $^{49}_{22}Ti$ by K-electron capture.

5. Proton emission, while possible, is a highly improbable process and therefore of no importance: it warrants no further comment.

6. Alpha-particle emission decreases the mass number by four units ($A \rightarrow A-4$) and the nuclear charge by two ($Z \rightarrow Z-2$). The emission of a particle with such a large mass and charge requires a very energetic nucleus. Only the heaviest nuclei can be expected to participate in this process.

The decay processes outlined above are complicated by a further factor, nuclear isomerism, where a given nucleus (fixed Z and A) may exist in more than one nuclear energy level and thus possess different properties. Figure 7.4 shows that the unstable nucleus X may exist in any one of several energy levels A, A' or A'', where A and A' are metastable with respect to A''. Such a nucleus could gain stability by losing energy through transition from A or A' to A'', the process of isomeric transition. Although no mass or charge alteration occurs, energy is lost either as short wavelength (γ) radiation or by transfer to orbital electrons. The latter process may cause one or more orbital electrons to be emitted from the atom in question. Such electrons are called internal conversion electrons and their emission is accompanied by X-radiation as with K-electron capture.

The stable nucleus X in Figure 7.4 may decay to the stable nucleus Y of energy C' by alternative paths. If the isomeric transition from A to A'' requires an appreciable time interval, decay may proceed by the path $A \rightarrow A'' \rightarrow C'$ as well as by the direct path $A \rightarrow C$ and to another nucleus Z by the path $A \rightarrow C$. This can only occur if C and C' are

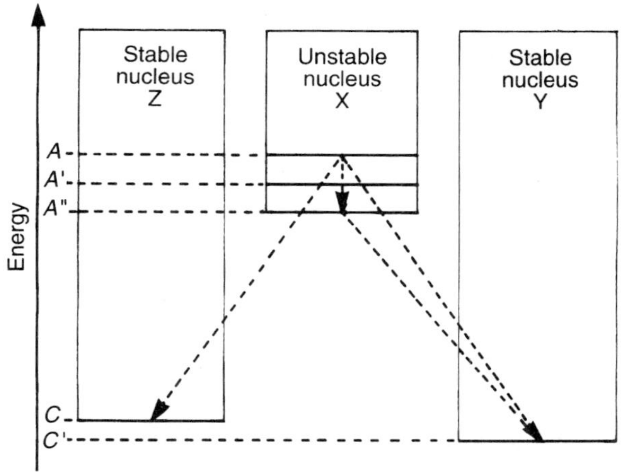

Figure 7.4 The processes involved in nuclear isomerism.

Table 7.1 *Summary of the nuclear decay process*

Decay process	*Designation*	*Change in Z*	*Change in A*	*Examples*
Beta emission	β^{-1}	+1	0	${}^{14}_{6}C \rightarrow {}^{14}_{7}N$ ${}^{99}_{42}Mo \rightarrow {}^{99}_{43}Tc^{m}$
Positron emission	β^{+}	−1	0	${}^{23}_{12}Mo \rightarrow {}^{23}_{11}Na$ ${}^{68}_{31}Ga \rightarrow {}^{68}_{30}Zn$
K-electron capture	EC	−1	0	${}^{55}_{26}Fe \rightarrow {}^{55}_{25}Mn$ ${}^{123}_{53}I \rightarrow {}^{123}_{52}Te$
Isomeric transition	IT	0	0	${}^{99}_{43}Tc^{m} \rightarrow {}^{99}_{43}Tc$ ${}^{52}_{25}Mn^{m} \rightarrow {}^{52}_{25}Mn$
Alpha emission	α	−2	−4	${}^{226}_{88}Ra \rightarrow {}^{222}_{86}Rn$

n → p + electron (neutron conversion)
p → n + positron (proton conversion)

nearly identical. An example of a nucleus which exhibits such behaviour is ${}^{60}_{27}Co$, which decays to ${}^{60}_{28}Ni$ by β-emission and isomeric transitions of the above kind. Table 7.1 summarizes the nuclear decay processes.

7.3 RADIONUCLIDE PRODUCTION

In order to create a radioactive element we must expect to have to create an imbalance of the neutron : proton ratio in the nucleus. There are four ways in which we can do this:

1. neutron capture (or activation)
2. nuclear fission
3. charged particle bombardment
4. the radionuclide generator.

Each of these routes is useful in the medical context.

Neutron capture is the absorption of a neutron by an atomic nucleus with the production of a new radionuclide, an isotope of the target material, through a reaction such as that depicted in equation 7.5. The two nuclei have the same number of protons.

$$ {}^{98}Mo + n \rightarrow {}^{99}Mo + \gamma \qquad (7.5)$$

For this process to work we need neutrons with energies in the range 0.03–100 eV, so-called ‘thermal’ neutrons, such as are conveniently produced in a nuclear reactor. The outcome of this process is a mixture of the original stable material plus the radionuclide, which decays initially by emission of a β^- particle according to equation 7.6, thus

$$^{99}\text{Mo} \xrightarrow{\beta^-} {}^{99}\text{Tc}^{\text{m}} \xrightarrow{\text{IT}} {}^{99}\text{Tc} + \gamma \qquad (7.6)$$

The parent material ^{99}Mo has a half-life of approximately 66 h and decays largely (85%) to the metastable form of ^{99}Tc^{m} (half-life 6.02 h) and thence by isomeric transition (IT) to ^{99}Tc with the emission of a 140 keV γ-ray. Often the radionuclide can be separated easily from the target radionuclide using physical or chemical techniques. This approach is used in radionuclide generators, the subject of section 7.7.

A more complex situation arises when radionuclides are produced by fission of heavy nuclei, since there will generally be a range of fission products from which the desired radionuclide must be harvested. Usually, fission products are rich in neutrons and decay primarily by β^- emission, which rules them out for medical imaging purposes. However, there are some useful radionuclides produced in this way and these include ^{99}Mo, ^{131}I and ^{133}Xe.

An alternative route for the production of radionuclides is through the interaction of charged particles such as H$^+$, D$^+$, ^{3}He^{2+} and ^{4}He^{2+} with the nuclei of stable atoms. In order to overcome the electrostatic repulsive forces of the positively charged nucleus it is necessary to give the charged particles energies in the range 1–100 MeV. This is done through the use of a particle accelerator, for example, a linear accelerator or a cyclotron. Both of these accelerators operate in much the same way. Charged particles are accelerated over short distances by the application of alternating potentials either between the drift tubes of a linear accelerator or between the 'dees' of a cyclotron (Figure 7.5). Charged particle bombardment has a major advantage when it comes to isotope production, since the desired isotope is almost always of different atomic number to the target material. The resultant radionuclide has very high specific activity and small radionuclide impurity level. A further advantage of charged particle bombardment as a means of radionuclide production is that the radionuclides are usually neutron-deficient and decay by β^+ (positron) emission or electron capture (EC). The latter type of radioisotope is particularly useful for clinical applications since there is no particulate emission and hence a reduced patient dose. Some of the radionuclides which find routine use in medicine are listed in Table 7.2.

We see from the preceding account that radioactive decay can lead to the generation of either a stable or a radioactive nuclide. This new nuclide may have the same, or different, atomic number according to

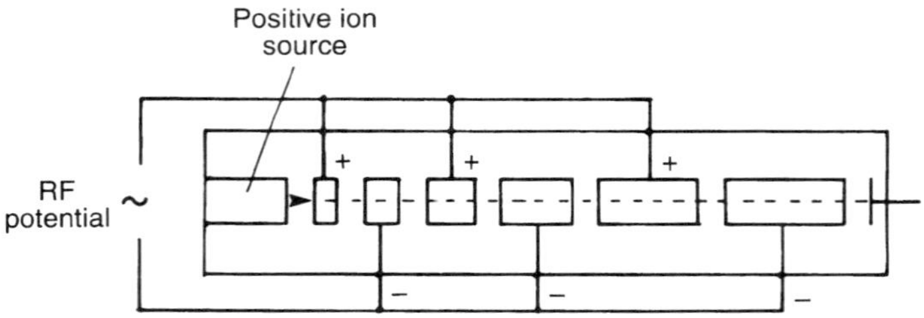

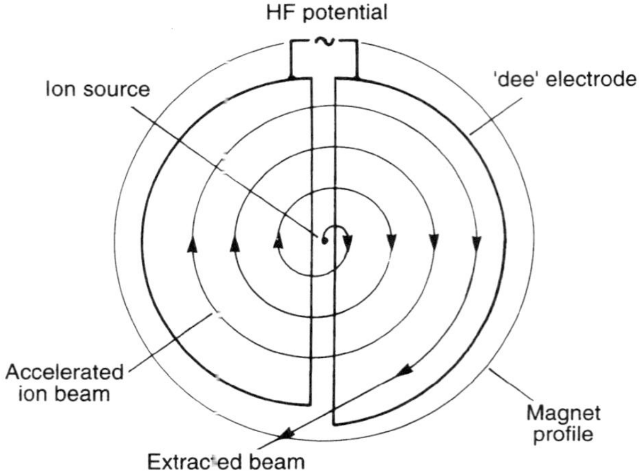

Figure 7.5 (a) Schematic outline of the structure of a linear accelerator used for producing high energy charged particles. (b) Section through a cyclotron showing the accelerator 'dees' and the charged particle trajectory.

the type of decay involved. Where the radioactive decay process leads to the production of a radioactive nuclide with a different *Z*, known as a daughter, then we can contemplate the possibility of simple chemical separation of the parent–daughter combination, particularly if the daughter radionuclide has good physical characteristics compatible with medical imaging, and the parent has a sufficiently long half-life to allow for production, processing and shipment. Remote parent–daughter separation provides a particularly convenient source of medically useful, short-lived radionuclides: this type of radionuclide production has been referred to already and is known as a radionuclide generator. The most widely used, generator-produced radionuclide in nuclear medicine is $^{99}Tc^{m}$, already mentioned in our discussion of isomeric

Table 7.2 *Characteristics of some 'medical' radionuclides*

Element	*Principal γ-ray energy* (keV)	*Half-life* $t_{1/2}$
^{11}C carbon	511	20.30 min
^{15}N nitrogen	511	9.97 min
^{15}O oxygen	511	122 s
^{18}F fluorine	511	109.80 min
^{24}Na sodium	1369	15.00 h
^{51}Cr chromium	322	27.70 d
^{59}Fe iron	1099	44.51 d
^{99}Mo molybdenum	740	65.94 h
^{99}Tc technetium	143	6.02 h
^{123}I iodine	159	13.1 h
^{138}Xe xenon	81	5.25 d
^{201}Tl thallium	68–80.3	3.05 d

transitions. It is produced from ^{99}Mo (equation 7.6). The types of processes involved in parent–daughter, generator-produced radionuclides are summarized in Figure 7.6.

7.4 RADIATION DETECTORS

The early radiation detectors, as depicted in Figure 7.2, find no use in modern radioisotope imaging. Instead arguably the most important detectors are based on materials which emit visible, or near visible, light when energy is absorbed from ionizing radiation. They are used both to count the output from, and also to, image radioisotopes. These materials are high *Z*, inorganic scintillators, having good stopping

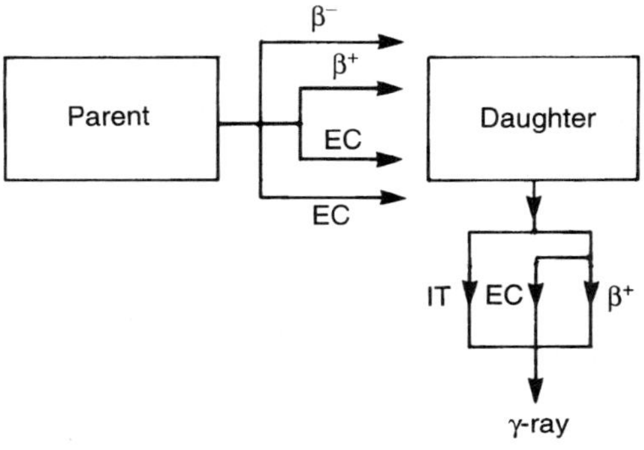

Figure 7.6 The processes involved in parent–daughter generation of radionuclides.

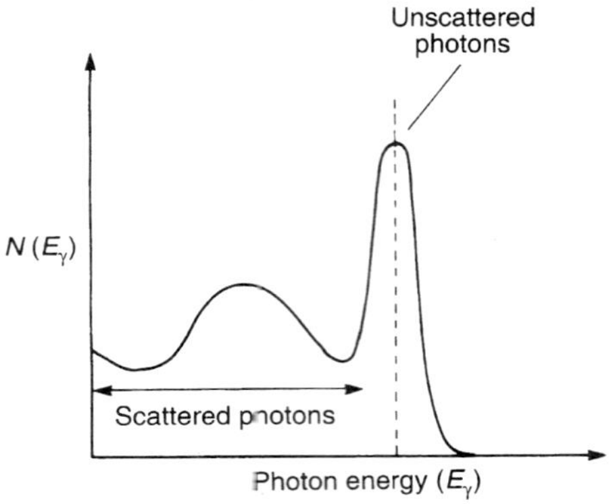

Figure 7.7 The pulse height spectrum for γ-rays detected by an inorganic scintillator showing the distinction which can be made between scattered and unscattered photons; the dashed line corresponds to the photon energy.

power. They are combined with a photomultiplier to constitute a radiation detector of high sensitivity. This will be the case particularly when the scintillator is transparent to its own light and has emission characteristics which match those of the photomultiplier. In addition, light emission by most inorganic scintillators is proportional to the energy deposited therein, so that the detector also serves as a means by which the detected photon energy may be measured. At 100–200 keV the energy resolution is about 10–15%, which is good enough to permit a distinction between γ-rays emitted unscattered and those which have lost energy by scattering. This distinction is made clearer in Figure 7.7. The materials which find use as inorganic scintillators are sodium iodide (thallium iodide doped), barium fluoride and bismuth germanate. Their properties are summarized in Table 7.3.

Inorganic scintillators operate by using the energy of the incoming γ-ray photon to generate excited atoms in the crystal lattice. Light will

Table 7.3 *Properties of inorganic scintillators*

Scintillator	*Effective Z*	*Relative light yield* (%)	*Decay time* (ns)	*Scintillation wavelength* (nm)
Sodium iodide (NaI)	50	100	230	410
Barium fluoride (BaF_2)	54	5	0.7	195, 220, 310
Bismuth germanate ($Bi_4Ge_3O_{12}$)	74	12	300	480

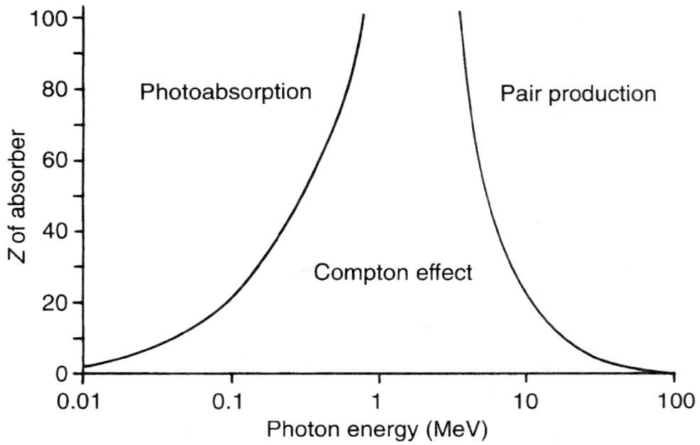

Figure 7.8 Domains in the photon energy versus atomic number (Z) of an absorber. The domains are labelled according to which energy loss mechanism is dominant.

be emitted from these as they de-excite. The intermediary for this generation of excited atoms is an energetic electron in all cases. There are three mechanisms associated with this process; the actual mechanism operating depends on the atomic number (Z) of the scintillation material and the energy of the incident γ-ray photon. Figure 7.8 shows the relationship between these quantities and the three operating regions:

1. photoabsorption
2. Compton effect
3. pair (electron–positron) production.

In the first region, the photoabsorption region, which operates at the low energy end of the γ-ray photon spectrum, we find that the photon energy is completely absorbed to yield energetic electrons which in turn excite the scintillator atoms. In fact, of course, energetic electrons are produced in all three regions but not always efficiently. In the case of pair production, the γ-ray photon energy must have a threshold value of 2×511 keV, the electron-positron pair energy (twice the rest mass equivalent energy of the electron). Although neither the electron nor the positron will itself escape because it loses energy too fast, the positron ultimately annihilates an electron to liberate two new γ-ray photons, each of energy 511 keV. To register the full energy, both photons must be contained within the detector. Reference to Figure 7.8 shows that desirable detector characteristics may be summarized as

1. high *Z*, enhancing the domain in which photoabsorption occurs;
2. reasonable size, giving photons a chance to convert energy to ionization in more than one encounter;
3. high density, which maximizes the number of electrons encountered in the distance travelled.

Crystals of sodium iodide, with thallium iodide doping, NaI(Tl), can be grown with diameters up to 813 mm, so that a large range of crystal diameters are readily available.

A second type of radiation detector is now important in medical imaging and that is based on silicon or germanium single crystals. The operating mechanism here is simply electron–hole pair production during γ-ray interaction, and results in a current flowing in a biased crystal or *p–i–n* (*p*-type intrinsic *n*-type) diode. Detectors of this sort have inherently higher energy resolution than their scintillator counterparts. Figure 7.9 contrasts the energy resolution of a lithium doped germanium crystal Ge(Li) against that of a NaI(Tl) detector for two different γ-ray emitters. The germanium detector has an energy resolution which can be a few keV compared with tens of keV for a NaI(Tl)

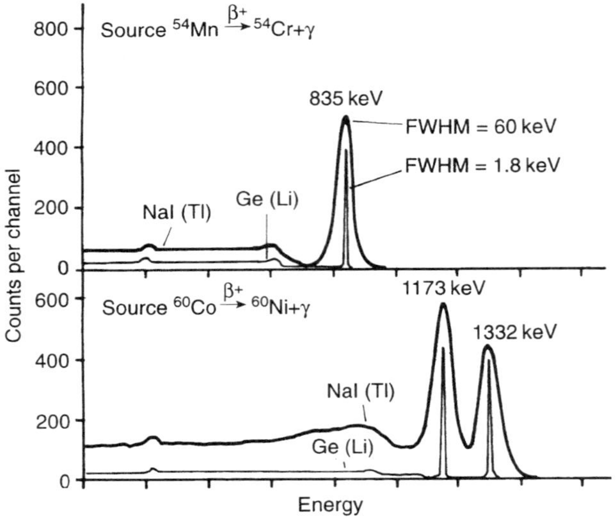

Figure 7.9 Gamma ray spectra from radiative sources as measured with both NaI(Tl) and Ge(Li) detectors. In (a) the source is ^{54}Mn in (b) the source is ^{60}Co. The relatively high energy resolution of Ge(Li) detectors is apparent here.

detector. In each case the background signal at low energies represents events in which a γ-ray photon lost some energy to the crystal, but eventually escaped; this is called the 'Compton platform'.

There remains a third general class of detectors which may be termed gas detectors, or gas-filled ionization chambers. These structures are very sensitive and accurate detectors of ionizing radiation, although they are relatively transparent to γ-rays or X-rays unless the gas filling has a high atomic number (Z); high pressure xenon is often used to meet this requirement. A schematic, simplified view of a gas detector is shown in Figure 7.10. The incident photon ionizes a gas atom and the electron thus released is available to cause further ionization as it gathers energy from the electric field between the anode and cathode electrodes. This process may be thought of as gas amplification, and amplification factors of approximately 10^8 are possible. The resultant current in the external circuit is proportional to the applied electrode voltage.

A variation on the gas-filled detector is the multiwire detector shown in Figure 7.11. It consists of a multilayer sandwich in which thin sheets of lead foil are interleaved with arrays of parallel-wire, planar anodes. Incident photons collide with the lead foil sheets where they are converted to electrons. These electrons cause ionization of the gas in the same fashion as described for the gas-filled detector. Although detectors of this sort have relatively low efficiency, they can readily be built to cover large areas and thus yield adequate sensitivity; their spatial resolution is very good.

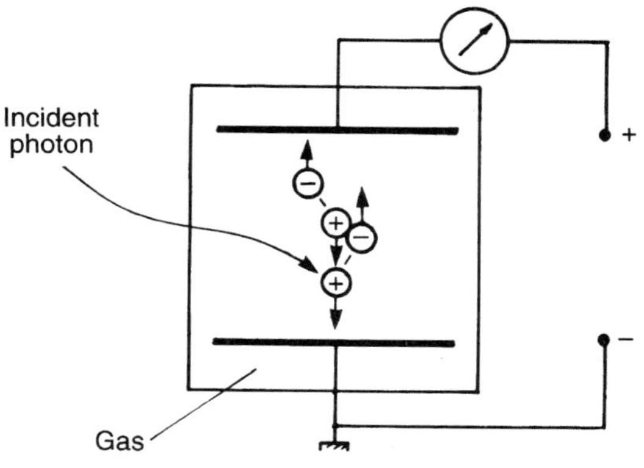

Figure 7.10 Schematic diagram of a simple gas ionization chamber. After the photon collision creating the first ion, the free electron is available to create further ions by collision.

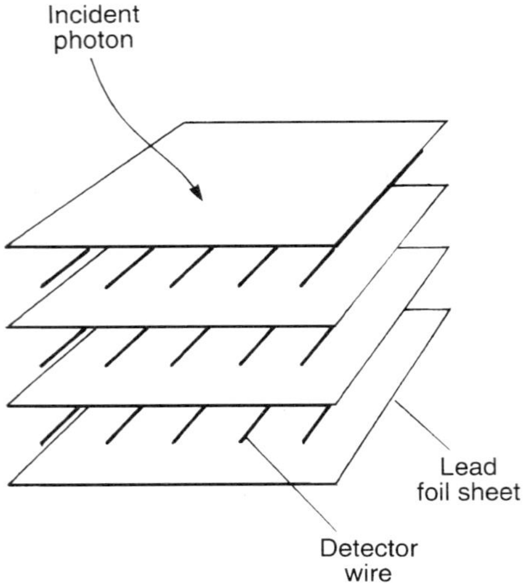

Figure 7.11 A multiwire proportional chamber stack showing the wire-plane anodes interleaved with thin lead foil cathode convertors.

7.5 RADIOISOTOPE IMAGING BASICS

Radioisotope imaging has already been distinguished from X-ray imaging by the fact that the point of origin of the γ-ray photon is not defined. In consequence, any radioisotope imaging system must incorporate a collimator which will only accept photons from a small region of the body before allowing them to impinge on the scintillator–detector combination. The basic idea underlying a focused collimator is shown in Figure 7.12, which depicts a section through such a structure. The collimator is a large, truncated lead cone, with multiple holes set out on a regular pitch. Each hole tapers towards the focal point and may have circular, square, hexagonal or triangular cross-section. Choice of cross-section depends largely on cost and sensitivity factors. Multihole collimators increase sensitivity and can be designed to focus at different distances within the body, depending on the organ being examined. If the energy of the emitter is low (<150 keV) a large number of holes can be used (100–200 holes for a 12.5 cm diameter scintillation crystal) since the septa between the holes can be made relatively thin, say less than 5 mm. However, for higher energy emitters, radiation will be transmitted across such septa, resulting in a loss of spatial resolution. Consequently there is an optimum thickness of septa for a given γ-ray energy at which the increase in spatial resolution, accompanying a thin

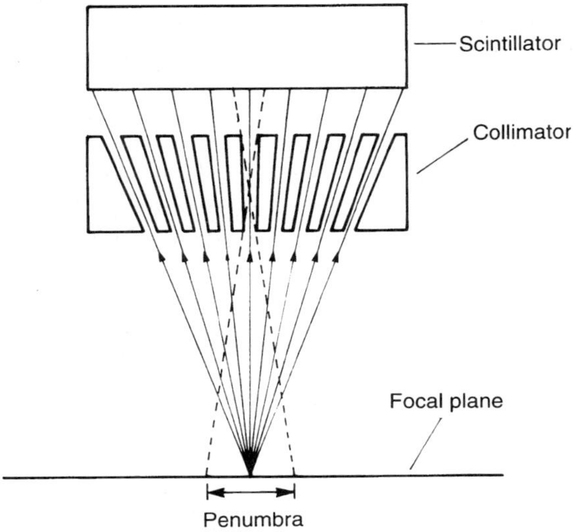

Figure 7.12 Cross-section through a focused lead collimator.

septa and narrow holes, is balanced against the loss in sensitivity, that is attenuation of the scintillator signal. A generally useful radioisotope for imaging purposes is ^{99}Tc with a γ-ray energy of 140 keV.

In order to construct an image from the radioisotope distributed throughout a body organ, the point of focus of the collimator has to be traversed, both laterally and longitudinally, across the patient. The earliest example of such a scanning system is the rectilinear scanner depicted schematically in Figure 7.13. Scanning systems of this type make poor use of the patient dose, spending very little time imaging any particular part of the body.

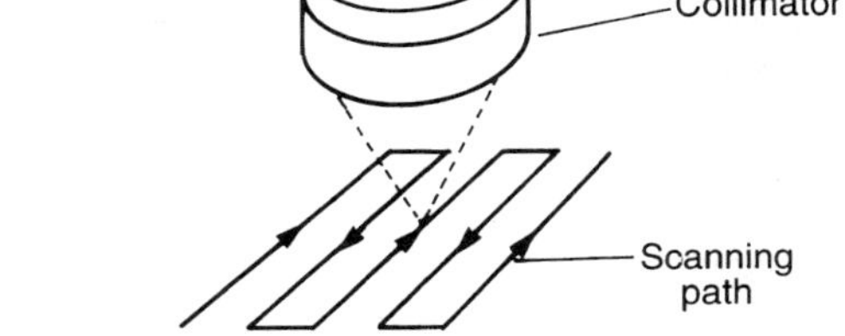

Figure 7.13 A rectilinear scanner.

An alternative approach to the imaging problem depended on the development and introduction of large-area single crystals of NaI(Tl). These could be coupled to the distribution of radionuclides in the patient's body by means of a large-area, parallel-hole lead collimator, giving a 1 : 1 image of the organ being viewed. This arrangement constitutes the gamma camera introduced in 1958 by H.O. Anger, and is shown in Figure 7.14. It enabled the position of the photon interaction in the scintillator to be determined using analogue electronics coupled to an array of photomultipliers attached to the rear surface of the scintillator. Crystal–collimator combinations up to 500 mm in diameter permit the imaging of large body organs and also imaging showing temporal changes, thus conveying dynamic information about the distribution of the radionuclide. The gamma camera can also be used for planar, dynamic and multiplane scintigraphy, as we shall see in section 7.11.

A particular category of radionuclide imaging is that based on radioisotopes which decay by positron emission. This process, as we have seen, results in the annihilation of an electron and the emission of two approximately antiparallel γ-rays, each of 511 keV. The most efficient method of imaging the positron-emitting radioisotope is to surround the patient with a large number of individual scintillation detectors, the detectors being paired off spatially so that antiparallel

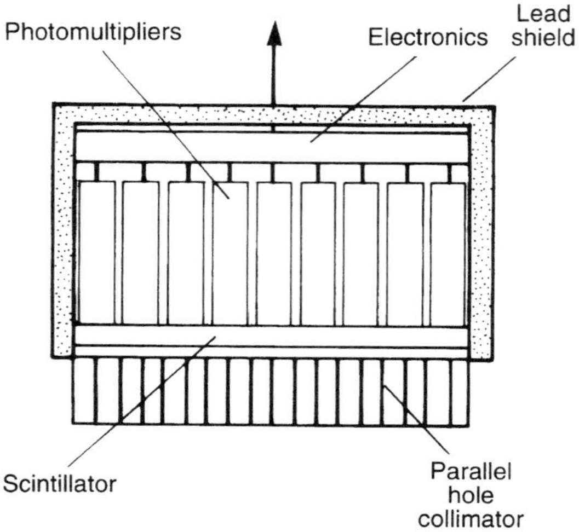

Figure 7.14 Section through a gamma camera.

γ-rays can be detected. Figure 7.15(a) shows the arrangement of detectors and collimators which, together with electronic coincidence detection, enable the position of positron emitters to be determined. This arrangement forms the basis of positron emission tomography (PET), which finds particular application in studies of brain metabolism and neurological problems. Figure 7.15(b), (c) shows the disposition of scintillators and photomultiplier tubes which constitute a typical PET system. Figure 7.16 shows a section through the arrangement utilising multiwire detectors.

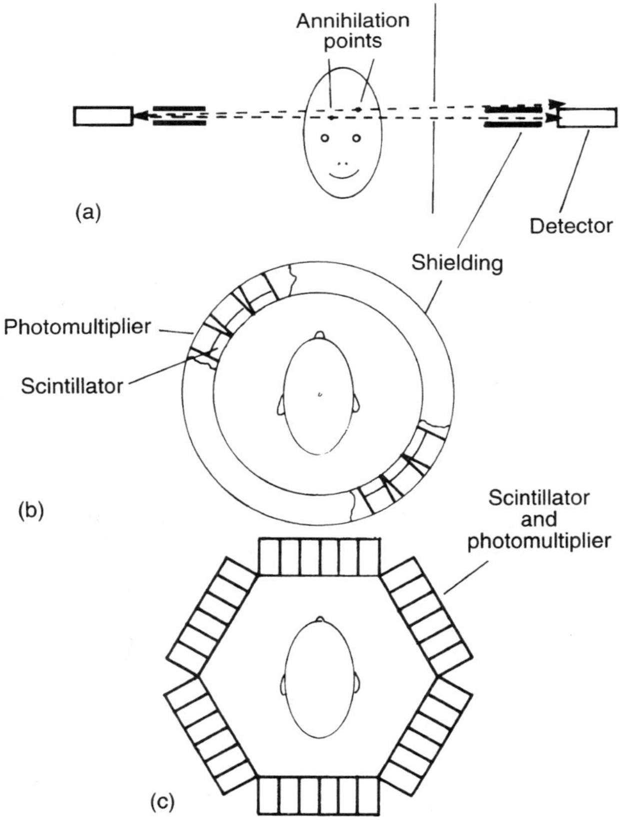

Figure 7.15 (a) Side view of single slice multicrystal PET camera, showing corresponding detectors. (b) Plan view of ring system of detectors. (c) Plan view of a hexagonal detector system.

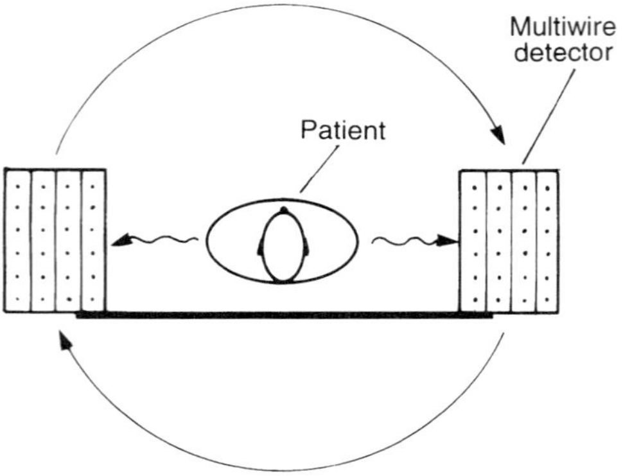

Figure 7.16 Schematic view through a multiwire, proportional chamber, PET imaging system. The pair of detectors are rotated around the patient.

Before making a more detailed consideration of the individual imaging systems it is worthwhile summarizing the characteristics which should be possessed by an 'ideal radionuclide' for use in an imaging system. Our ideal radionuclide should decay via photon emission, but with an absence of particulate emission, thus reducing patient dose. The emitted photon should possess sufficient energy to penetrate body tissue without significant attenuation. This photon energy is in fact critical for two reasons. We need, as we have seen, for the photon to escape from the body efficiently, at the same time it is desirable that the photopeak be readily distinguished from any scattered radiation (Figure 7.7). These requirements are easily met by high energy photons. Unfortunately, when using a conventional gamma camera, the septa thickness would have to be increased significantly, thus decreasing the sensitivity; consequently there has to be a trade-off. A further consideration with high energy photons is the shielding needed in order to protect staff. Finally, the effective half-life t_e of a radoipharmaceutical is actually a combination of the physical half-life t_p and the biological half-life t_b, where the biological half-life is the time required to eliminate half of the radiopharmaceutical by the regular elimination process of the body, thus

$$\frac{1}{t_e} = \frac{1}{t_p} + \frac{1}{t_b} \tag{7.7}$$

Equation 7.7 shows that the effective half-life must be matched to the experiment or study duration, purely from dosimetric considerations.

The radioisotope which comes nearest to fulfilling the above requirements is $^{99}Tc^{m}$. In the following section we look more closely at the way in which radioactive elements are used and the manner in which they are incorporated into compounds which are accepted and metabolized by the human body, and then move on to a more detailed examination of the imaging systems.

7.6 RADIOPHARMACEUTICALS

The particular importance of imaging with radioisotopes is that it delivers both qualitative and quantitative information on the physiological performance of the body. This information is important in that it is not supplied by any other imaging technique, and yet can yield indications of the physiological or biochemical changes which are precursors of the changes accompanying pathological conditions. In other words, we are able to assess biological function or localize specific disease. However, in order to do this, special regard must be given to the characteristics of the radiopharmaceutical itself, since if this does not meet certain criteria, then the imaging process will be largely a waste of time.

The properties which an effective radiopharmaceutical must possess can be roughly grouped into three categories: pharmacological, physical and chemical. Further subdivision of these categories is possible so that within the pharmacological category we must consider sterility, specificity and pyrogenic character; physical properties are concentration, molecular size, half-life, molecular conformation (shape), solubility in lipids (ability to penetrate phospholipid structures) and general membrane penetration ability. In chemical terms a radiopharmaceutical has no measurable effect of its own. Molecular size is important since, in general, only low molecular weight compounds (less than 500 daltons*) can penetrate the body's membranes. Thus radiopharmaceuticals intended to gain access to the brain through the blood–brain barrier must meet this latter requirement, in addition to having suitable lipid solubility. It will be recalled from the discussion in Chapter 2 (Figure 2.2(b)) that cell membranes comprise water-soluble phosphate groups sandwiching a lipid, hydrophobic, central layer. Membrane penetration can be classified as passive – simply a matter of diffusion – or active, and involving either a carrier, receptor

* The dalton is a unit of mass which equals the atomic weight of a hydrogen atom, i.e. 1.657×10^{-21} kg. It is named after John Dalton and is widely used in biochemistry.

recognition or some energy dependent mechanism. Receptor recognition relies on highly selective receptors which recognize a particular molecule and actively accumulate it.

The most satisfactory approach to radiopharmaceutical design is to incorporate or insert the radionuclide atom into a molecule which the body already accepts. If this is a material important in the body's metabolic processes, so much the better. There are some general rules to be observed in carrying out this sort of insertion and these relate to molecular conformation and polar character. Thus we find that radioisotopes of iodine may be substituted for methyl (CH_3) or ethyl (C_2H_5) groups in an organic molecule since the volumes of an iodine atom and the methyl or ethyl groups are comparable. With regard to polar character, the electronegativity of iodine, that is the tendency of an iodine atom to attract electrons to itself, has been established (by L. Pauling) as 2.4. On the same scale the electronegativity of carbon is 2.5, essentially the same as that of iodine. We can therefore expect to be able to replace an ethyl or methyl group with an iodine atom without altering, in any significant way, the molecular shape (conformation), or the polar character of the host molecule. Finally, regard has to be had to the radionuclide half-life as summarized in equation 7.7.

An attractive approach to creating organic pharmaceuticals is to exchange a naturally occurring atom of the parent organic molecule with its radioactive equivalent. In principle, we can exchange C, N and O atoms, or even an H or OH group of a naturally occurring organic molecule. For the latter two entities we would use ^{18}F, which may be included in this group of natural substitutes since its insertion into a host molecule does not lead to polar or conformational perturbation. Examination of Table 7.2 shows that all of the corresponding radionuclides are positron emitters and are therefore restricted to PET systems. Clearly, radiopharmaceuticals for general imaging must be synthesized from other, non-organic radionuclides. In this connection the chemical properties of what will be an entirely 'unnatural' molecule become significant.

As we have already seen, it is possible to insert radioisotopes of iodine (in place of methyl or ethyl groups) in organic molecules to construct iodine-containing analogues. For metallic radionuclides the position is generally more difficult, and the metal has to be bound up or stabilized with organic groups to form a complex, or co-ordination compound. The metals which are most active in the formation of complexes are the transition metals, although this property is not exclusive to them. Thus we find that, of the metals used in radioimaging which are stabilized by organic complexes, or chelating agents, the most important is the transition metal technetium (Tc); three members of the boron family (group IIIb elements) also find widespread use, namely gallium (Ga), indium (In) and thallium (Tl). Table 7.4 attempts to classify some of the common radionuclides, both 'organic' and metallic.

Technetium, probably one of the most widely used radionuclides, is usually presented to the body as an organometallic compound or metal-chelate complex. In this form its behaviour in the body can be described according to whether the properties of the complex depend explicitly on the presence of technetium or not. We have therefore 'Tc-labelled' pharmaceuticals. To make this concept clearer, if Tc is complexed with hexamethylpropylene amine oxime to form Tc-HMPAO, it can cross the blood–brain barrier and become localized in brain tissue, in amounts dependent on the blood flow. HMPAO on its own, however, does not cross the blood–brain barrier, making this a Tc-essential radiopharmaceutical. The presence of the transition metal atom influences, in some way, the properties of the chelate group. The word chelate is taken from the Greek *chele* meaning crab's claw, and refers to the ring-forming characteristics of the co-ordinating or complexing agent. The chelating group will have two or more atoms which can serve simultaneously as donor atoms, the presence of the accepting

Table 7.4 *Medical uses of some common radionuclides*

Radionuclide	*Use*	*Notes*
^{11}C	PET	^{11}C-glucose for brain imaging, positron emitter
^{13}N	PET	Positron emitter
^{15}O	PET	Positron emitter
^{18}F	PET	^{18}F-fluorodeoxyglucose, substitute for glucose in brain imaging
^{133}Xe	Cerebral blood flow imaging	
^{67}Ga	Tumour imaging	Strong affinity for the plasma protein transferrin
^{111}In	Imaging of inflammatory processes	Strong affinity for the plasma protein transferrin
^{201}Tl	Cardiac imaging, myocardial perfusion	Substitutes of K in cardiac muscle
^{123}I	Thyroid studies	
$^{99}Tc^{m}$	Widespread	Can be organ specific
^{82}Rb	Myocardial perfusion studies	Potassium analogue in cardial muscle

metal atom causing the ring to close. Before considering the application of these radionuclides it is necessary to review the basis for, and features of, the last remaining radionuclide source, the radionuclide generator.

7.7 RADIONUCLIDE GENERATORS

The fourth, and last remaining source of clinically useful radionuclides is the radionuclide generator; it is an important source of some positron emitters. Before considering the physical structure of a generator it will be helpful to review the relationships which exist between parent and daughter activities in a generator system. Figure 7.17 summarizes the genetic relationships in a nuclear decay series for a possible generator system. Using equations 7.1, 7.2 and 7.3 as a basis we can write that the change in the number of 'daughter' nuclei n_2 as a function of time is the difference between the formation rate $\lambda_1 n_1$ and the decay rate $\lambda_2 n_2$, thus

$$\frac{dn_2}{dt} = \lambda_1 n_1 - \lambda_2 n_2 \tag{7.8}$$

Also, we can rewrite this using equation 7.2 to obtain

$$\frac{dn_2}{dt} + \lambda_2 n_2 - \lambda_1 n_1^0 \exp(-\lambda_1 T) = 0 \tag{7.9}$$

where n_1^0 is the number of parent species at time zero.

The solution of this differential equation is

$$n_2 = \left(\frac{\lambda_1}{\lambda_2 - \lambda_1}\right) n_1^0(\exp(-\lambda_1 t) - \exp(-\lambda_2 t)) + n_2^0 \exp(-\lambda_2 t) \tag{7.10}$$

where n_2^0 is the number of daughter species at time zero.

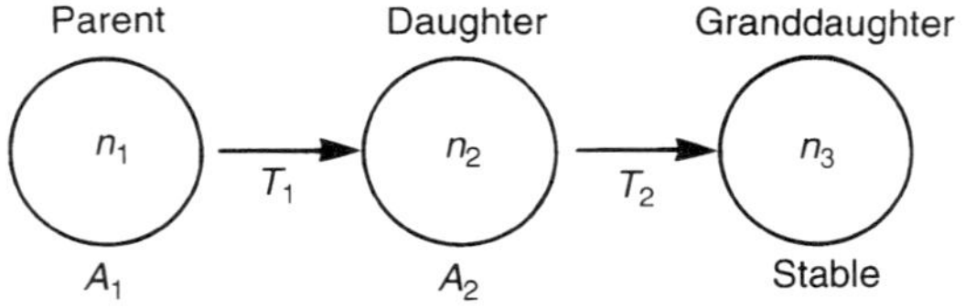

Figure 7.17 Nuclear decay series relationships for a possible generator system.

If the decay rate $-dn/dt$ is written more compactly as A, or the activity, then equation 7.10 may be rewritten to show the general relationship between parent and daughter activities as a function of time:

$$A_2 = A_1\left(\frac{T_1}{T_1 - T_2}\right)\left\{1 - \exp\left[-0.693\left(\frac{T_1 - T_2}{T_1 T_2}\right)t\right]\right\} \qquad (7.11)$$

where T_1 is a more compact representation of $t_{1/2}$ in equation 7.3 and is the half-life of the parent, T_2 is the half-life of the daughter, and both of these quantities are defined by equation 7.3. The activity of a sample is defined as the number of decays per second, which at $t = 0$ will just be $n_1^0 \lambda_1$. The units of activity are *curies* (abbreviated Ci), where 1 Ci = 3.70×10^{10} disintegrations per second. The SI unit of activity is the *becquerel* (abbreviated Bq), where 1 Bq = 1 disintegration per second.

In a system such as this we can state that the parent/daughter activities have come to equilibrium when the ratio of activities becomes a constant, that is A_2/A_1 is constant. By a somewhat arbitrary convention we state that the genetic series depicted in Figure 7.17 is capable of transient equilibrium if $100 > T_1/T_2$. On the other hand if $T_1/T_2 > 100$ the system is said to be capable of reaching secular equilibrium.

Transient equilibrium is shown schematically in Figure 7.18. Here, if the pure parent is isolated in an ideal generator, growth of the daughter activity is described by equation 7.11. Equilibrium is reached when the time since separation t becomes large compared with T_2. At this time

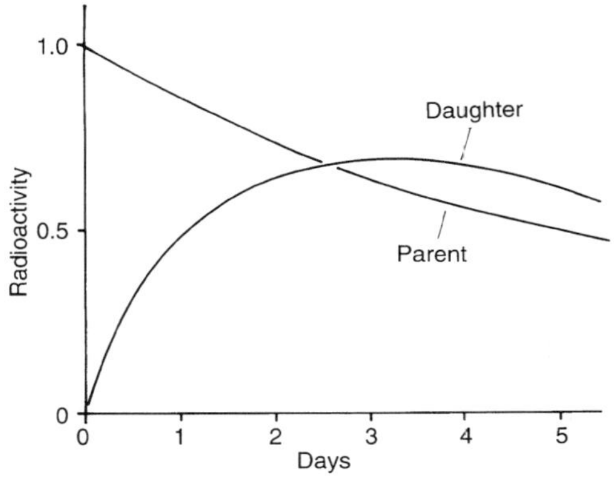

Figure 7.18 Relationship between 'parent' and 'daughter' activity for a radionuclide exhibiting transient equilibrium.

the exponential term in equation 7.11 tends to zero, so that, when the system is in equilibrium, equation 7.11 becomes

$$A_2 = A_1\left(\frac{T_1}{T_1 - T_2}\right) \tag{7.12}$$

or put another way, the activity of the daughter exceeds that of the parent by a factor of $T_1/(T_1 - T_2)$. If the difference between the parent and daughter half-lives is small, then the greater will be the ratio of activities at equilibrium. If the difference between parent and daughter half-lives is large, that is T_2 is small compared with T_1, then equation 7.11 can be reduced to the form

$$\begin{aligned} A_2 &= A_1[1 - \exp(-0.693t/T_2)] \\ &= A_1[1 - \exp(-\lambda_2 t)] \end{aligned} \tag{7.13}$$

This is the secular equilibrium relationship (Figure 7.19), with the further simplification that at equilibrium $\exp(\lambda_2 t)$ is small compared with unity and equation 7.13 reduces to $A_2 = A_1$. We could have obtained this result from equation 7.12 by putting $T_1 \gg T_2$ and it shows the development of parent/daughter activity with time in the approach to secular equilibrium. A typical example of a genetic relationship exhibiting secular equilibrium is ^{68}Ge to ^{68}Ga, while transient equilibrium is demonstrated by ^{100}Pd to ^{100}Rh. Figure 7.20 shows, schematically, the essential features of a radionuclide generator which would be used, for example, to produce perhaps the most important and widely

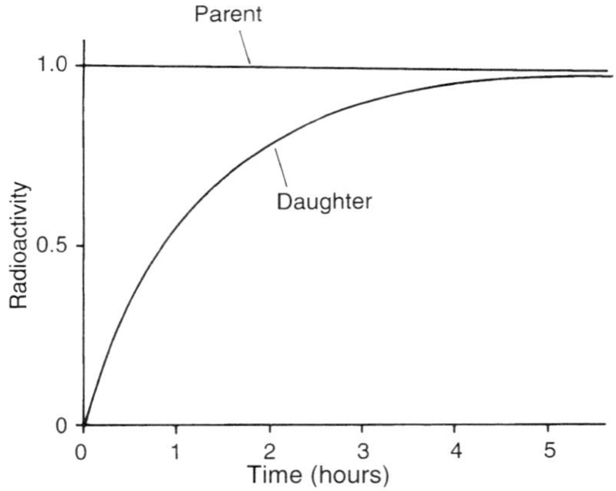

Figure 7.19 Relationship between 'parent' and 'daughter' activity for a radionuclide exhibiting secular equilibrium.

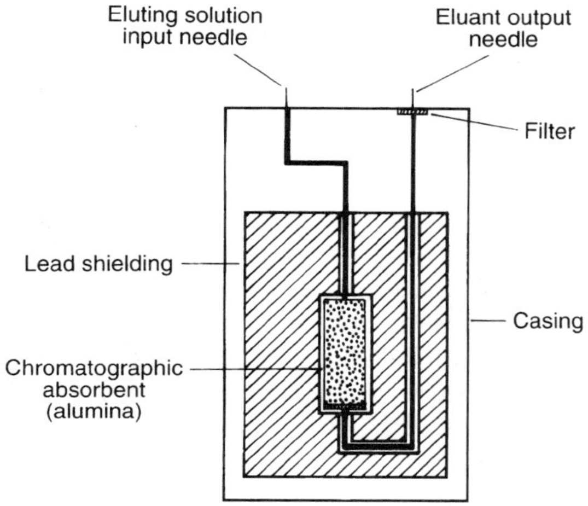

Figure 7.20 Schematic diagram for a radionuclide generator based on chromatographic separation of parent and daughter.

used radioisotope of all, namely $^{99}Tc^{m}$. The daughter is separated from the parent within the generator by a variety of techniques of which chromatography is most commonly used, and is the method illustrated in Figure 7.20. Here the chromatographic medium is alumina and the $^{99}Tc^{m}$ is eluted from this by passing a solution of isotonic saline (0.9% NaCl) through the column. The $^{99}Tc^{m}$ can be separated almost totally from the parent ^{99}Mo. The eluted $^{99}Tc^{m}$ is then combined chemically with compounds chosen according to the application. We shall consider within this chapter its combination with HMPAO as a particular example. It should be noted that $^{99}Tc^{m}$ is not a positron emitter; some generator produced positron emitters are listed in Table 7.5.

We can now proceed to consider the generation of cross-section images of tissue function, in a manner analogous to that employed with X-rays, only now using radioisotopes. This technique is emission computed tomography (ECT) and is conveniently split into two separate modalities, positron emission tomography (PET) and single photon emission computed tomography (SPECT).

7.8 POSITRON EMISSION TOMOGRAPHY (PET)

An important technique, particularly for imaging the brain, is that of positron emission tomography (PET). It is based upon the use of radionuclides which decay by positron emission. The final stage of this decay process is the annihilation of a positron–electron pair to produce two photons, each of approximately 511 keV, travelling essentially collinearly. The imaging system depends on coincidental detection of the photon pair and Figure 7.15(a) will make this idea clearer. Here only one of the annihilation events is registered by the two collimated detectors. The other photon pair registers its presence by interacting with only one of the collimated detectors and is excluded from the measuring process as a consequence.

Where coincidental reception of photons is registered we have, in effect, defined a line passing through the point of origin of the annihilation process. We have at the same time measured a line integral for the annihilation process, since all annihilation events along that line will be measured and recorded. This measurement is therefore exactly analogous to that carried out for transmitted X-rays in X-ray computed tomography. The decision that two photons have actually arrived simultaneously, and are therefore connected with the same annihilation event, is made by triggering an electronic gating signal of appropriate duration. If a photon is registered on an opposing detector within this time, then a coincidence is said to have occurred. The gating time is usually set in the range 4–10 ns. The above approach can be extended to produce time-of-flight information, as reference to Figure 7.21 shows. An annihilation event occurring at point P, displaced from the midpoint by Δd, produces two photons which have different pathlengths. The photon travelling to detector A has a path of length $d + \Delta d$ while that travelling to detector B has a pathlength $d - \Delta d$; The pathlength difference is $2\Delta d$. If we set a limit on the arrival time differences which can be recorded, say 1 ns, then the smallest displacement of the point of annihilation from the midpoint is 150 mm. In practice the time resolution can be reduced to about 500 ps, simultaneously improving the spatial resolution to around 75 mm. Imaging

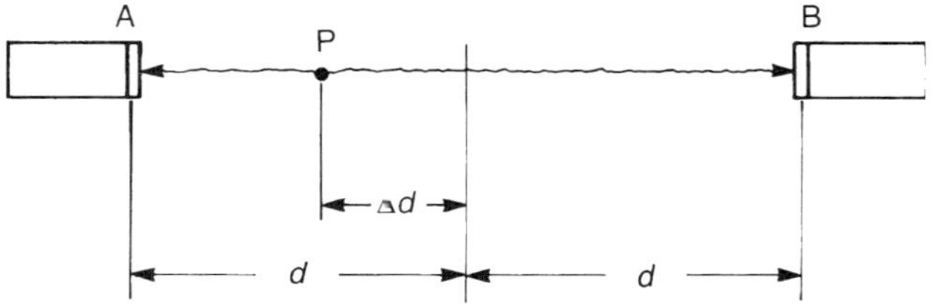

Figure 7.21 Basis for utilization of photon time-of-flight information for the localization of a positron annihilation event along the coincidence line for two radiation detectors operating in the coincidence mode.

systems utilizing the time-of-flight approach are generally referred to as time-of-flight positron emission. Although the spatial resolution figures quoted above seem far from impressive in an imaging context, the time-of-flight technique has a less obvious advantage which is that it improves the signal-to-noise ratio. The gain in signal-to-noise ratio resulting from the use of time-of-flight information depends on the size of the object being viewed and the time resolution of the system. An improvement by a factor of 5–10 would be typical. If all the possible line integrals through a given plane of the object are measured, then the distribution of radioactivity within the plane may be determined using the techniques outlined in Chapter 6. So, for example, filtered backprojection can be used to reconstruct an image of the distribution. The usual practical arrangement is shown in Figure 7.15(b), (c) or Figure 7.16.

The data obtained from annihilation coincidence detection is a better approximation to the actual line integrals than that obtained in a single photon/collimator combination. The linespread functions for annihilation coincidence detection are almost constant over much of the distance between the detector because of the geometric 'rigidity' imposed by the detector arrangement. In addition, the fraction of photons detected at coincidence is dependent on the thickness of the object only, rather than the depth within the object at which annihilation took place. This fraction can be measured and corrected for. The correction is usually made, with an external ring source, by making a transmission scan with and without the object in place, the ratio of the two signals gives the attenuation factor. Figure 7.22(a), (b) gives an idea of the contrast between single photon detector and coincidence detection systems.

A further advantage possessed by PET imaging systems over single photon based systems is the absence of a bulky collimator, so that the detection sensitivity is nearly two orders of magnitude higher. This has the direct consequence that the reconstructed image is more accurate.

7.9 APPLICATIONS OF PET

A particularly important arena in which PET finds application is in studies of brain function and metabolism. In 1902 P. Ehrlich suggested that there must exist a barrier which prevents many molecules present in the bloodstream from gaining access to the brain; the so-called blood–brain barrier (BBB). Although this title implies total impermeability, this is not actually the case and what we are confronted with is

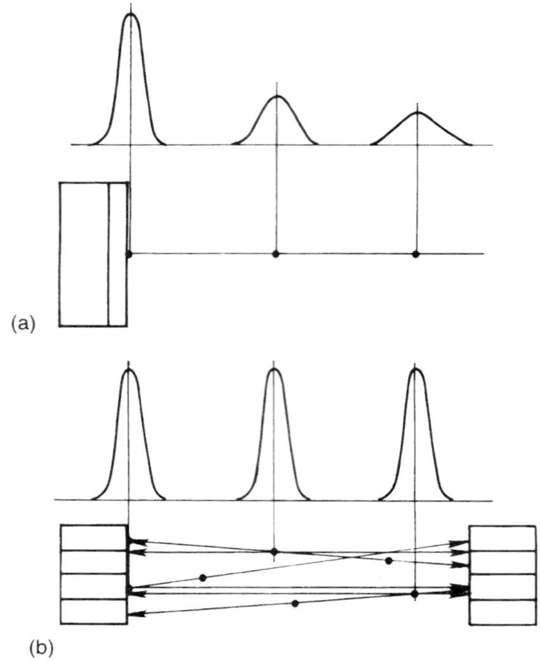

Figure 7.22 (a) Spatial signal resolution with position for a single photon tomographic system, with (b) positron annihilation photon system signal for comparison. The size of the region of object space 'seen' by the collimator varies with distance from the collimator. This is not the case for a PET system.

selective permeability. Indeed, whenever brain function is sufficiently abnormal the BBB is lost altogether.

Many molecules routinely pass through the blood–brain barrier, since they are vital to brain function. A good example would be β-D-glucose, which is the brain's 'fuel'. The molecular structure of β-D-glucose is shown in Figure 7.23(a). On the other hand a molecule which is present in many people's bloodstream, but which is not a normal constituent of the bloodstream, is nicotine. Nicotine is able to penetrate the BBB. Molecules which are not too large (less than 500 daltons) and posses high lipid solubility are also capable of traversing the BBB; a good example would be the technetium complex $^{99}Tc^{m}$ HMPAO, the structure of which is shown in Figure 7.23(b). The use of this complex forms the subject of section 7.10 on SPECT.

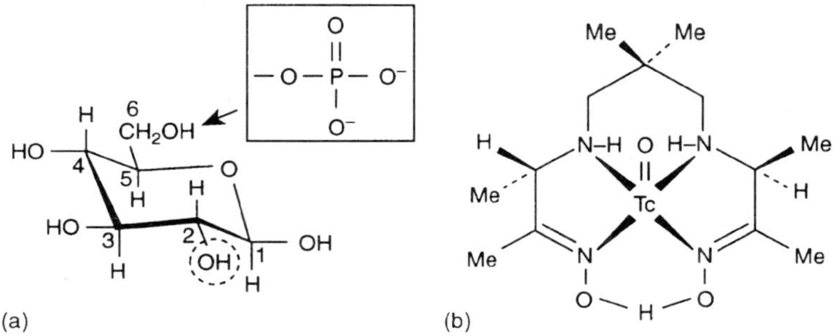

Figure 7.23 (a) Basic structure of the brain 'fuel' D-glucose. The molecule has a 'chair' structure with the hydroxyl groups in equatorial positions. The hydroxyl group ringed at the 2 position may be replaced by hydrogen or ^{18}F for brain studies. (b) The structure of ^{99}Tc^{m} HMPAO. Me indicates a methyl group; dashed bonds lie below the plane of the diagram. During complex formation, the HMPAO molecule wraps around the Tc atom and loses three protons to close the ring by a hydrogen bond at the lowest point of the figure. The Tc atom sits within a tetrahedron composed of four nitrogen atoms and one oxygen atom. This is a ^{99}Tc-essential radionuclide.

An elegant, but not exclusive, example of the way in which positron emission tomography may be applied clinically is to be found in studies of the brain's metabolism. The fact that the brain's energy consumption is prodigious, and that its fuel is D-glucose, permits the exploration of the regional distribution of brain activity via 'labelled' glucose utilization. In order for this approach to work the glucose must not only be suitably 'labelled' with a positron emitter, but it must also be modified structurally, so that it is eventually trapped in the brain. A simple structural modification which meets the above requirements is the replacement of the hydroxyl group, circled in Figure 7.23(a), by a hydrogen atom. If the carbon atoms in the ring are now replaced by ^{11}C, then we have ^{11}C-2-deoxy-D-glucose, which is as readily accepted by the brain as if it were D-glucose.

Alternatively, we can simply exchange the encircled hydroxyl group in Figure 7.23(a) for ^{18}F to give ^{18}F-2-deoxy-D-glucose (^{18}F 2 DOG). This molecule is also accepted as 'fuel' by the brain. Intracellular hexokinase normally phosphorylates D-glucose to glucose-6-phosphate; it also phosphorylates ^{11}C-2-deoxy-D-glucose and ^{18}F-2-deoxy-D-glucose, replacing the arrowed hydroxyl group in Figure 7.23(a) by the phosphate OPO_3^{2-} group. However, the next glycolytic enzyme does not 'recognize' either of these 'labelled' compounds so that they are trapped intracellularly in the brain. The amount of these compounds generated within the cells of the brain in any given region is an indication of that region's flux of glucose in the glycolytic pathway at that time.

7.10 SINGLE PHOTON EMISSION COMPUTED TOMOGRAPHY (SPECT)

Single photon emission computed tomography dates from the early 1960s when the first transverse section tomographs were presented by D.E. Kuhl and R.Q. Edwards, who used a rectilinear scanner (Figure 7.13) and analogue backprojection techniques. The reconstruction algorithms employed in SPECT are similar to those for X-rays with the exception that attenuation compensation methods are incorporated.

For some situations, such as brain imaging, the attenuation coefficient is essentially constant so that efficient compensation methods have been devised. In most cases the attenuation is variable and a weighted least-squares technique is employed. The problems associated with correction for the attenuation error are easily implemented in positron tomography because only line integrals of the attenuation coefficients are needed and the acquisition of the transmission data is easy compared with that for single photon systems.

The size of the region of object space 'seen' by the collimator varies with distance so that the point spread function, or resolution, varies with distance in the manner shown in Figure 7.22(a). This effect has to be compensated for computationally. The spatial resolution FWHM for a single photon device can be as little as 11 mm (Anger camera) – contrast this with the 1 mm attainable with X-ray tomography. The radiation dose associated with ECT is, of course, distributed throughout much of the body for the lifetime of the radiopharmaceutical. In contrast, the dose delivered in an X-ray tomographic measurement is confined to the duration of the X-ray exposure and to the section of the body exposed. If we were to match the radiation dose presented to the patient by the two systems then we would find that in emission computed tomography we have used 10^4 times more photons to produce an image.

Within emission computed tomography we can distinguish two general types – limited angle or transaxial. In limited angle, also known as longitudinal tomography, photons are detected within a limited angular range from several sections of the body simultaneously. The reconstructed image planes are those parallel to the face of the detector.

In transaxial tomography the detectors move around, or actually surround, the body to achieve complete 360° angular sampling of photons from sections of the body. In contrast to limited angle tomography, the sections taken in transaxial tomography are perpendicular to the face of the detector(s). Figure 7.24 makes these differences clearer.

In limited angle emission tomography, the radioactive distribution is viewed from several different angles so that information about the depth of the radioactive sources can be obtained. One way in which this objective can be achieved is by means of a rectilinear scanner with a highly focused collimator such as is depicted in Figure 7.12. The image from this scanning system represents the radioactive distribution of the plane 'in focus', together with blurred information from the planes above and below the focal plane. Methods of deblurring a limited angle tomogram have been developed using either deconvolution or iterative techniques, but the depth resolution or effective slice thickness is still poor compared with the in-slice resolution. This difficulty is exacerbated with increasing depth. In consequence, these tomographic images are essentially qualitative, but they do have the advantage that dynamic studies are possible and the image times are less than those required for transaxial tomography.

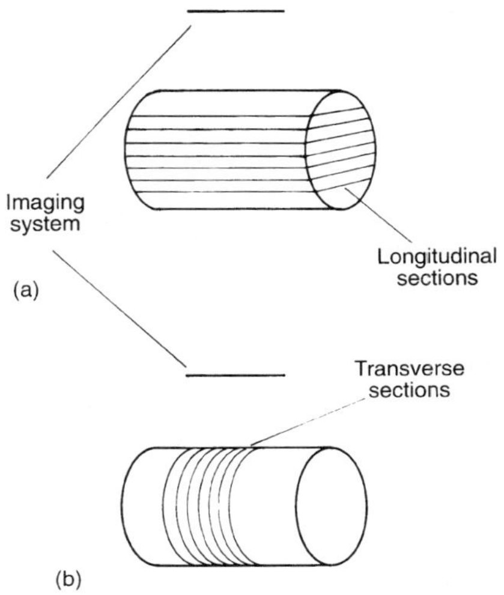

Figure 7.24 Two types of single photon emission computed tomography (SPECT). (a) Scanning basis for limited angle tomography. (b) Scanning basis for transaxial tomography.

Transaxial tomography is conducted using a rotating gamma camera/computer system. As the detector is rotated around the patient, multiple views or two-dimensional projections of the three-dimensional distribution of radioactivity are acquired. The system is exactly analogous to that used in the later developments of X-ray computed tomography, Figure 6.6 for example. Each 'view' represents the linear sum (neglecting attenuation) of the radioactive elements along the line of view of the detector, as determined by the collimator. In this way it is possible to reconstruct cross-sectional images that represent the distribution of radiopharmaceutical within the body.

7.11 RADIOISOTOPE IMAGES

We end this examination of the use of radiation emitted from within the body as a means of body imaging by considering two examples, using $^{99}Tc^{m}$ and ^{201}Tl, respectively.

Calcium does not possess a suitable radioisotope for bone imaging. However, this problem is overcome by using phosphonate complexes labelled with $^{99}Tc^{m}$. These complexes diffuse from the blood into the extracellular fluid that surrounds the porous mineralized surface of the bones, and rapidly fix to the solid phase of the bone by ion-exchange with the metabolically active bone mineral, hydroxyapatite. The concentration of radioactivity in the bones will then be highest where there is increased bone activity and blood flow, for example in areas of healing bone or bone stress. Figure 7.25 shows whole-body skeletal images, obtained with a scanning gamma camera, from a patient who had recently received $^{99}Tc^{m}$ labelled diphosphonate complex (HDP). The images show small areas of increased counts over the midshaft of the right fibia, and the left ankle. The active area of the right tibia corresponds to a stress fracture while the increased counts over the left ankle are due to the reactive changes as the patient 'protects' his right leg. The image should be compared with that in Figure 1.21 where a stress fracture of the same bone is revealed thermographically, that is without the injection of a radiopharmaceutical!

The function of the heart is central to the functioning of all other body organs. Cardiac muscle tone and contractility is maintained by the body's sodium–potassium pump which we have already met in Chapter 2, section 2.14. Blood flow is critical for cardiac muscle function and potassium is extracted from this blood with an efficiency of some 70%. Again, as with calcium, no useful potassium isotopes exist, but we can use the monovalent cation thallium, which has a similar ionic radius to

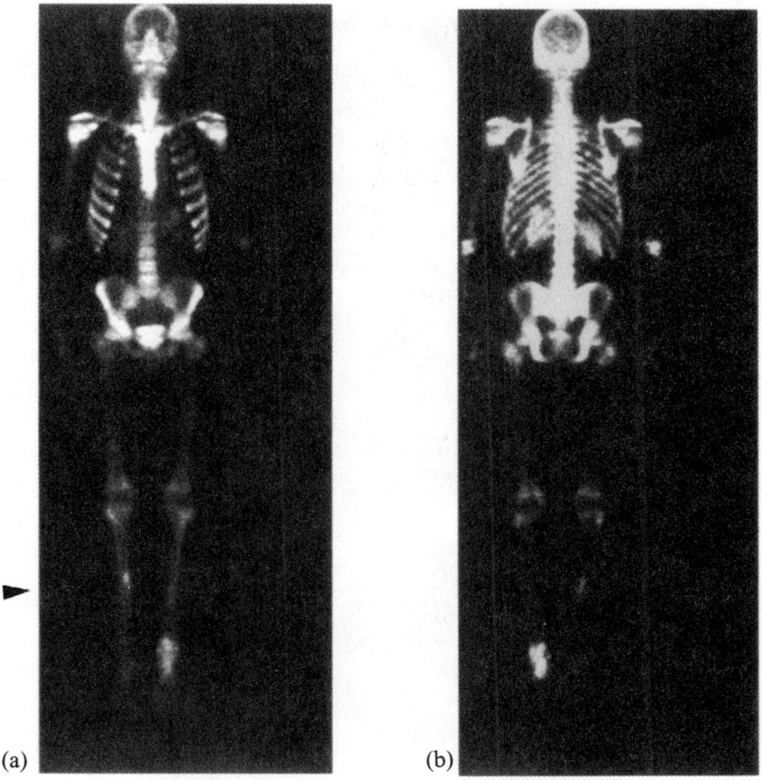

Figure 7.25 (a) Anterior and (b) posterior bone scan images of the human skeleton, taken with a scanning gamma camera, following the injection of 600 MBq of a $^{99}Tc^{m}$ labelled diphosphate (HDP). The increased counts over the midshaft of the right **tibia** represent a stress fracture. (Courtesy of Dr D. Taylor.)

potassium, 0.147 nm compared with 0.133 nm, respectively. Although the thallium ion is not a true potassium analogue, its biological distribution is very similar to potassium. Consequently, the monovalent ^{201}Tl has been used to study the blood flow associated with the heart. Figure 7.26 shows SPECT images of the heart which have been assembled into a three-dimensional image, looking down through the mitral valve. The topmost image shows the distribution of ^{201}Tl under heart stress while the lower image represents the distribution under resting conditions some 4 h later. The maximum concentration of ^{201}Tl redistributes from the nine o'clock position to the eleven o'clock position in the

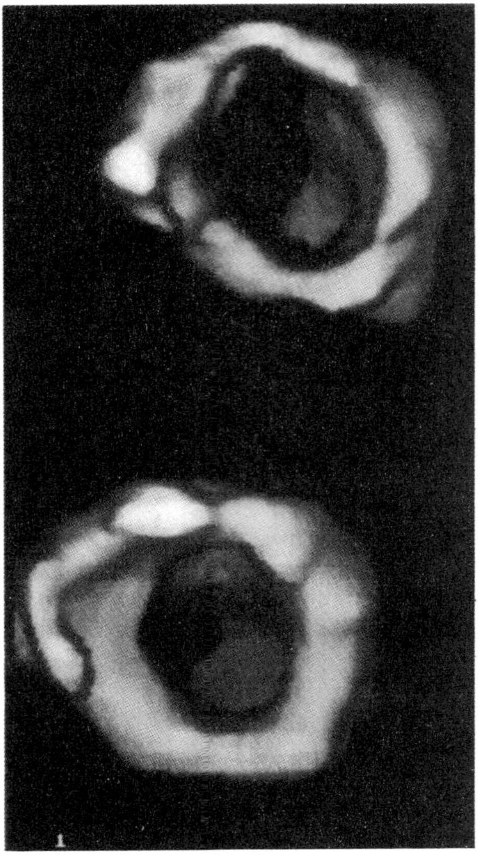

Figure 7.26 Top, three-dimensional display of SPECT images following stress injection of ^{201}Tl. Bottom, injection of ^{201}Tl 4 h later under rest conditions. (Courtesy of Dr D. Taylor.)

process, owing to a defect in the blood supply to the anterior wall of the heart.

PROBLEMS

1. ^{201}Tl has a half-life of 3.05 days. What is (a) the activity of 1 g of pure ^{201}Tl? (b) What will the activity be at the end of 5 days?
2. The isotope ^{99}Mo decays to the daughter ^{99}Tc^{m} in the scheme

$$^{99}\text{Mo} \xrightarrow{\beta^-} {}^{99}\text{Tc}^{m} \longrightarrow {}^{99}\text{Tc} + \gamma$$

The atomic mass of the parent and daughter are 98.9077 and 98.9062 atomic mass units (amu). Calculate the energy of the electron emitted in this transition. (1 atomic mass unit = 1.66×10^{-27} kg.)

3. ^{68}Ge decays to ^{68}Ga by electron capture. The atomic mass of ^{68}Ge is 67.928 amu with a half-life of 288 days, while the atomic mass of ^{68}Ga is 67.927 amu with a half-life of 68.1 min. (a) What type of equilibrium is represented by this genetic relationship? If 1 g of ^{68}Ge decays according to the above scheme, what will be the activity of ^{68}Ga (b) at equilibrium and (c) after 10 min?
4. A positron emitter produces 511 keV γ-rays which are incident on NaI(Tl) and bismuth germanate (BGO) radiation detectors. What thickness of each of these detector materials is required to absorb 80% of the incident radiation? The mass attenuation coefficients are 9.36×10^{-3} m^2 kg^{-1} and 1.38×10^{-2} m^2 kg^{-1}, with densities of 3.67×10^3 kg m^{-3} and 7.13×10^3 kg m^{-3}, respectively.

8

Magnetic resonance imaging: the last window

8.1 INTRODUCTION

In the diagnosis of the human condition we have seen that an important role is played by techniques which provide images of internal organs or structures of the body. The human body is opaque within the optical region of the electromagnetic spectrum, and we must be grateful that we are unable to see inside it directly. Instead, we have to resort to the use of X-rays, γ-rays or even longitudinal pressure waves (ultrasound), all of which are able to penetrate human tissues to a significant extent. It is convenient to think of the body as if it possessed an absorption 'window' for radiation at these frequencies.

There remains one 'window' into the body which has not yet been exploited, namely the 'window' beyond the infrared, in the low energy radio region, the radiofrequency sector of the electromagnetic spectrum. Mention has already been made of this 'window' in our discussion of thermal mapping (Chapter 1), where the innate radiative output of the human body is used to provide diagnostic information. The technique which utilizes this final imaging 'window' is magnetic resonance imaging (MRI). It depends upon the phenomenon of nuclear magnetic resonance (NMR). While the imaging methods we have considered so far depend on the absorption, emission or reflection of a beam of energy, and the principles governing image formation are easily understood, this is not the case for magnetic resonance imaging. Although we still irradiate the subject with electromagnetic radiation, the absorption which occurs is now a resonance phenomenon occurring at a single frequency, or discrete set of frequencies. Also, in contrast to X-ray or γ-ray systems, NMR preserves phase as well as intensity information. Furthermore, in its imaging role the physical property which is displayed is to some extent a matter of choice, and may be spin density, one or other of the two principal relaxation times, or local diffusion or flow effects. Consequently, MRI is also probably the most complex diagnostic technique with which we have to deal.

Nuclear magnetic resonance as a spectroscopic method of chemical analysis dates back to 1946 and the work of F. Bloch and co-workers on the one hand, and E. Purcell and co-workers on the other. Bloch and Purcell shared the 1952 Nobel Prize for physics in recognition of their work. Soon after the introduction of NMR spectroscopy the 'chemical shift' was discovered, which allowed nuclei in different chemical environments to be identified. The physical phenomenon on which both NMR spectroscopy and MR imaging depend is the magnetic resonance exhibited when atoms with nuclear spin respond to the application of radiofrequency magnetic fields, by absorbing or emitting electromagnetic radiation.

In contrast to X-ray imaging, which provides a map of the electron density (related to the physical density) of body structures, MRI, in its simplest manifestation, provides a mapping of the proton density. Since all tissues have about the same NMR observable density of hydrogen, this type of image is generally rather uninteresting. However, hydrogen is the most abundant element in the human body and, more significantly, MRI can distinguish between protons in different environments by the 'chemical shift'. Furthermore, as bone is transparent to radiofrequency radiation, the interiors of bone cavities are as accessible for imaging as other regions of the body. It is properties such as these that make MRI extremely important; it is able to produce information not obtainable by X-ray tomography. Magnetic resonance imaging has an additional tremendous advantage in that it does not subject the body to biologically harmful radiation. The energy of a quantum of RF radiation, suitable for exciting a ^{1}H nucleus for example, is about 10^{-7} eV. This is orders of magnitude less than the energy required to break chemical bonds, and twelve orders of magnitude less than the maximum energy of X-ray photons used in diagnosis, at around 10^{5} eV. So far as is known, neither the RF intensities, nor the magnetic field strengths customarily used, are in any way hazardous to the patient. In addition, the resolution provided by MRI images surpasses that of X-ray CT images, since MRI can be 'tuned' to different elements.

Magnetic resonance imaging offers considerable potential for the imaging of elements other than hydrogen; phosphorus, for example, is of particular interest due to its role in metabolic processes. The molecules of various phosphates, especially adenosine triphosphate (ATP), mediate the storage and transfer of energy in the living cell. Non-invasive measurements of phosphate levels provide unprecedented insight into the condition of the body's vital organs. The NMR signals from phosphorus atoms in different chemical environments, that is different biological phosphates, can be collected separately. Again, this is possible because the resonances are modified by the chemical environments of the nuclei, the so-called chemical shifts already alluded to. Other nuclei, apart from hydrogen and phosphorus, are important in MRI, and some of these are listed in Table 8.1. We can regard MRI as a multipurpose technique in the sense that it can provide both anatomical and metabolic information, in contrast to techniques such as X-ray computed tomography or emission computed tomography. In addition, for reasons which are not yet understood, the NMR signal characteristics of diseased tissue are significantly different from those of healthy tissue.

Table 8.1 *NMR characteristics of selected nuclides. The nuclei* ^{12}C *and* ^{16}O *have* $I = 0$, *consequently they have no spin angular momentum, no magnetic moment and no NMR spectrum. The nuclides with negative gyromagnetic ratios have magnetic moments which align antiparallel to the spin angular momentum*

Nucleus	*Nuclear spin quantum number* I	*Gyromagnetic ratio* $\gamma/2\pi$ (MHz T^{-1})	*Approximate chemical shift range* (ppm)	*Relative sensitivity (to* ^1H*) at constant field**
^{1}H	$\frac{1}{2}$	42.57	10	1
^{2}H	1	6.54	–	9.64×10^{-3}
^{12}C	0	–	–	–
^{13}C	$\frac{1}{2}$	10.71	250	1.59×10^{-2}
^{14}N	1	3.07	–	1.01×10^{-3}
^{15}N	$\frac{1}{2}$	−4.31	1000	1.04×10^{-3}
^{16}O	0	–	–	–
^{17}O	$\frac{5}{2}$	−5.77	650	2.91×10^{-2}
^{19}F	$\frac{1}{2}$	40.06	500–1000	8.34×10^{-1}
^{23}Na	$\frac{3}{2}$	17.91	–	9.27×10^{-2}
^{31}P	$\frac{1}{2}$	27.41	700	6.64×10^{-2}
^{129}Xe	$\frac{1}{2}$	11.78	185	2.12×10^{-2}
^{129}X		Hyperpolarized to 25%		2×10^{3}

* For identical volumes and species concentration sensitivity is proportional to $\gamma^3 I(I + 1)$ and this is the quantity calculated here. These values must be modified, according to the species' physiological concentration, in order to determine the relative *imaging* sensitivity.

The exploitation of NMR as a means for imaging the body's anatomical and metabolic detail did not occur until nearly 30 years after the discovery of NMR itself. In 1973 P.C. Lauterbur, and independently P. Mansfield and P.K. Grannell, produced the first images of small objects. Coincidentally this occurred at essentially the same time as the introduction of X-ray CT. In 1977 R. Damadian and co-workers produced the first whole-body image. Since that time it has become possible to produce not only proton images, but sodium and phosphorus images. Methods have been developed to separate water and fat in proton images and measure blood flow *in vivo*. Even spatially localized NMR spectroscopy can be performed. Unfortunately, MRI systems remain relatively expensive and therefore not as widely available as X-ray or gamma camera systems. In order to understand the basis of MRI we have first to treat the classical description of nuclear magnetic resonance, in which we account for the interaction of a magnetic nucleus with an external magnetic field. It transpires that nuclear magnetic moments are exquisitely sensitive to their surroundings and yet interact very weakly with them.

8.2* PRINCIPLES OF NUCLEAR MAGNETIC RESONANCE

We are interested in the origins and properties of nuclear magnetism. Magnetic nuclei possess an intrinsic angular momentum known as *spin*, whose magnitude is quantized in units of $\hbar$ ($h/2\pi$, where h is Planck's constant). The overall spin angular momentum depends upon the spin quantum number of the nucleus concerned, where the latter quantity is the net result of the addition of the intrinsic spin quantum numbers of the neutrons and protons. This addition must be conducted according to the rules of quantum mechanics. The magnitude of the spin angular momentum $\boldsymbol{S}$, a vector quantity, is given by

$$|\boldsymbol{S}| = \hbar\sqrt{I(I+1)} \tag{8.1}$$

where I is the nuclear spin quantum number which may take the values $0, \frac{1}{2}, 1, \frac{3}{2}, 2 \ldots$, with quantum numbers greater than four being rather rare (Table 8.2). Within a quantum mechanical viewpoint the direction of $\boldsymbol{S}$ cannot be established unless we define an axis of quantization by applying an external magnetic field $\boldsymbol{B}_0$ in a particular direction, for example the z direction. The z components S_z will be $m_I\hbar$, where m_I is the magnetic quantum number; it may take the values $\pm I, \pm(I-1), \ldots$, yielding $2I+1$ possible orientations of angular momentum, or projections onto our arbitrarily chosen axis.

Table 8.2 *Guide for predicting nuclear spin quantum numbers I from the numbers of protons and neutrons in a nucleus*

Number of protons	*Number of neutrons*	*I*
Even	Even	0
Odd	Odd	1 or 2 or 3 or …
Even	Odd	$\frac{1}{2}$ or $\frac{3}{2}$ or $\frac{5}{2}$ or …
Odd	Even	$\frac{1}{2}$ or $\frac{3}{2}$ or $\frac{5}{2}$ or …

If $I = \frac{3}{2}$, such as is the case for ^{23}Na, then $m_l = -\frac{3}{2}, -\frac{1}{2}, \frac{1}{2}, \frac{3}{2}$. This situation is shown in Figure 8.1 where $\boldsymbol{S}$ is orientated at a semi-angle β to the z axis such that

$$\cos\beta = m_l\hbar/|\boldsymbol{S}| = m_l/[I(I+1)]^{1/2} \tag{8.2}$$

For the proton $I = \frac{1}{2}$ and m_l takes the values $\pm\frac{1}{2}$ corresponding to spin-up (parallel) and spin-down (antiparallel) states. If we consider an isolated proton of charge $+e$ and intrinsic spin angular momentum I, then from a purely classical point of view, we can imagine that the charge is distributed and rotating about a centre axis as a consequence of its angular momentum. This gives rise to a magnetic field $\boldsymbol{B}$ and a magnetic dipole moment $\boldsymbol{\mu}_p$ (Figure 8.2(a)). An alternative point of

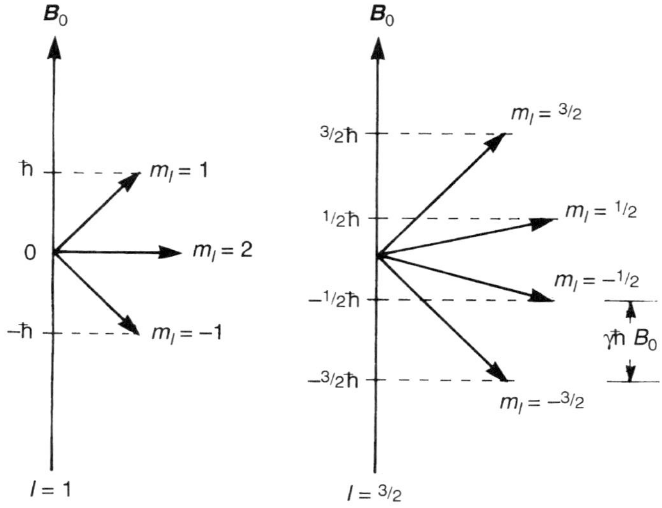

Figure 8.1 The permitted orientations of I for nuclei of spin 1 and $\frac{3}{2}$ showing the corresponding values of $m_l\hbar$

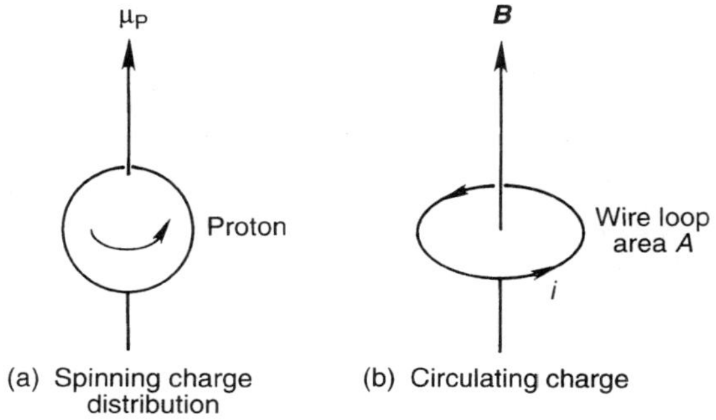

Figure 8.2 (a) The spinning electric charge distribution of a proton generates a magnetic moment $\boldsymbol{\mu}_P$. (b) The equivalent circulating charge model of a proton where a wire loop of area A carrying a current i generates a magnetic field $\boldsymbol{B}$ perpendicular to the plane of the one-turn coil, and a magnetic moment with magnitude $\mu_P = iA$.

view is that of Figure 8.2(b), where the circulating charge is imagined flowing in a conducting loop. From a classical point of view a current i flowing in a loop of area A generates a classical magnetic moment with magnitude given by

$$\mu_p = iA \tag{8.3}$$

The orientation of the magnetic moment vector is perpendicular to the plane of the loop and along the direction of the $\boldsymbol{S}$ vector. In fact, the magnetic moment is directly proportional to $\boldsymbol{S}$ so that, for a proton of mass m_p, we can write

$$\boldsymbol{\mu}_p = \gamma \boldsymbol{S} = \frac{e}{2m_p}\boldsymbol{S} \tag{8.4}$$

The constant of proportionality γ is known as the gyromagnetic (or magnetogyric) ratio and has the units Hz T^{-1}. For nuclei generally, the appropriate constant of proportionality would be $Ze/2m_N$, where m_N is the mass of the nucleus and Z its charge number. Usually though, the nuclear magnetic moment is written

$$\mu = g_N \mu_N S/\hbar = g_N \mu_N [I(I+1)]^{1/2} \tag{8.5}$$

where $\mu_N = e\hbar/2m_p$ (the nuclear magneton), and the difference in mass from that of the proton, together with quantum mechanical effects, are dealt with by including them in the nuclear g factor, g_N. The magnitude of μ_N is 5.0505×10^{-27} J T^{-1}. Electron magnetic moments are also quantized, but now in units of $e\hbar/2m_e$, where m_e is the electron mass.

This unit, which will be written μ_B, is an important quantity known as the Bohr magneton, and equals 9.2732×10^{-24} J T^{-1}.

It might be thought that, since all nuclei are positively charged, all nuclear g factors (g_N) must be positive. However, this is not so, and the reasons for this may be traced to the structure of the nucleus. The neutron itself has $I = 1/2$, but zero charge, and we might imagine that its magnetic moment would therefore be zero. However the neutron has structure (quarks) and we actually find a relatively large magnetic moment μ_n, where $\mu_n = -1.91\mu_N$. The negative g factor here implies that the spin magnetic moment is antiparallel to the angular momentum (as for the electron); positive values indicate that $\boldsymbol{\mu}$ and $\boldsymbol{S}$ are parallel. Inspection of Table 8.1 shows that ^{15}N and ^{17}O have negative γ, and hence negative nuclear g factors.

The energy of the interaction between a magnetic moment $\boldsymbol{\mu}$ and a magnetic field $\boldsymbol{B}_0$ is given by

$$E = -\boldsymbol{\mu} \cdot \boldsymbol{B}_0 \tag{6.6}$$

where E is the energy of the resulting state. If we identify the z axis with the direction of the static magnetic field, then the allowed energy levels of the system are

$$E = -\gamma S_z B_0 = -\gamma m_I \hbar B_0 \tag{8.7}$$

The number of these states is equal to $2I + 1$ and their separation is

$$\Delta E = \gamma \hbar B_0 = \mu_B B_0 \tag{8.8}$$

We know that for a system of atoms, the presence of well-defined energy levels, such as are depicted in Figure 8.1, can be detected by the resonant absorption of radiation. We expect, therefore, that if the nuclear system is subject to a time-varying electromagnetic field, it should be possible to induce transitions between these energy levels. For radiation of angular frequency ω_0 we have the relationships

$$\hbar\omega_0 = \Delta E = \gamma \hbar B_0 \tag{8.9}$$

or

$$\omega_0 = \gamma B_0 \tag{8.10}$$

We note that Planck's constant has disappeared from the final result, indicating that it is accessible via a classical route, we shall see later that this is so. All $2I$ allowed transitions for a spin I nucleus have the same energy, but the local magnetic field experienced by a nucleus in a molecule differs slightly from the external field so that the exact resonance frequency is characteristic of the chemical environment of the nucleus.

Figure 8.1 illustrates the situation for nuclei of spin 1 and spin $\frac{3}{2}$. A proton (spin $\frac{1}{2}$) is permitted to occupy only two states, spin-up or spin-down, and it can occupy either of these states with a probability which varies from observation to observation. In the absence of a magnetic field $\boldsymbol{B}_0$, spin-up and spin-down cannot be defined. In the presence of a magnetic field $\boldsymbol{B}_0$, the spins will be distributed between the two states and, at equilibrium, the spin-up state, the lower energy state, will be favoured (Figure 8.3). The population distribution can be described by the Boltzmann distribution, where

$$\frac{\text{spin-up}}{\text{spin-down}} = \frac{n_{+1/2}}{n_{-1/2}} = \exp(\Delta E/k_\text{B} T_\text{s}) \tag{8.11}$$

Here k_B is the Boltzmann constant and T_s is the temperature of the spin system. At equilibrium, as in Figure 8.3(a), T_s is just the sample temperature and a relatively small excess of protons are in the spin-up state. This situation means that NMR has a relatively low sensitivity when compared, say, with radioisotope measurements. Inspection of equation 8.11 shows that since ΔE is fixed, changing the population difference between two states must be equivalent to a change in T_s. Thus T_s will increase if the number of spin-down $n(-\frac{1}{2})$ states is increased: T_s will return to its equilibrium value once the equilibrium population distribution is restored. For large assemblies of independent nuclei we can treat their behaviour classically and it is convenient to do this. We have already seen from equation 8.10 that the radiation angular frequency required to cause transitions between spin states

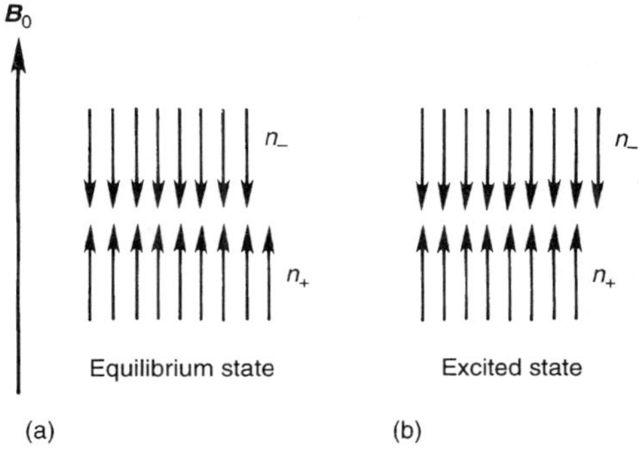

Figure 8.3 The distribution of spin $\frac{1}{2}$ nuclei between two states (a) at equilibrium and (b) following an RF pulse which results in an excess of nuclei in spin-down states.

must be accessible through a classical model. We will now develop the principles of NMR using a classical viewpoint.

8.3* NUCLEAR MAGNETIC MOMENT BEHAVIOUR

In an external magnetic field $\boldsymbol{B}_0$, a couple $\boldsymbol{C}$ will be exerted on a magnetic dipole moment causing the angular momentum to change at a rate equal to the couple or torque,

$$\boldsymbol{C} = \boldsymbol{\mu}_p \times \boldsymbol{B}_0 = d\boldsymbol{S}/dt \tag{8.12}$$

This effect is analogous to that of a gravitational field on a spinning top, and the magnetic dipole precesses about the applied magnetic field. If we substitute for $\boldsymbol{S}$ from equation 8.4 we obtain

$$\frac{d\boldsymbol{\mu}_p}{dt} = \gamma\boldsymbol{\mu}_p \times \boldsymbol{B}_0 \tag{8.13}$$

Equation 8.13 is the Larmor equation which describes the precession of $\boldsymbol{\mu}_p$ about $\boldsymbol{B}_0$ with angular velocity ω_0, the Larmor angular frequency, where $\omega_0 = \gamma B_0$ as in equation 8.10. If the magnetic dipole is orientated in the same direction as $\boldsymbol{B}_0$, no couple will be exerted on it and precession will not occur. On the other hand a magnetic dipole orientated at an angle ϕ to the magnetic field will precess, maintaining the angle ϕ. Figure 8.4 depicts the situation diagrammatically.

The discussion above is focused upon a single spinning proton; a typical sample will contain many nuclei so that we have to work with the net magnetic moment, comprising the vector sum of all the nuclear magnetic moments. While a full description of single spins should really be couched in terms of quantum mechanics, classical theory is adequate to describe fully the behaviour of a large ensemble.

In an applied magnetic flux density $\boldsymbol{B}_0$, the net magnetic moment $\boldsymbol{M}$ at equilibrium will be aligned with $\boldsymbol{B}_0$ and cannot be measured directly. In order to measure $\boldsymbol{M}$ it must be tilted away from the z axis, or $\boldsymbol{B}_0$ direction, to yield a measurable component in the x–y plane, and this is the basis of NMR measurements.

Tilting is achieved by applying a magnetic flux density $\boldsymbol{B}_1$ orientated in the x–y plane, and rotating at the Larmor angular frequency ω_0. The magnetic moment $\boldsymbol{M}$ will then experience a torque moving it through an angle α from the $\boldsymbol{B}_0$ axis. The value of α will depend upon the magnitude and duration of $\boldsymbol{B}_1$; the flux density $\boldsymbol{B}_1$ will have a frequency in the radiofrequency range. Figure 8.5(a) shows how the magnetic moment spirals away from the $\boldsymbol{B}_0$, or z axis, on application of a transverse flux

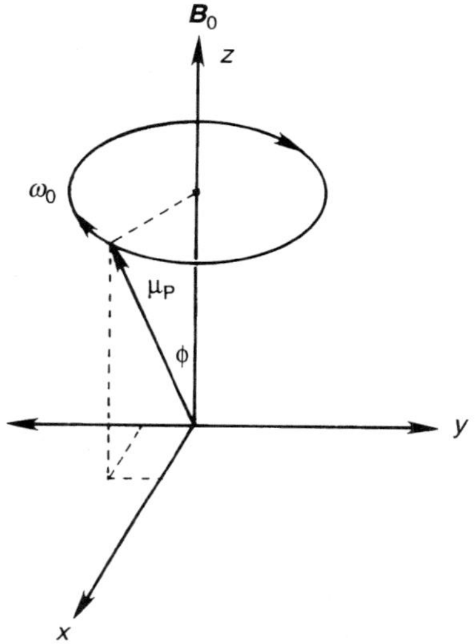

Figure 8.4 Interaction of a nuclear magnetic moment $\boldsymbol{\mu}_P$ with a magnetic flux density $\boldsymbol{B}_0$ results in a couple acting to produce precession of the moment about $\boldsymbol{B}_0$ with angular velocity ω_0.

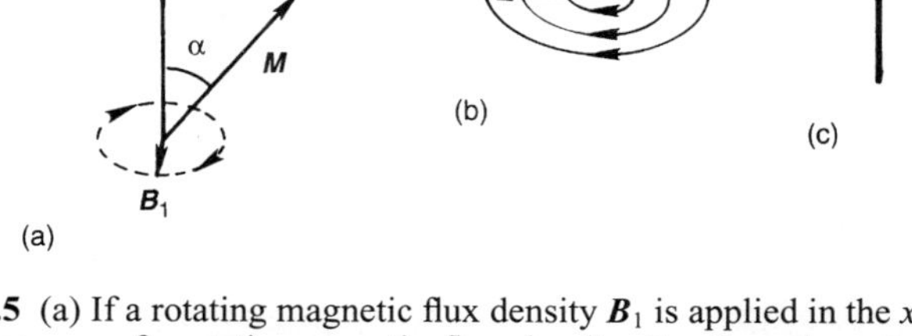

Figure 8.5 (a) If a rotating magnetic flux density $\boldsymbol{B}_1$ is applied in the x–y plane, in the presence of a static magnetic flux density B_0, then the net magnetic moment experiences a couple which gives rise to precession at a final angle α. (b) The horizontal component of the net magnetic moment builds up as shown. (c) In the absence of the rotating magnetic flux density, the net magnetic moment is directed essentially along $\boldsymbol{B}_0$.

density, $\boldsymbol{B}_1$, yielding a horizontal component $\boldsymbol{M}\sin\alpha$ which builds up with time in the fashion shown in Figure 8.5(b). If the rotating magnetic field is applied for a time t, $\boldsymbol{M}$ will rotate through an angle

$$\alpha = \omega_0 B_1 t \tag{8.14}$$

When $\alpha = \pi/2$, $\boldsymbol{M}$ will be exactly in the x–y plane and we have applied what is known as a 90° pulse. Clearly α depends on both $\boldsymbol{B}_1$ and t. In principle we can choose to vary either quantity to give the desired angle. Since α also depends on the gyromagnetic ratio γ, the pulse power or duration required to give a 90° tilt will also depend on the particular nucleus under observation. Figure 8.6(a) shows the effect of a 90° pulse. When $\boldsymbol{M}$ has been turned through 90°, the magnetization vector will continue to precess about the z axis in the x–y plane, so that the component of $\boldsymbol{M}$ in the z direction, M_z, will be zero. The components M_x and M_y in the x and y directions will oscillate about zero at the Larmor frequency with amplitude M. This rotating magnetization will induce an RF signal (at the Larmor frequency) in a detector coil whose axis lies in the x–y plane. The resultant signal, known as the free induction signal, does not persist indefinitely.

8.4* RELAXATION EFFECTS

It might be thought that the horizontal component, $M\sin\alpha$, would simply decrease to zero as the z component of $\boldsymbol{M}$ recovers its original value. In fact the situation is much more complex than that and the horizontal component decays through field inhomogeneities and spin–spin interaction. These have the effect of dephasing individual magnetic moments so that the net horizontal component of the magnetic moment decreases with time, the so-called free induction decay (FID), the decay rate being characterized by the transverse relaxation time T_2. Figure 8.6(b) shows the nature of this free induction decay signal and its decay envelope. It depends on the local environment of the nuclei. Figure 8.6(c) shows the dephasing of individual magnetic moments, the dashed moments moving more slowly than ω_0, the dot-dashed moments moving faster.

During the decay of the horizontal component of $\boldsymbol{M}$, the vertical component M_z grows from zero to its maximum value. This process occurs with a different (generally longer) time constant T_1 and by a different mechanism, which simply involves the transfer of the energy, originally supplied by the $\boldsymbol{B}_1$ pulse, from the precessing magnetic moments to the surrounding lattice. The heating effect of this transfer is

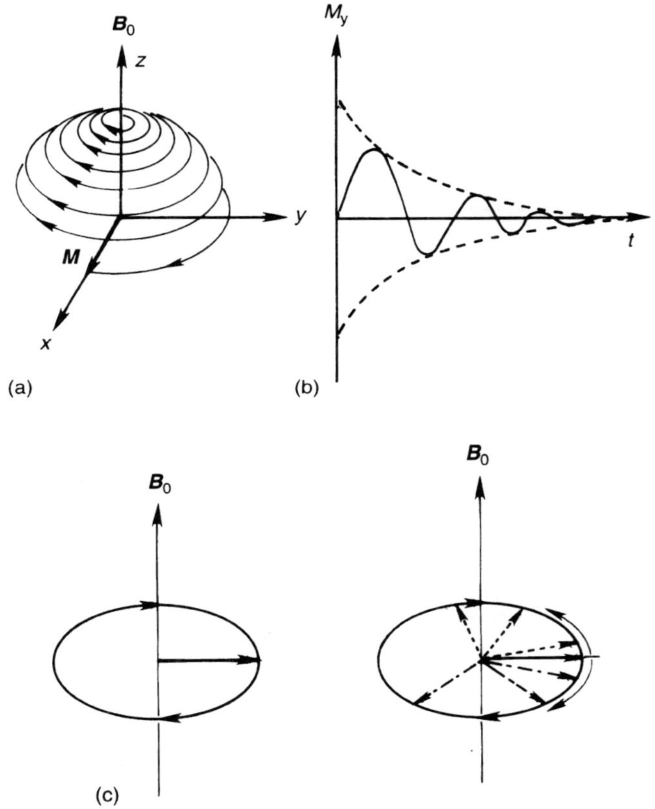

Figure 8.6 (a) Effect of a 90° pulse, whereby the rotating magnetic field $\boldsymbol{B}_1$ causes the net magnetic moment to spiral down the surface of an imaginary sphere during the pulse. (b) The resulting free induction decay (FID) signal measured here in the y direction, the sample magnetization following termination of the 90° pulse. The decay results from the dephasing of the individual horizontal components of $\boldsymbol{M}$. (c) Dephasing of spins with time showing the transition to spins rotating more slowly than ω_0 (dashed lines), and more rapidly (dot-dashed lines).

negligible. In consequence T_2 is known as the spin–spin relaxation time, while T_1 is the spin–lattice relaxation time or transverse relaxation time. Figure 8.7 illustrates the different decay characteristics displayed by these two relaxation mechanisms.

An alternative way of understanding or viewing these processes is to transform to a co-ordinate system that rotates about the z axis at the

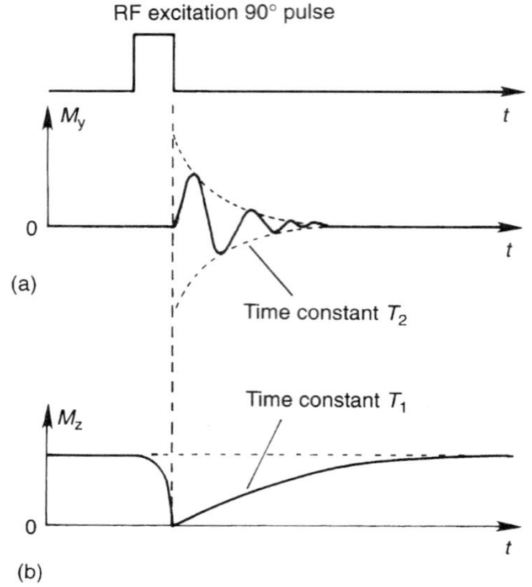

Figure 8.7 Contrast between the free induction decay (FID) following a 90° pulse in the x direction, which decays with a characteristic time constant T_2, and the restoration of the full vertical component of $\boldsymbol{M}$, which occurs with a time constant T_1. The mechanisms responsible for the two processes are different.

Larmor frequency. In the rotating reference frame, the magnetisation vector 'sees' a constant field $\boldsymbol{B}_1$ pointing in some fixed direction in the new co-ordinate system rotating in the x–y plane. In the rotating co-ordinate system the magnetization vector now precesses about this field. As it does so its z component gradually falls to zero, becomes negative, and then increases again to its original value. The rate of this precession depends on the strength of $\boldsymbol{B}_1$ through the equation $\omega_1 = \gamma B_1$. Figure 8.8 shows the relationship between the laboratory reference frame and the rotating reference frame $x'y'z'$. In the rotating frame $\boldsymbol{B}_1$ is arbitrarily aligned with the y' direction.

8.5* THE BLOCH EQUATIONS

For our system the net magnetization is due to the imbalance of up and down spins, where up and down refer to the z component of the spin vector. Since each spin carries an intrinsic magnetic moment $\mu = \gamma\hbar m_l$, the z component of magnetization is obtained by summing over all the nuclear moments, n, with the result that

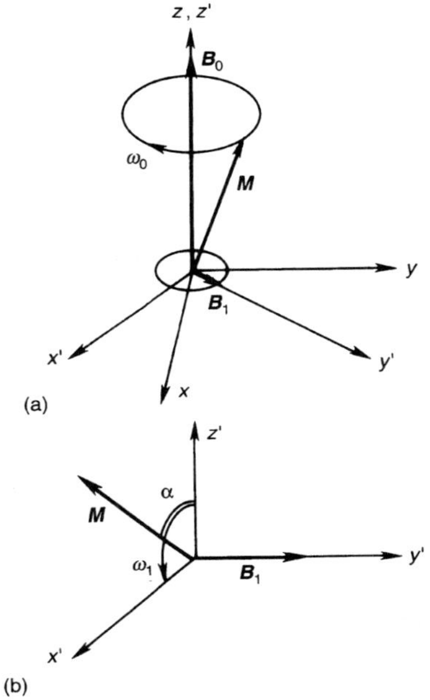

Figure 8.8 (a) Effect of a rotating magnetic field $\boldsymbol{B}_1$ on the magnetization vector $\boldsymbol{M}$ in the laboratory reference frame where it precesses about $\boldsymbol{B}_0$ with angular frequency ω_0. (b) In the rotating frame of reference, $\boldsymbol{B}_1$ is stationary and does not precess about the $z = z'$ axis. It does precess about $\boldsymbol{B}_1$, i.e. the y' axis, at a rate ω_1 which depends on the strength of $\boldsymbol{B}_1$. The tilt angle α depends on $\boldsymbol{B}_1$ and the time t for which it is applied.

$$M_z = \gamma \hbar m_l \tag{8.15}$$

We now follow Bloch and co-workers (1946) and write the appropriate version of the Larmor equation as

$$\frac{\mathrm{d}\boldsymbol{M}}{\mathrm{d}t} = \gamma \boldsymbol{M} \times \boldsymbol{B}_0 \tag{8.16}$$

Under the action of spin–lattice relaxation, that is the transfer of the magnetic excitation energy to the surrounding lattice, the rate of change of n is given by

$$\frac{\mathrm{d}n}{\mathrm{d}t} = \frac{n_0 - n}{T_1} \tag{8.17}$$

where n_0 is the value of n at thermal equilibrium, so that the z

component of $\boldsymbol{M}$ must obey a similar equation, namely

$$\frac{\mathrm{d}M_z}{\mathrm{d}t} = \frac{M_0 - M_z}{T_1} = -\frac{(M_z - M_0)}{T_1} \tag{8.18}$$

If the net nuclear magnetization has a component in the x–y plane, that is perpendicular to the z direction, this transverse magnetization will decay due to interactions with local spins which cause the magnetic moments to dephase according to the equations

$$\frac{\mathrm{d}M_x}{\mathrm{d}t} = -M_x/T_2 \tag{8.19}$$

$$\frac{\mathrm{d}M_y}{\mathrm{d}t} = -M_y/T_2 \tag{8.20}$$

where T_2 is the transverse relaxation time. We can combine the above equations for motion due to relaxation and motion due to the free spins to yield

$$\frac{\mathrm{d}\boldsymbol{M}}{\mathrm{d}t} = \gamma \boldsymbol{M} \times \boldsymbol{B}_0 - \frac{(M_x \boldsymbol{i} + M_y \boldsymbol{j})}{T_2} - \frac{(M_z - M_0)\boldsymbol{k}}{T_1} \tag{8.21}$$

where $\boldsymbol{i}, \boldsymbol{j}$ and $\boldsymbol{k}$ are unit vectors in the laboratory frame. We have here assumed that the motion due to relaxation can be superimposed on the motion of the free spins under the influence of a static field and a much smaller RF field. Before looking in greater detail at T_1 and T_2, it is instructive to break off briefly to examine the basic requirements for NMR measurements.

8.6 OBSERVATION OF NMR

The basic arrangement required for observing NMR is shown schematically in Figure 8.9. It comprises a main magnet (which may be a large electromagnet or, more likely, a superconducting solenoid), the poles of which carry modulation coils. Within the magnet gap is placed an RF coil surrounding the sample. Searching for resonance can be accomplished either by varying the frequency of the RF current that generates the magnetic field $\boldsymbol{B}_1$, while holding $\boldsymbol{B}_0$ fixed, or by varying $\boldsymbol{B}_0$ while keeping $\boldsymbol{B}_1$ fixed. The latter approach can be implemented by augmenting the steady magnetic field with a smaller magnetic field due to current passed through modulation coils wound around the main magnet poles. The RF coil around the sample may then be used, both to excite the resonance condition, and also to detect the resonance. Alternatively, a separate sensing coil, usually orientated in the x–y plane, at 90° to the

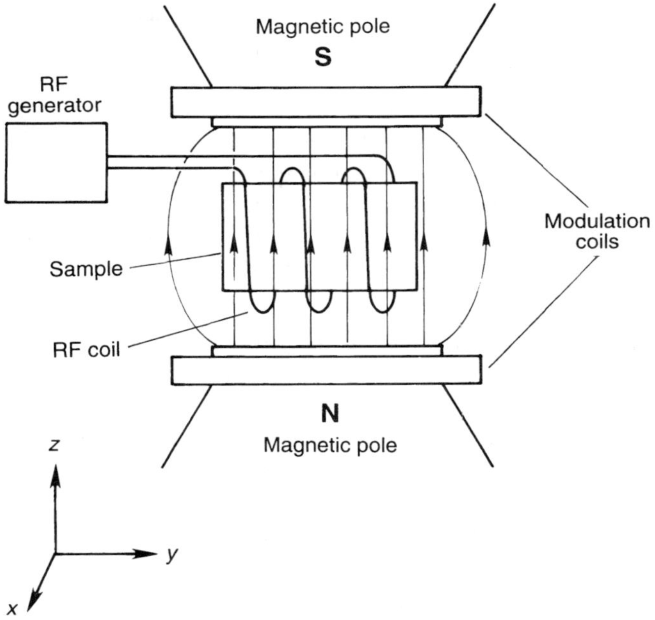

Figure 8.9 Schematic diagram showing the arrangement of sample, main magnet, modulation coils and RF coil for NMR measurements.

RF coil, may be used. Before we consider these techniques more fully we need to return to the processes occurring during the resonance condition, in particular we need to explore T_1 and T_2 in more detail.

8.7* NUCLEAR MAGNETIC RESONANCE PULSE SEQUENCES

If we start with a nuclear spin system in thermal equilibrium in a field $\boldsymbol{B}_0$, the magnetization vector $\boldsymbol{M}$ will be almost exactly along the z direction, precessing about $\boldsymbol{B}_0$ at the Larmor frequency. An RF field of frequency $\omega_0 = \gamma B_1$, applied in the x–y plane, will rotate the magnetization vector away from the z direction, as we have seen previously. If we choose to turn off this RF field when $\boldsymbol{M}$ has been turned through 90°, the magnetization vector will continue to precess about the z axis. When this condition has been established, the component of $\boldsymbol{M}$ in the z direction, M_z, will be zero. The components M_x and M_y in the x and y directions will oscillate about zero at the Larmor frequency with amplitude M. This rotating magnetization will induce an RF electric field in a detector coil whose axis lies in the x–y plane. The resulting signal, the free induction signal, does not persist indefinitely, but decays because of field inhomogeneities and spin-spin interactions. The decay envelope is described by T_2^*, the effective spin–spin relaxation time. It turns out that we can decompose this quantity to distinguish between the 'true' spin–spin relaxation time T_2, the contribution T_{2f} due to field inhomogeneities, and that due to magnetic susceptibility effects T_{2m}. The term T_{2f} will be given by $\gamma \Delta B_0/2$ where ΔB_0 is the variation in the magnetic field due to inhomogeneities. We can write

$$1/T_2^* = 1/T_2 + 1/T_{2f} + 1/T_{2m} \tag{8.22}$$

Following a 90° pulse the nuclear magnetic moments will dephase as they precess about the z axis owing to local static field inhomogeneities. Let us assume that $T_1 \gg T_2$ and that the free induction signal decays to zero in a time Δt. If we apply a 180° pulse, that is a pulse which inverts every magnetic moment, then the nuclear magnetic moments will continue to precess about the z axis, in the same direction. Those spins which precessed too rapidly during the time interval between $t = 0$ and $t = \Delta t$ because their local magnetic field was greater than $\boldsymbol{B}_0$ will still precess too rapidly, only now starting from behind the other magnetic moments. Thus, after an additional time Δt, all spins will be back in phase. An identical argument applies to those spins which precessed too slowly (Figure 8.10(a)). Experimentally we will see an exponential increase in the free induction signal, followed by an identical, or mirroring, decay; this signal is the *spin-echo*. Figure 8.10(b) depicts this process, while Figures 8.10(b) and (c) show the individual spin behaviour. In the limit that T_1 and T_{2s} are infinitely large, the peak amplitude of the spin-echo will be the same as that of the original signal. The spin-echo procedure allows us to account for the spatial inhomogeneity of the static magnetic field $\boldsymbol{B}_0$, while the second half of the echo, which has the same form as the free induction decay, can be detected and Fourier transformed to give a spectrum containing NMR lines whose amplitudes are independent of field inhomogeneities. It should be noted that the spin-echo procedure does not counter the rephasing which arises from the fluctuating magnetic fields associated with molecular motion, the third term in equation 8.22, since these are random and not 'refocused' by the 180° pulse. We can determine T_2 since the peak value $I(\Delta t)$ of each successive pulse in the pulse train will follow the equation.

$$I(2\,\Delta t) = I(0)\exp(-2\,\Delta t/T_2) \tag{8.23}$$

This is the large dashed profile in Figure 8.10(a).

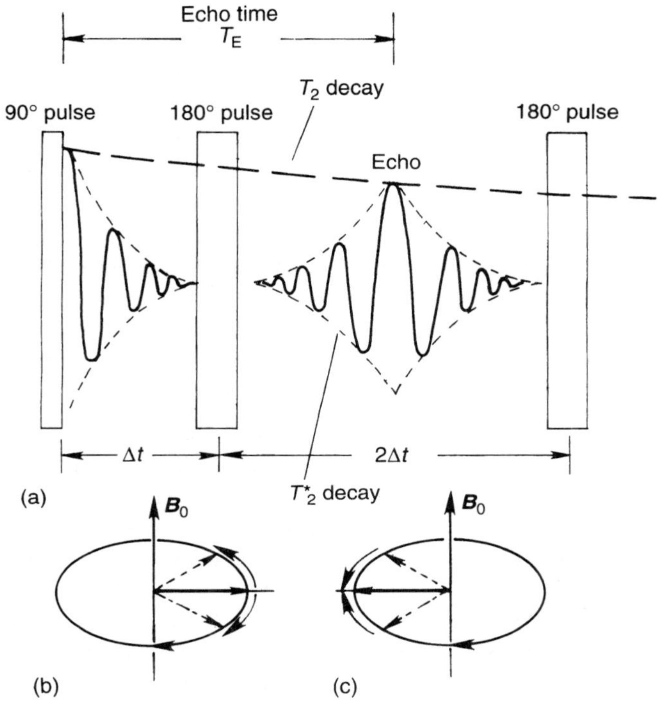

Figure 8.10 (a) Measurement of the true spin–spin relaxation time T_{2s} using a 90° pulse followed by 180° pulses to generate a spin-echo (only one 180° pulse is shown here). The true decay constant T_2 is obtained from the overall pulse train envelope (large dashes) while the true spin–spin relaxation time T_{2s} is obtained from the individual spin-echo envelopes in the frequency domain. (b) Dephasing of spins after 90° pulse. (c) Rephasing of spins by 180° pulse.

The spin–lattice time constant T_1 can also be determined using pulse sequences, only now it is sufficient to apply 90° pulses at a time interval Δt, where $\Delta t \simeq T_1$. The free induction signals following two 90° pulses are shown in Figure 8.11. Adjustment of the value of Δt allows one to determine the dependence of the amplitude of the free induction signal on time, since if $\Delta t \gg T_1$, M_z will have recovered from zero to its full value. As Δt is shortened the extent of recovery of M_z will be reduced, enabling the dashed curve of Figure 8.11 to be produced. If a series of pulses are applied at equal intervals, a steady state will be set up. The signal in this steady state is given by

$$M_z = M_0[1 - \exp(-\Delta t/T_1)] \tag{8.24}$$

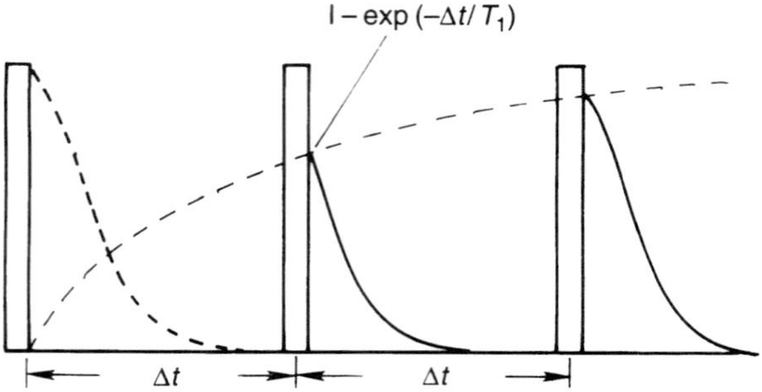

Figure 8.11 The free induction signals following three 90° pulses. The amplitude of succeeding pulses is proportional to the recovery of M_z due to spin–lattice relaxation in the interval between pulses, leading to a value for T_1.

where M_0 is the maximum value of $\boldsymbol{M}$ and Δt is the interval between pulses. This effect is known as saturation and can be used as a source of contrast in MRI. The spin–spin relaxation process can also be used as a source of contrast in MRI. The pulse-echo sequence leading to a spin-echo is the basis for many of the current imaging sequences and is thus important. It should not be assumed that this discussion of pulse sequences is exhaustive. To give an idea of the size of these time constants, for protons in biological materials T_1 is approximately 300–1000 ms and T_2 is approximately 10–500 ms.

8.8 THE CHEMICAL SHIFT AND NMR SPECTRA

So far we have treated the nuclei as isolated systems, although we did recognize the need for some interaction with the environment to allow for the attainment of thermal equilibrium. In fact, each nucleus is surrounded by atomic electrons, and the configuration of the valence electrons depends critically on the chemical composition. These same electrons are affected by the applied magnetic field and, in turn, their response influences the magnetic field at the nucleus. In saturated chemical compounds electron spins are paired off; in this case the application of a magnetic field modifies the electronic wavefunctions, giving rise to diamagnetism. From a classical point of view, we say that the magnetic field induces currents according to Faraday's law, which act to oppose that field. This magnetic coupling of the nucleus with the atomic electrons produces the chemical shift. Expressed slightly differently, we can say that the field $\boldsymbol{B}_{\mathrm{n}}$ actually experienced by a nucleus in an atom or molecule differs slightly from the field $\boldsymbol{B}_0$, that is the field which would be felt by a bare nucleus, stripped of its electrons. In an

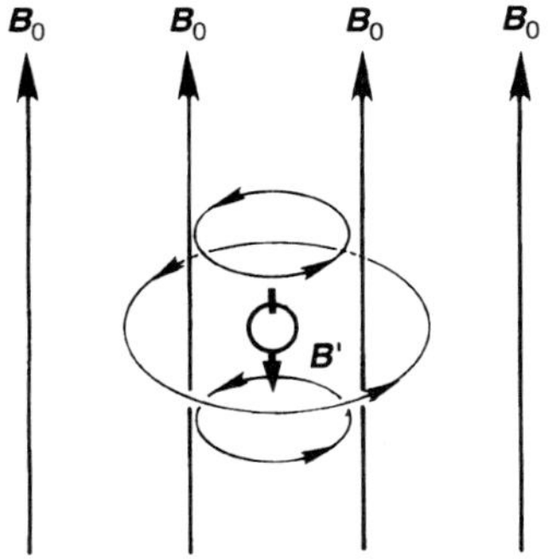

Figure 8.12 An applied magnetic field $\boldsymbol{B}_0$ causes the electrons in an atom to circulate within their orbitals. This motion generates an additional field $\boldsymbol{B}'$ at the nucleus which opposes $\boldsymbol{B}_0$. (The arrows on the loops indicate the direction of electron motion, opposite to the current directions.)

atom the field is slightly smaller than $\boldsymbol{B}_0$ because the external field causes the electrons to circulate within their atomic orbitals. This induced motion, much like an electric current passing through a coil of wire, generates a small magnetic field $\boldsymbol{B}'$ aligned in the opposite direction to $\boldsymbol{B}_0$. Figure 8.12 shows this effect. We can say that the nucleus is *shielded* from the external field by its surrounding electrons, so that

$$\boldsymbol{B}_{\mathrm{n}} = \boldsymbol{B}_0 - \boldsymbol{B}' = \boldsymbol{B}_0(1 - \chi) \qquad (8.25)$$

where χ is the constant of proportionality between $\boldsymbol{B}'$ and $\boldsymbol{B}$, known as the *shielding constant* or *screening constant*; it is a tensor quantity. $\boldsymbol{B}'$ is proportional to $\boldsymbol{B}_0$ since the stronger the external field, the more the electrons are impelled to move. $\boldsymbol{B}'$ is typically 10^4–10^5 times smaller than $\boldsymbol{B}_0$ and as a result of this nuclear shielding, the resonance condition of equation 8.10 is modified to

$$\omega_0 = \gamma B_0(1 - \chi) \qquad (8.26)$$

This tells us that the resonance frequency of a nucleus in an atom is slightly lower than that of a bare nucleus, stripped of all its electrons.

Similar effects occur for nuclei in molecules, except that the motion of the electrons is rather more complicated than that in atoms and the induced fields may *oppose* or *augment* the external field. The size and sign of the shielding constant in equation 8.25 is determined by the electronic structure of the atoms surrounding the nucleus; the resonant frequency of a nucleus is therefore characteristic of its environment. The shielding constant is actually an inconvenient measure of the 'chemical shift', and we shall adopt a different measure.

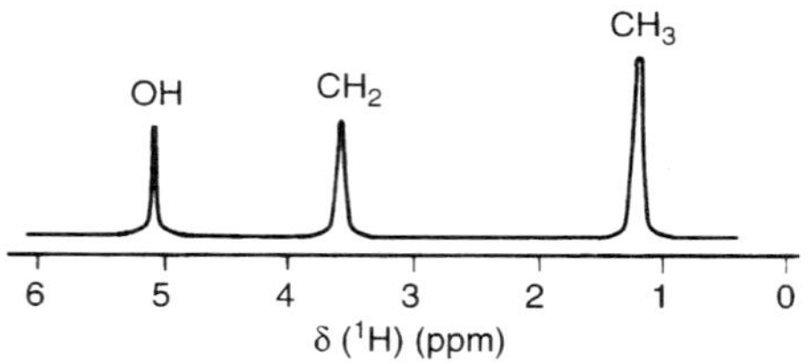

Figure 8.13 Chemical shift spectrum for ethanol showing the effect of the different environments on the three types of proton.

A simple, and often quoted, example of chemical shifts is the proton resonance pattern in ethyl alcohol, CH_3CH_2OH. Figure 8.13 shows the resonance lines obtained from this material, where the integrated areas under the peaks are in the ratio 3 : 2 : 1. The strongest line is due to the three equivalent protons of CH_3, the next strongest line is due to the two equivalent protons of CH_2 and the weakest line is due to the single proton of the OH group. These shifts in resonance frequency reflect the local conditions where the electron cloud of the molecule produces a small additional magnetic field. We shall see in a later section that we can develop a spatially localized spectroscopy based on the chemical shift so it is important to quantify it. The chemical shift δ (a tensor) is defined with respect to a reference frequency f_{ref}, and is a dimensionless quantity. It is given by

$$\delta = \frac{(f_{\text{sample}} - f_{\text{ref}}) \times 10^6}{f_{\text{ref}}} \tag{8.27}$$

δ is expressed in parts per million (ppm) and is independent of frequency. The chemical shift characteristics of atoms in many molecules have been tabulated and in some instances calculated; Table 8.1 records some values for nuclei of medical interest.

To see how δ relates to χ, equations 8.26 and 8.27 can be combined to give

$$\delta = 10^6(\chi_{\text{ref}} - \chi)/(1 - \chi_{\text{ref}}) \tag{8.28}$$

We see from this that δ is actually a deshielding parameter.

The NMR signal detected following an excitation pulse is a function of time. In the simple case of one type of nucleus in a uniform magnetic field, it is a single, exponentially decaying signal (the FID) whose frequency depends on the nucleus being studied. If there are several groups of nuclei within the sample in different local magnetic fields, as will be the case for MRI, then the signal is more complex. In this case

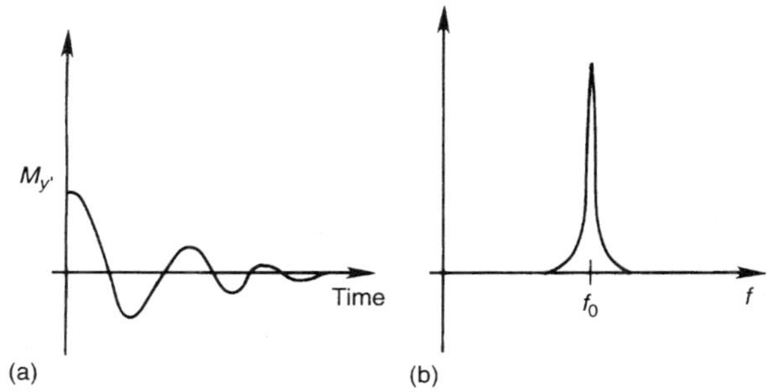

Figure 8.14 Fourier transform of the real component (a) of the FID signal in the time domain, leading to a single Lorentzian peak (b) in the frequency domain.

the signal is the sum of the signals from the nuclei in these different fields precessing at their own frequency and decaying with their own relaxation rate. Such a complex signal is difficult, if not impossible, to interpret. However, if we transform the FID signal to the frequency domain by means of a Fourier transform, a great simplification is achieved. The exponential FID signal transforms to a Lorentzian line; the slower the decay the sharper the line. Figure 8.14 shows this transformation diagrammatically for the real part of the FID signal. Of course, for a more complex FID signal a family of lines would be obtained.

The NMR signal is also complex in the mathematical sense with real and imaginary components which can be thought of as the x and y components of the transverse magnetization. In order to detect the NMR signal fully we must use coil structures which will measure the two orthogonal components of the NMR signal separately, that is, we must record the signal in quadrature. If we use just one coil and detect only one component, we will still obtain a spectrum (and in MRI, an image), but we are not making the most effective use of the available data.

8.9 PRINCIPLES OF MAGNETIC RESONANCE IMAGING

We now turn to the practical details of what is probably the best medical imaging modality currently available. In MRI it is possible to observe not only the distribution of static parameters within a sample, such as the proton density, but also the dynamic parameters T_1 and T_2 which provide information about the environment of the nuclei.

Nothing that we have considered so far provides the localization necessary for imaging the actual spatial distribution of nuclei. Indeed, with the uniform flux densities implicit in our description up to now, there would be no spatial discrimination at all in the detected signal: nuclei would respond, regardless of their position, in the entire volume exposed to the magnetic flux density. Since the radiofrequency signals employed correspond to wavelengths in the range 10–100 m, we cannot contemplate any form of collimation such as is used in X-ray tomography. The key to imaging the magnetic resonance process was found by P.C. Lauterbur and independently by P. Mansfield and P.K. Grannell in 1973. They had the ingenious idea of superimposing linear flux density gradients on the main flux density $\boldsymbol{B}_0$. Then, for a given pulse frequency, only nuclei in a plane at a particular value of x, and perpendicular to x, will have the correct precession frequency to resonate with the pulsed radiation (the resonance frequency being proportional to field strength). A flux density gradient applied in different directions in the selected plane (a slice of finite width Δx, say) then gives a set of projections which can be used to reconstruct an image of the proton density in the slice. For a field gradient G_x superimposed on the main magnetic field, the Larmor frequency will depend on position along the x axis as shown by equation 8.29

$$\omega_\mathrm{a} = \gamma(B_0 + xG_x) \tag{8.29}$$

If we consider a phantom structure comprising the two tubes of water shown in section in Figure 8.15, then the signal obtained, after detection and Fourier transformation to the frequency domain, is a single line at the Larmor frequency. In the presence of a field gradient G_x, however, each tube will have a different Larmor frequency

$$\omega_\mathrm{a} = \gamma(B_0 + x_\mathrm{a} G_x) \tag{8.30}$$

$$\omega_\mathrm{b} = \gamma(B_0 + x_\mathrm{b} G_x) \tag{8.31}$$

and the spectrum will consist of two lines, separated in frequency by $\gamma(x_\mathrm{a} - x_\mathrm{b})G_x$.

In order to construct an image of the two tubes it is necessary, of course, to have more information than a single projection onto the x axis. We can obtain a second projection by imposing a y direction gradient. However, for complex structures, more information is necessary. One way round this problem, the way originally used by Lauterbur, is to use a combination of x and y gradients ($G_x \cos\theta + G_y \sin\theta$) to produce a series of projections as a function of θ, and to reconstruct an image by the filtered backprojection approach used in X-ray computed tomography (CT). This approach is rarely used now.

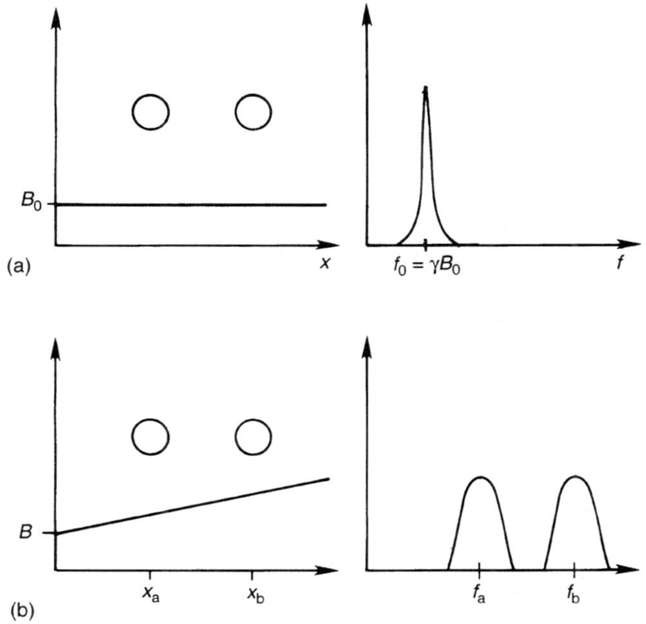

Figure 8.15 (a) The single Larmor frequency obtained from two tubes of water (viewed end on) in a uniform magnetic field $\boldsymbol{B}_0$. (b) The two Larmor frequencies obtained when a field gradient is superposed on the static field.

The imaging methods used currently originate from the ideas of R.R. Ernst and co-workers (1975) on NMR Fourier zeugmatography and, as the name implies, are based on the properties of Fourier transforms. If we collect a series of n data points in the presence of an x gradient the resulting frequency encoded signals may be transformed to produce a series of n spatially encoded data points. For a planar image we also need spatial information concerning the y axis. We can do this using the signal phase ϕ, which we can express mathematically as

$$\phi = \gamma y \int_0^t G_y(t)\,\mathrm{d}t \tag{8.32}$$

where G_y is the y direction magnetic field gradient. If we consider two cosine waves of equal amplitude, but slightly different frequency, that is $\cos(\omega_1 t)$ and $\cos(\omega_2 t)$, then at time zero we have both signals in phase and $\phi = 0$. After a time interval τ the higher frequency signal will have 'got ahead' of the lower frequency signal and the two signals will be out of phase by

$$\phi = (\omega_1 - \omega_2)\gamma \tag{8.33}$$

or

$$\phi = \gamma G_y y \tau \tag{8.34}$$

and we have another variable which is a function of position. The way in which this technique is applied may be seen from the simplified waveforms of Figure 8.16, which illustrates phase encoding. The information-gathering period is split into two parts, relating to the y and x axes respectively. At the end of the first part, lasting for t_1 seconds, the relative phase of the signal depends on the gradient strength G_y and its

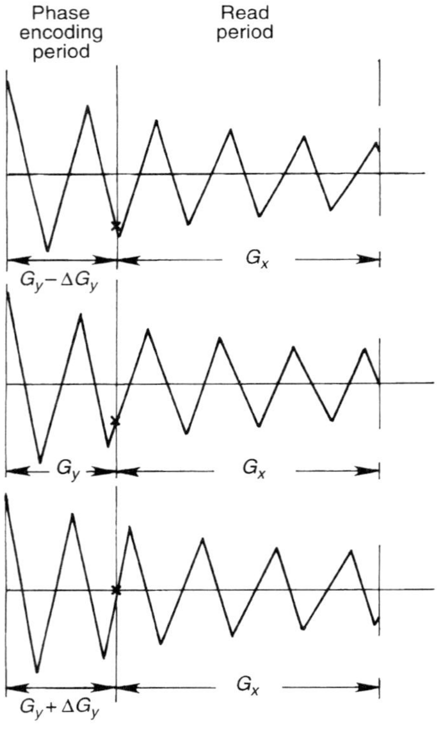

Figure 8.16 The principle of phase encoding. The value of the phase encoding gradient is incremented in each successive pulse sequence during the phase encoding interval before the start of the read period. The phase at the start of the read period is marked by a cross in each case.

position in the y direction. Thus the signal at the start of the data collection period, which follows immediately, is phase-shifted by an amount dependent on the position of the spins with respect to the y gradient. If we repeat the procedure for a different value of G_y (or equivalently for a different duration of the same gradient) we then get a set of n time domain points which differ from the first set in their initial phases. If we repeat the procedure for m different values of G_y, then we obtain a data set comprising m rows and n columns. The variable along the rows is the time for which the variable down the rows is in phase. A two-dimensional Fourier transform, both along the rows and down the columns, produces an image within the two axes defined by the gradients used.

To summarize, we have a system of 2-D Fourier imaging in which two magnetic field gradients are used but for different purposes, namely phase and frequency encoding. In this instance we have equated phase with the y axis and frequency with the x axis, although this assignment is not mandatory. Choice of variables depends on the technique used and the imaging plane. The alternative possibilities are summarized in Table 8.3. It will be convenient, in view of this lack of specificity, to refer in future to the phase encoding gradient as G_ϕ and the frequency encoding gradient, or read gradient, as G_ω.

8.10 GENERATING AN MR IMAGE

The basic procedure for generating an MR image may be summarized as:

1. excite the spin system;
2. apply a phase encoding gradient G_ϕ for a fixed time;
3. apply a read gradient G_ω and collect n data points;

Table 8.3 *Allocation of variables in Fourier transform MRI. Variables in brackets will often have very limited digitization*

Technique	*Selection*	*Phase*	*Frequency*
2-D Fourier transform (spin warp)	Slice (z)	y	x
Chemical shift selective	(z)	y, (δ)	x
Chemical shift specific	(z), (δ)	y	x
Chemical shift imaging	(z)	x, y	δ
Flow	(z)	y, (flow)	x
3-D Fourier transform	–	x, y	z

4. increment the value of G_ϕ and repeat steps 1, 2 and 3 m times (commonly 256);
5. perform a 2-D Fourier transform on the data to produce an $m \times n$ image.

It should be noted that all axes are equal in this procedure, since axes are defined by the orientation of the gradient coils. Consequently **sagittal** and **coronal** images, as well as any oblique set of axes, can be selected just as easily as the more familiar axial images we are accustomed to with X-ray CT.

The time required to obtain an MR image is largely determined by the repetition frequency adopted for the 90° pulse. In practice a few seconds must elapse between 90° pulses in order to avoid complete saturation, although steps 1, 2 and 3 themselves take only a few tens of milliseconds. For m phase encoding steps, steps 1, 2 and 3 are repeated m times and the total time required is mt_r, where t_r is the time between successive excitations. Of course, if averaging is used, this time will be increased.

In order to construct an image, a complex pulse sequence is required where the imaging pulse can be built up from a series of building blocks. These are combined together to produce a signal at the desired time which is weighted by the appropriate spin properties, T_1 and T_2. These blocks are:

1. a non-selective RF pulse
2. a selective RF pulse
3. a field gradient
4. the data collection period.

In fact, the two types of pulse, indicated under (1) and (2) above, form a Fourier transform pair and are depicted in Figure 8.17. The non-selective pulse is designed to excite all nuclei equally.

Field gradients are used to encode spatial information onto the NMR signal, but they actually have two effects on an NMR signal, the desired one stated above and an unwanted one of dephasing the signal; the two effects are intimately related. Following a 90° pulse the vector $\boldsymbol{M}$ representing the nuclear spins is pointing, say, along the y axis. The non-uniform field which is, of necessity, associated with a field gradient, will cause this vector to fan out in the x–y plane. The degree of dephasing gives the detectable spatial information, that is the pixels in the final image. If this dephasing is too great, the signal just decays beyond usefulness. If dephasing with a field gradient occurs early in a pulse sequence, we can restore the unwanted dephasing by a 'time

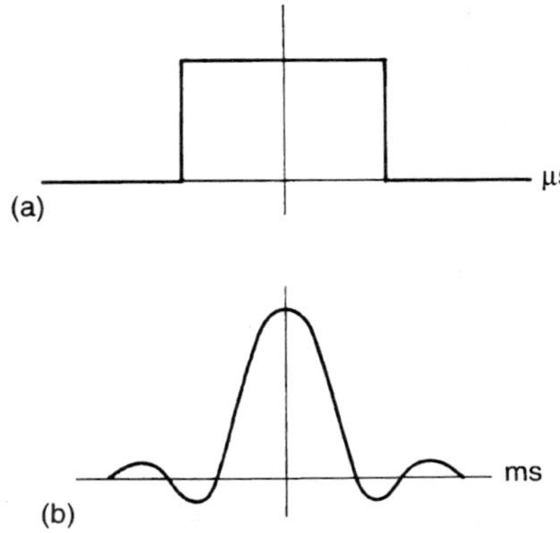

Figure 8.17 The Fourier transform pair which typify the RF pulse types used in MRI. (a) Rectangular non-selective pulse with short (μs) duration and large bandwidth. (b) Relatively long (ms), selective, sinc-function pulse with narrow bandwidth.

reversal' gradient, that is a gradient of equal magnitude, but opposite sign. Under the influence of a time reversal gradient the dephasing caused by the first gradient is reversed and the signal grows again; the net phase is zero, as the direction of dephasing with respect to time is reversed. The final element in the imaging sequence is the period when the signal is detected in the presence of a read gradient G_x. The NMR signal will now contain a mixture of frequencies whose range will depend on the size of the object and the amplitudes of the gradients used. If an object with a 200 mm diameter is located in the centre of a field having a gradient of 5×10^{-3} T, then equations 8.30 and 8.31 tell us that the frequencies in the NMR signal will be approximately ± 21 kHz about the basic frequency of the pulse. Nyquist's theorem tells us that we must sample this signal at approximately 40 kHz if we wish to record the signal without aliasing. The rate of data acquisition will be proportional to the product of the field gradient (T m^{-1}) and the field of view (m); the duration of the NMR signal will be of the order of ms. In practice, the data acquisition time is usually kept constant and the gradients adjusted to match the field of view. The pulse sequences employed in 2-D Fourier imaging are shown in Figure 8.18.

8.11 LOCALIZED SPECTROSCOPY (MRS)

In the early sections of this chapter we have examined the physics which forms the basis for NMR spectroscopy, and identified the sort of

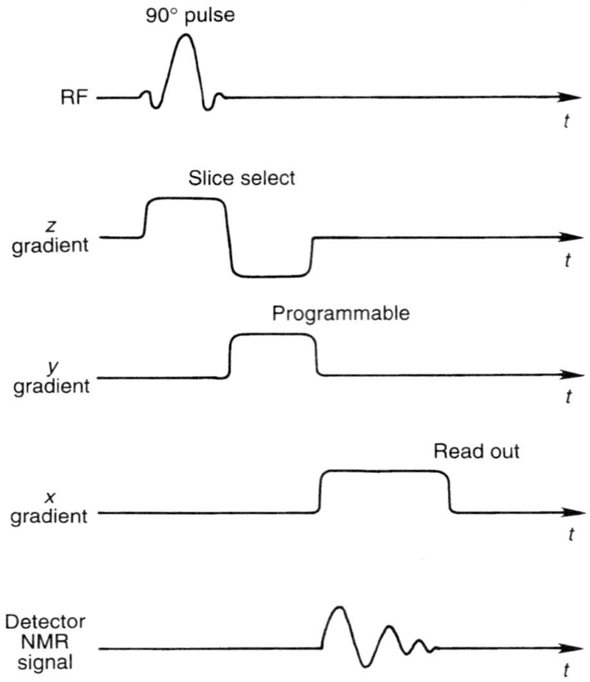

Figure 8.18 Two-dimensional Fourier imaging: the FID is acquired in the presence of a read gradient (in this case the x gradient) after a period of phase encoding in the presence of a y gradient. The duration of the phase encoding period is incremented at each repetition of the sequence.

information which may be obtained from these spectra, in particular the chemical shift. The chemical applications of NMR involve small, homogeneous samples, usually liquid; in contrast human beings are very heterogeneous in structure and large in extent. Consequently we are faced with the problems of:

1. localizing the source of the signal to a volume small enough to be considered homogeneous; and
2. establishing just where this volume is.

In principle, we can use a small surface coil to localize the region from which the signal originates; this has the advantage of simplicity. However, its volume selection capability is rather primitive and is restricted to regions near the surface. Despite these drawbacks it can be used to assist in localizing a source before other methods are applied. These other methods involve electronic localization and utilize three properties of the NMR system, the fields $\boldsymbol{B}_0$ and $\boldsymbol{B}_1$ and the field gradients.

By appropriate magnet design and the use of shim coils, it is possible to produce a base magnetic field which has at its geometric centre a spherical volume where the field is highly homogeneous, surrounded by a very inhomogeneous region. The NMR signals produced by such a magnet are confined to the homogeneous region since elsewhere the tissue signal will decay rapidly and can be filtered out. The limitation with this method is that the sensitive volume is only at the geometric centre of the magnet, so that positional freedom is largely removed. We may think of this technique as $\boldsymbol{B}_0$ field shaping.

The surface coil's apparent disadvantage of producing a $\boldsymbol{B}_1$ field which is inhomogeneous can be used as an additional method of localization if the coil is used in a transmit–receive mode, as opposed to the receive-only mode. The $\boldsymbol{B}_1$ field decreases with distance from the plane of the coil and so consequently does the flip angle produced by an excitation pulse.

The final method of localization utilizes field gradients, along with frequency selective pulses. This combination can be used in two ways. First, a specific volume may be excited without a field gradient, or second, multidimensional imaging may be used. For example, the simplest of the gradient localization techniques is depth-resolved spectroscopy (DRESS), where a plane parallel to the face of a surface coil is excited with a selective pulse in the presence of a gradient. This excitation, along with the approximately hemispheric sensitivity region of the surface coil, results in a spectrum originating from an approximately disk-shaped planar volume. The location of this plane can easily be established by obtaining a conventional image using the same coil.

An alternative way to use field gradients for localized spectroscopy is through multidimensional Fourier techniques where x, y and z form the three variables. A straightforward way to achieve this is through a chemical shift imaging sequence, where a slice is selected by the normal means, x and y information is phase encoded, and chemical shift information is frequency encoded. From this procedure images can be produced for a given chemical shift or spectra. Figure 8.19 shows the type of information which may be gathered in this way, together with a spectrum obtained from a region of interest (ROI).

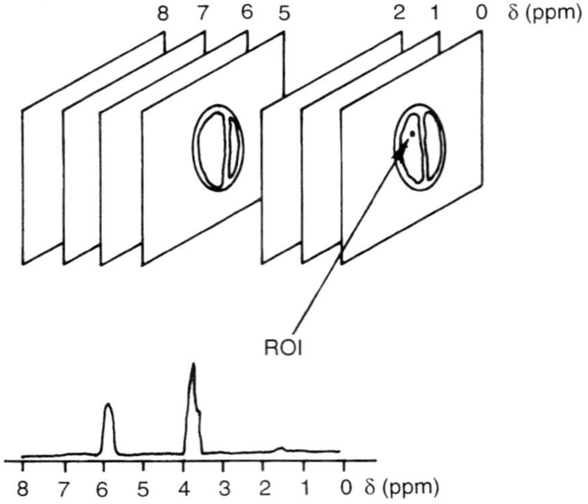

Figure 8.19 Chemical-shift imaging where an array of images is generated, each corresponding to a particular chemical shift. Alternatively, the data can be presented as spectrum from a particular region of interest.

8.12 THE MRI MAGNET

We now turn to the details of the instrumentation required to implement magnetic resonance imaging and magnetic resonance spectroscopy; it will comprise four main parts, of which the most important part is the magnet assembly. The interrelationship of the four parts is shown in Figure 8.20.

The magnet design must meet several criteria, the most important of which is field uniformity. The most convenient way of generating the magnetic field shape required for MRI is the use of a simple solenoid. Unfortunately, this does not, of itself, meet the criteria for field uniformity since the field decreases towards the solenoid ends. The usual solution to this problem is to wind additional coils at each end of the solenoid to provide end-correction; the final structure will involve four to six coils.

A major classification of MRI magnets results from the nature of the conductor, that is whether it is resistive or superconducting. Although magnets using superconductors are more difficult to construct, they have the advantage that since they possess no resistance they can use high currents with very low power consumption, yielding large volumes at high fields. They have come to dominate MRI. Figure 8.21 shows a simplified cross-section through a superconducting magnet

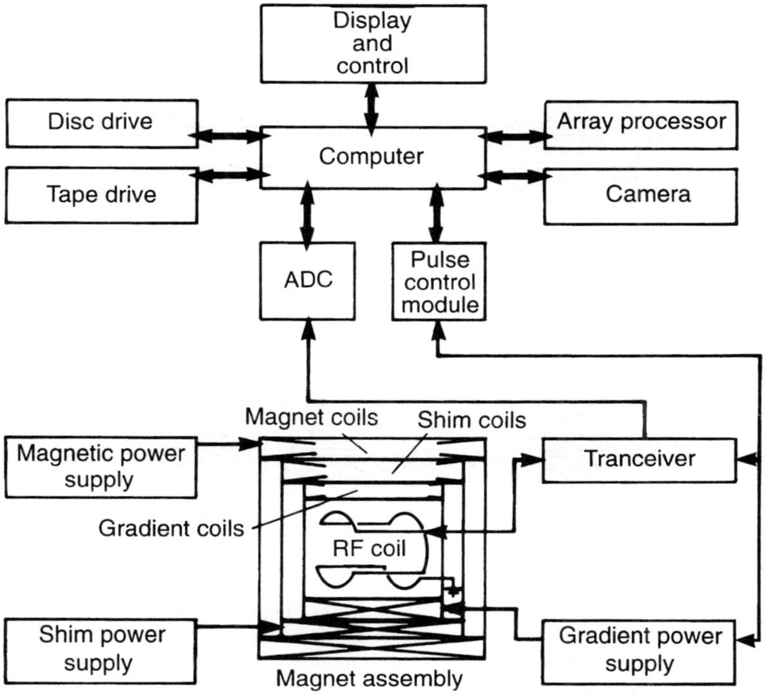

Figure 8.20 Block diagram of a typical MRI system showing the interrelationship of the constituent parts.

assembly for human imaging. Because these magnets must be capable of accepting a human body they need a large bore, of the order of 1m in diameter, yet must provide a magnetic field in the range 0.05–2 T with a uniformity over the 0.5 m central region of 20 ppm. If it is intended to carry out spectroscopy in addition to imaging, the field uniformity over the central 0.1 m must be improved to 0.1 ppm, at a flux density of at least 1.5 T. Usually, the magnet is designed so that the major axis of the patient, that is head to toes, can lie along the $\boldsymbol{B}_0$ direction. The superconducting coils are constructed from niobium–titanium alloy strands, embedded within a copper matrix. The superconducting state is only achieved at liquid helium temperature (4.2 K), so that particular care has to be exercised to minimize heat loss through conduction, convection and radiation. This is done through the use of an outer vacuum case and secondary evacuated shields, cooled by liquid nitrogen and liquid helium, which minimize the heat load on the

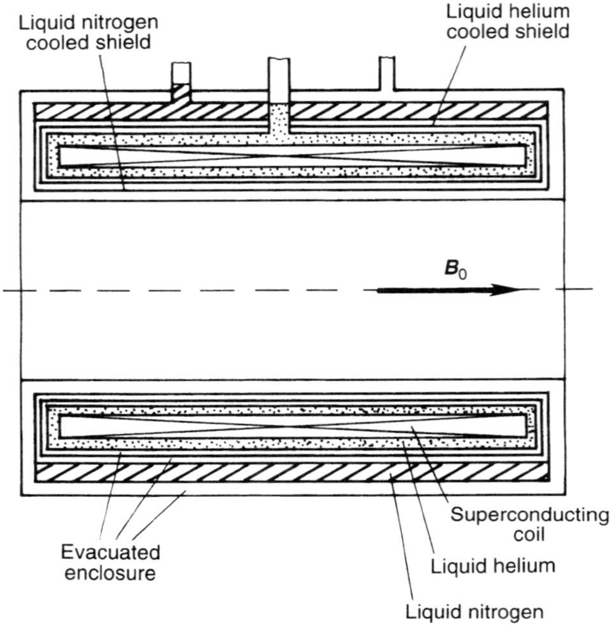

Figure 8.21 Section (simplified) through a superconducting magnet showing how the superconducting coil is isolated thermally from conduction, convection and radiation heat loads.

liquid helium cooled coils. The whole assembly is enclosed by a vacuum envelope. Ultimately, with the advent of high temperature superconductors, we can expect to see coils becoming available constructed from these new materials. This will lead to changes in coil design, since the need for liquid helium disappears and perhaps, one day, the need for liquid nitrogen also.

As we have seen, the essence of MR imaging is the ability to establish appropriate magnetic field gradients in the x, y or z directions. Accordingly, three sets of gradient coils are used to generate magnetic fields directed along z axis, but varying linearly in magnitude in the x, y and z directions. As a result they always add to, or subtract from, the main $\boldsymbol{B}_0$ field. The exact way in which these fields are used depends upon the application. For most imaging requirements, highly linear static gradients which can be switched on and off, as required, with only the overall amplitude being changed, are sufficient. Fast imaging requires the gradients to be switched very rapidly though. In all cases the maximum gradient amplitude is important since this affects the degree of chemical shift experienced in selective excitation and in the read-out direction. Gradients must exceed magnet inhomogeneities and strengths of 3–6 mT m^{-1} are usual, while values twice this magnitude are not uncommon. Another design consideration is the time required for the gradients to establish their maximum value when switched on, and to return to zero when switched off. These times can be a limitation in fast imaging or spectroscopic applications. In order to produce the magnetic field gradient we require power supplies capable of driving very high currents and providing rapid switching. Nowadays the gradient itself is determined by a digital-to-analogue convertor (DAC), driven by a microprocessor.

It is important for MRI, and essential for MRS, to have the most homogeneous magnetic field possible. This is achieved to first order by optimizing the design and construction of the basic magnet. Additional improvements to the overall homogeneity are obtained by using small correcting or shim coils (Figure 8.20). Adjustment of the shim coils is performed at the time of system installation or, in the case of MRS, on a patient-by-patient basis.

8.13 THE RADIOFREQUENCY COILS

Radiofrequency coils generate the oscillating, or in the case of quadrature detection, the rotating magnetic fields which are used both to excite the nuclei and detect the signal. There are four main types of RF coil; these are shown in Figure 8.22. With the exception of the surface coil (Figure 8.22(a)), these coils are designed to produce a magnetic field $\boldsymbol{B}_1$ which is as uniform as possible, since any non-uniformity in $\boldsymbol{B}_1$ affects the pulse angle induced in the nuclei, producing a shadowing in the final image. The most efficient design is the solenoid of Figure 8.22(b), but it can only be used in systems where the patient is at right angles to the main magnetic field, since it generates and detects magnetic fields parallel to its axis. A more useful design, at least for frequencies below 15 MHz is the Helmholtz coil of Figure 8.22(c), while the distributed capacitance coil (Figure 8.22(d)) can generate a rotating magnetic field.

The surface coil is essentially a one-turn solenoid, although its practical conception is usually much more complex. It 'sees' approximately a hemisphere, with a radius equal to the coil radius, although its sensitivity falls off with distance from the coil centre. Usually a surface coil operates simply as a receiver with a separate coil providing the

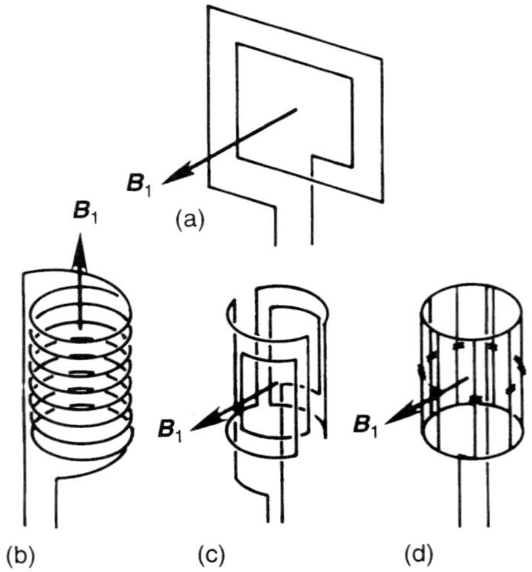

Figure 8.22 The four types of RF coil used in MRI (a) surface (b) solenoid (c) Helmoltz (d) distributed capacitance or 'bird cage' coil. The direction of the B_1 field is shown by a heavy arrow in each case.

uniform $\boldsymbol{B}_1$ excitation. In MRS the surface coil is frequently used as both a transmitter and receiver so that the inhomogeneity in $\boldsymbol{B}_1$ can be exploited as an additional method of localization.

8.14 THE TRANSCEIVER

The transceiver is central to the operation of an MRI system, since it is responsible for generating the correct RF voltage to be applied to the transmitter coil, as well as detecting the induced signal. The frequency is produced by standard frequency synthesis techniques and shaped, in the case of selective pulses, before being connected to the RF coil via a power amplifier. Since the RF coil acts as a receiving coil as well as a transmitting coil it is connected to a very sensitive low-noise preamplifier and then onto the main amplifier. Careful design is required to protect the preamplifier from the main high power excitation pulse. The amplified RF signal is demodulated with respect to the basic transmitter frequency to produce a signal in the audio frequency range, which is the direct equivalent of converting to the rotating reference frame described in section 8.4. The signal is then digitized in the ADC before processing by the computer.

8.15 IMAGING REFINEMENTS

The pulse sequence illustrated in Figure 8.18, although simple, has some unwanted side effects. The selection of an axial slice is achieved by applying a field gradient on the *z* axis and a selective 90° pulse. This results in the excitation of all the nuclei in the appropriate *x–z* plane. If a narrow slice is required, a high gradient and a long selection pulse are required. During the selection period a large amount of unwanted dephasing occurs and, as a result, the ultimate NMR signal is rather small. The gradient is present only to permit slice selection and once that is accomplished it is no longer required. We may say that dephasing is an unwanted consequence of slice selection.

The removal of this unwanted dephasing is achieved by modifying the *z* gradient waveform to include a time reversal gradient (Figure 8.23(a)). Now, at the end of the time reversal period, an NMR signal is obtained which is much larger than before. Once slice selection has occurred, spatial information must be encoded onto it and this is achieved, for one axis, by applying a phase encoding gradient after excitation, and then a read gradient during the collection of the free induction decay (FID) (lines three and four of Figure 8.18).

There are, though, further problems associated with the simple sequence shown in Figure 8.18, and these are connected with the fact that the field gradients do not switch as quickly as is implied in the drawing. There is a short, but significant period before the field gradients stabilize, and accurate data can be collected. This problem can be overcome by applying a time reversal gradient before the read gradient, so arranged as to be synchronous with the time reversal gradient, to correct the dephasing caused by G_z; this gradient is shown in Figure 8.23(b). Signal 'recovery' from the initial dephasing occurs during the initial part of the read period and we record an NMR signal containing what is known as a gradient echo (Figure 8.23(b)). We can consider this to be two FID signals back-to-back which, when Fourier transformed, produce a projection in the usual way, since it occurs as a time in the sequence which is well away from the switching transients.

The final refinement to the basic imaging process is to use a 180° pulse to produce a spin-echo (Figure 8.23(c)). The use of a spin-echo refocuses the unwanted defocusing, caused by magnet inhomogeneities, before the time of data collection. The time of data collection is defined by the timing of the field echo produced by the time reversal gradient period. If the spin and gradient echoes do not coincide, chemical shifts are introduced. It should be noted that the sign of the

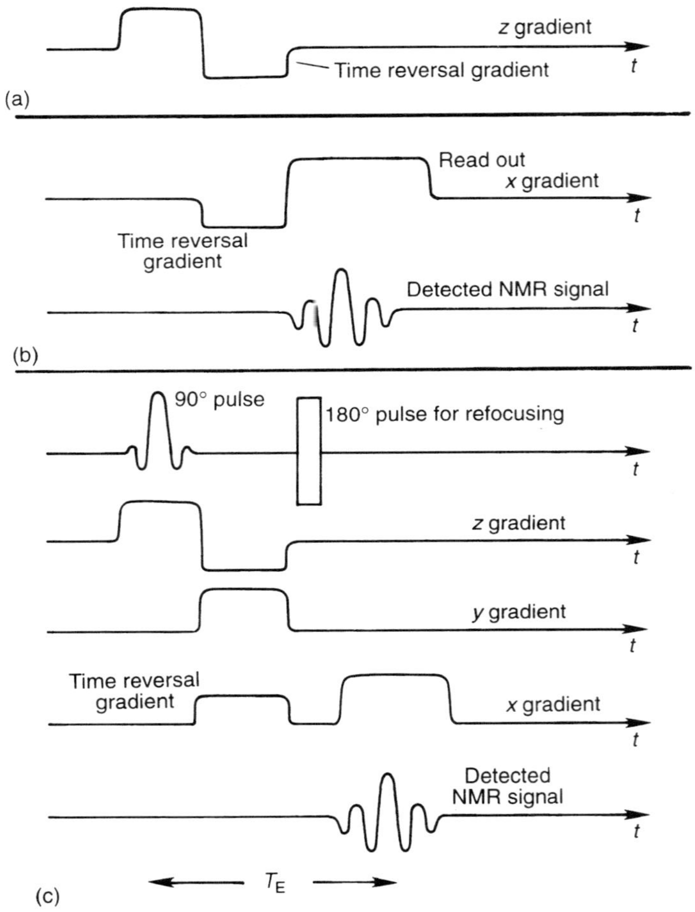

Figure 8.23 Pulse sequences used for image refinement: (a) *z* axis time reversal gradient; (b) *x* axis time reversal gradient; (c) use of 180° pulse to produce a spin-echo and refocus unwanted defocusing.

time reversal gradient is inverted to allow for the effect of the 180° pulse.

The improvements to the basic pulse sequence outlined above form the basis for the 2-D Fourier transform technique. It may be considered to contain three time periods: the slice selection period, the phase encoding period and the frequency encoding period. The phase encoding period is the most complex part of the pulse sequence, since in this period one or more variables may be encoded onto the signal. If there is only one variable, it is incremented during each pass and the whole image takes mt_r, where m is the number of values of the variables required in the final image and t_r is the repetition time. If more than one variable is applied in this period, the image takes lmt_r, where l is the number of data points required to define the new variable

8.16 CHEMICAL SHIFT SELECTIVE IMAGING

An interesting case of multivariable imaging occurs when the image is made chemical shift selective. Here l has the value 2. One image is taken, with the field and spin-echoes coinciding, so that the protons from water and fat yield coincident components (a normal 2-D Fourier transform). The second image is taken with the spin and gradient echoes adjusted so that the water and fat vectors are 180° out of phase. To achieve this, the timing of the 180° pulse is shifted by an amount equal to 1/(4 chemical shift differences). Addition and subtraction of these images produce separate 'fat' and 'water' images. The pulse sequence for chemical shift selective imaging is shown in Figure 8.24.

8.17 MULTISLICE IMAGING

In simple 2-D Fourier transform imaging, one slice is interrogated every t_r seconds, where t_r is the repetition time. This is clearly a very inefficient process, since large periods of time are wasted waiting for the spins to relax. A significant improvement is obtained if a multislice approach is adopted. Here the 'dead time' between the end of data collection and the next 90° pulse, which is dictated by relaxation effects, is used to interrogate other slices. The principle is shown diagrammatically in Figure 8.25. Once slice number 1 has been excited and data collected, slice 2 is excited and so on until the time t_r has elapsed, when slice 1 is returned to. The end of the data collection period is basically dictated by the time between the initial 90° pulse and the echo (this is the echo time T_E), plus a small margin of a few milliseconds for instrumental and software reasons. Very simply, the gain in efficiency is determined by t_r/T_E, and is usually of the order of 10–20.

Clearly, an alternative approach which would constitute true 3-D or volume imaging would be to encode the additional spatial variable in the fashion described previously. Although this procedure has the

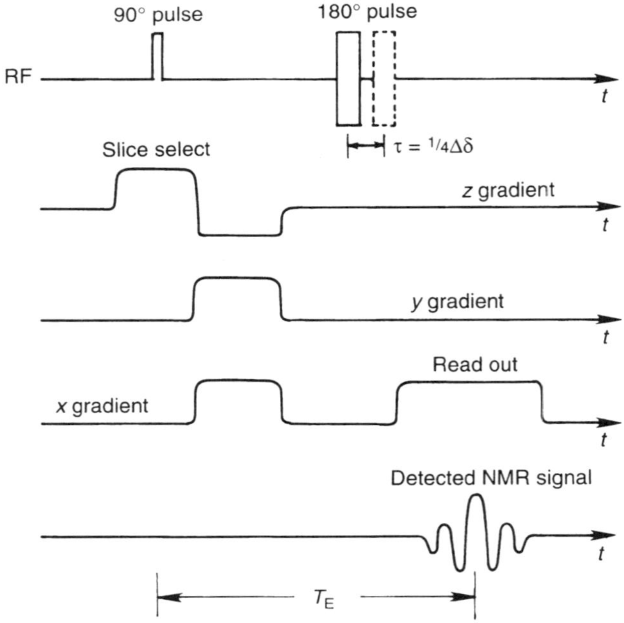

Figure 8.24 Pulse sequence for chemical shift selective imaging. The chemical shift selective sequence induces a 180° phase shift between two spectral components.

disadvantage that it takes l times longer, it is more efficient in that it has a higher signal-to-noise ratio, since it collects data from the whole volume and not just from a single slice, while the noise signal arises from the whole volume no matter how the signal volume is selected.

8.18 MAGNETIC RESONANCE ANGIOGRAPHY

Although the title magnetic resonance angiography (MRA) implies a single technique, it is actually a collection of methods in which flowing blood is imaged selectively. Within this collection of methods we find two categories: time-of-flight methods and phase-sensitive methods. These methods exploit different physical phenomena and each has unique advantages and disadvantages. The advantages, limitations and indeed physical principles of MRA, however, are fundamentally different from those arising in ultrasound, CT and X-ray angiography. Thus we find that MRA provides not only a morphological representation of

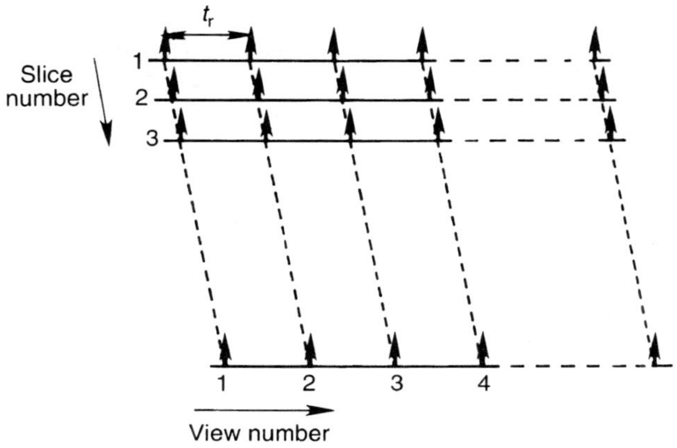

Figure 8.25 Multislice imaging routine which uses the 'dead time' between the end of data collection and the next 90° pulse.

blood vessels, but also details of blood motion. Physiological parameters such as blood velocity, velocity distributions and even vessel wall compliance can be measured, although we will only consider how blood velocity is measured.

There are two basic flow phenomena in magnetic resonance which alter the appearance of moving blood. Both of these phenomena were discovered at the end of the 1950s, years before the advent of MRI. The first of these, 'the time-of-flight effect', arises from the movement of blood during a relatively long period of time. By means of spatially selective RF pulses, blood in one region of the body is made to have an amount of longitudinal magnetization which is different from that found within the field-of-view of the image. As this blood moves into the imaged region, its longitudinal spin magnetization causes the blood to image with a different intensity from that of the surrounding tissue. In time-of-flight angiography methods blood enters the imaged region fully relaxed, that is with a substantial amount of longitudinal magnetization, and appears bright because the signal from stationary tissue within the imaged region is suppressed by saturation effects. This situation is illustrated in Figure 8.26. The partial spin saturation is due to strong RF pulses and short repetition time t_r. Alternatively, the blood may be intentionally saturated and then appears dark within the imaged region.

The second magnetic resonance flow phenomenon occurs wherever transverse magnetization is created in blood and the flow moves in the

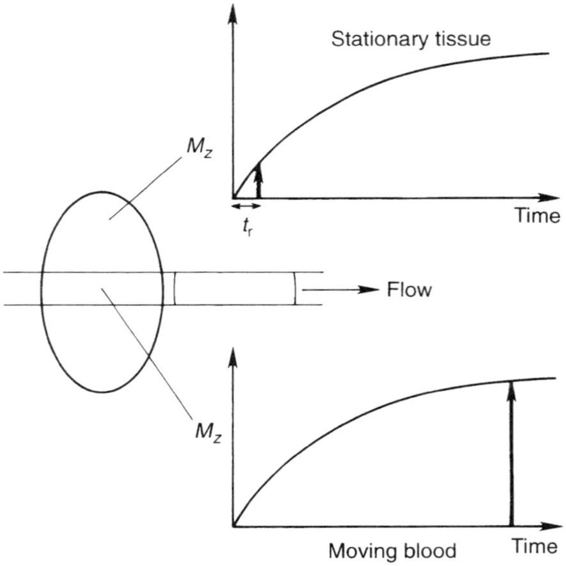

Figure 8.26 The time-of-flight effect. Blood in the region outside the imaged volume is given a different longitudinal magnetization M_z from the stationary tissue within the region of interest. Typically the blood is fully relaxed, that is it has recovered the full value of M_z, while the stationary tissue is still in a state of partial spin saturation due to a relatively short repetition time t_r and strong RF pulses.

presence of a magnetic field gradient. If this gradient has a time dependency, as in Figure 8.27, the transverse spin magnetization in the blood acquires a phase shift which is directly proportional to velocity. Since all MRI sequences generate transverse magnetization, and since most have bipolar gradient pulses, this effect leads to image artefacts in conventional MRI.

In the presence of a magnetic field gradient, the nuclei at different locations along the gradient resonate at different frequencies and consequently acquire phase shifts with respect to each other which are directly proportional to the position of each nucleus. The position-dependent phase shifts generated by a magnetic field gradient can be removed or anulled by applying a second magnetic field gradient of equal duration, but opposite polarity. This is illustrated in Figure 8.27 by the phase shifts ϕ_{S1} and ϕ_{S2}. If, however, the nuclei move between the first and second gradient pulses, the phase shift ϕ_{M1}, introduced by the first pulse, will not be exactly cancelled by that of the second pulse

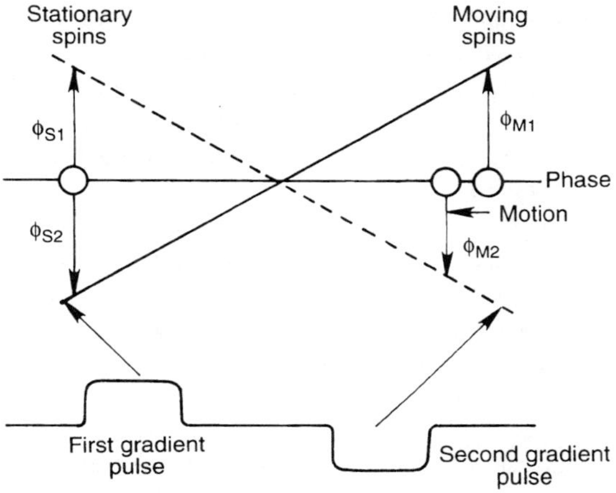

Figure 8.27 The effect of a bipolar gradient pulse on the phase of stationary and moving spin magnetization. The first lobe of the magnetic field gradient, bipolar pulse introduces a phase shift which is proportional to position. The second lobe induces a phase shift of identical magnitude, but opposite sign. For stationary nuclei these phase shifts cancel exactly. The phase shifts for moving nuclei, on the other hand, are not equal and do not cancel. The residual phase shift is proportional to the speed of movement.

ϕ_{M2}, and the residual phase shift will be directly proportional to the distance which the nuclei have travelled in the time interval between the gradient pulses. It will also be proportional to the flow velocity. The most common flow encoding gradient waveform is the simple bipolar gradient pulse described above. This induces a phase shift which is independent of position, but is proportional to velocity.

8.19 SPECIAL IMAGING

Imaging the lungs has always been a particular challenge for clinicians, since although X-ray pictures of the lungs can reveal fluid accumulation, **neoplasms** and calcified scars, they show hardly at all the very porous lung tissue or the inhaled gas. Magnetic resonance imaging, on the other hand, usually looks at the free protons, that is to say the ^{1}H nuclei, and even though the 'thermal' polarization of the proton spins in a 1.5 T static field is very small (around 3 in 10^6), there is so much hydrogen in the human body that a satisfactory image can still be obtained. Coupled with this, hydrogen has the highest spin moment of any spin $\frac{1}{2}$ nuclei. High resolution magnetic resonance imaging of a spin

$\frac{1}{2}$ nucleus, present only in low concentration, requires some means of increasing the polarization level since the magnetic resonance signal strength of a given nuclear species depends on its total magnetization in a given sample volume element, that is, on the product of the species concentration, the excess spin density per nucleus (polarization) and the volume of the element.

The lungs, of course, represent largely empty space, apart from inhaled air, so that the requirements for MRI cannot be met, even with ^{1}H. On the face of it this appears an impossible imaging situation, since the lungs would need to be filled with a non-toxic gas of spin $\frac{1}{2}$ nuclei polarized to greater than 10%, if a satisfactory image is to be formed. Interestingly, it is now possible to meet the above requirements by using the rare gases ^{3}He or ^{129}Xe, which can be polarized to around 50% (hyperpolarization) and maintain this state for days.

The method by which hyperpolarization of these rare gases is achieved involves an intermediate excited atom, an alkali metal atom. If circularly polarized laser light is shone on a vapour of alkali atoms, using a wavelength which is just right for exciting the single valence electron from its *s*-wave ground state to the first excited *p*-wave state, then conservation of the polarized photon's angular momentum imposes the selection rule $m_l = \frac{1}{2}$ for return to the ground state. This optical pumping process is extremely efficient, requiring on average around two photons per transition.

Optical pumping, or hyperpolarization, involves atomic electrons. What we are interested in though is polarized nuclei, and the question is: How do we obtain them? The answer lies in the fact that not only can an alkali metal, such as rubidium, be polarized optically, but it can also transfer its valence electron polarization to a rare gas nucleus by spin exchange in collisions. Moreover, because of the perfectly symmetrical, closed electron shell possessed by the rare gas atoms, the nuclear polarization can persist for hours. With appropriate storage conditions, for example freezing to 4.2 K, it can last for days.

Both the rare gas atoms ^{3}He and ^{129}Xe, which have spin $\frac{1}{2}$ nuclei, can be polarized to between 20 and 50% by this method. Both gases are entirely harmless if inhaled; indeed, ^{129}Xe has anaesthetic properties as well as exhibiting a large chemical shift (Table 8.1). These gases have been used to demonstrate MRI of the lungs of a guinea pig, and may well prove useful for imaging human lungs in the future. It is also conceivable that is will be possible to 'visualize' the anaesthetic action of ^{129}Xe using MRI, in view of the very large hyperpolarization which can be achieved.

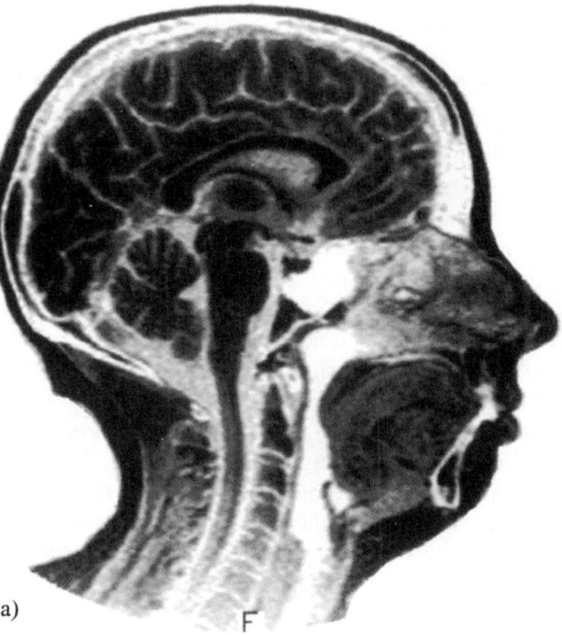

(a)

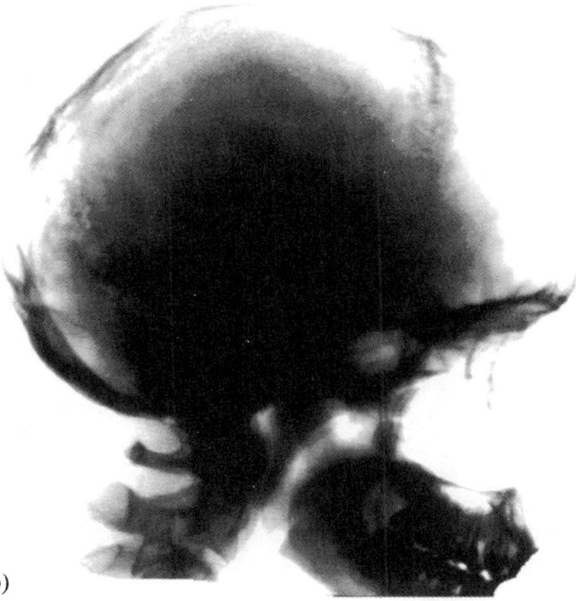

(b)

Figure 8.28 (a) Sagittal view of the human skull and neck obtained using MRI. (b) For contrast, a corresponding view obtained by means of standard X-ray projection radiography.

8.20 CLINICAL IMAGES FROM MAGNETIC RESONANCE

We close this chapter and finish this survey of physics in medical diagnosis by looking at some of the images obtainable by means of the magnetic resonance imaging technique. Figure 8.28(a) shows a sagittal view of the human skull and neck, with a standard projection X-ray image for contrast. (Figure 8.28(b)). These pictures reflect the improvement in imaging performance that has been achieved after some 90 years of effort.

Figure 8.28(a) was recorded at a magnetic field of 0.5 T, using a surface coil, and reveals considerable detail concerning brain structures of all sorts, details which are simply not obtainable by conventional X-ray techniques. The identity of the structures recorded can be determined by comparison with Figure 1.3.

Some alternative views of body structure are shown in Figure 8.29, where the knee and ankle joint are imaged with great clarity. Figures 8.28 and 8.29 show something of the scope of a technique which has the ability to demonstrate substantially greater tissue contrast than is achievable by any other imaging modality. This arises from the fact

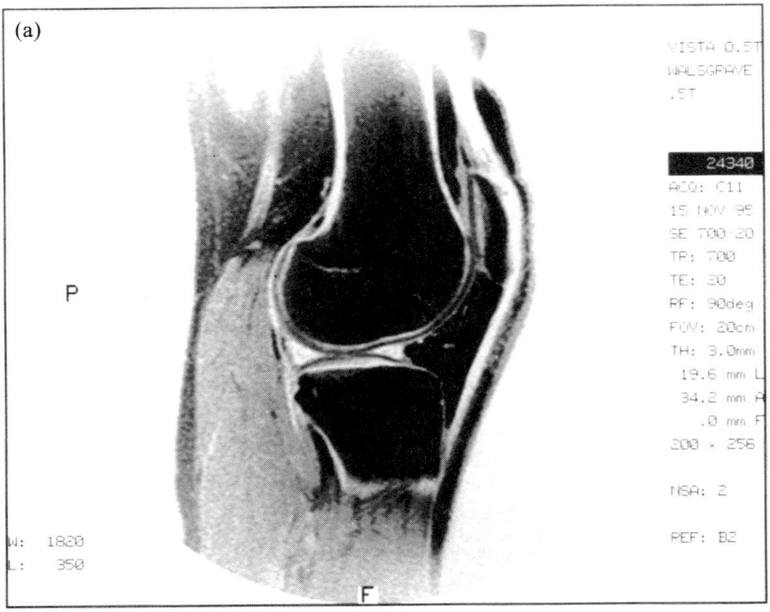

Figure 8.29 Sagittal sections through (a) the knee and (b) the ankle joint, obtained using MRI at 0.5 T.

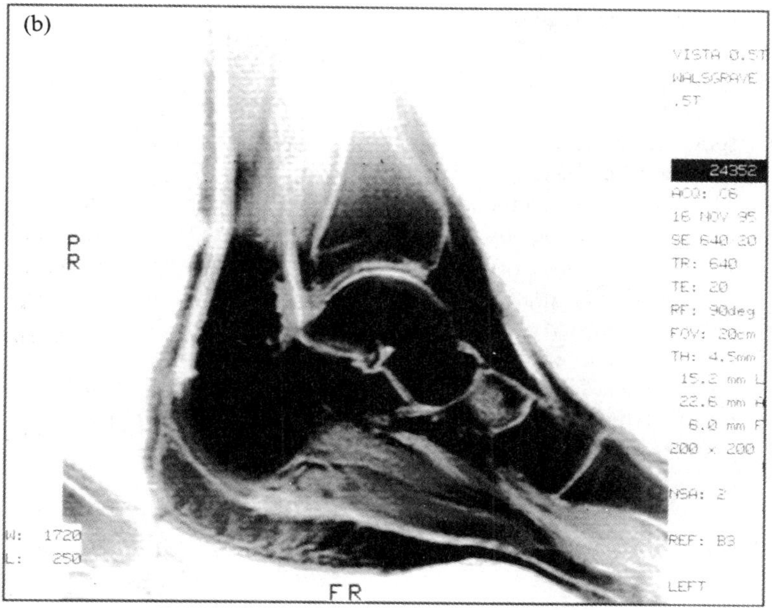

Figure 8.29 *Continued*

that there are more intrinsic tissue parameters available with which to determine MRI signal intensity than the simple parameter (X-ray attenuation coefficient) utilized in CT. The intricate, yet user controllable, interdependence of these parameters has a profound effect on the image characteristics such as signal-to-noise ratio, image contrast, spatial resolution and total exposure time. Consequently the user faces an heuristic problem, since there is typically no single, optimal set of scan parameters.

With the continued development of new pulse sequence techniques, it is probably safe to say that there remains further scope for expansion of the role of MRI in clinical diagnosis. We can confidently expect that MRI will ultimately become the application of physics in medical diagnosis which surpasses all others.

PROBLEMS

1. A ^{1}H nucleus is excited by RF radiation in a magnetic flux density of 0.5 T. Calculate the energy of the quantum of radiation responsible for this excitation.

2. Calculate the frequency of precession of a proton in a magnetic flux density of 10^{-4} T. The g factor for a proton is 2.79.
3. NMR is a relatively insensitive technique compared with other spectroscopic techniques. Calculate the excess of spin-up states (polarization), normalized to the total population, for protons in a magnetic flux density of 4 T at 300 K.
4. The population difference between spin-up and spin-down states can be increased, either by lowering the temperature, or by increasing the magnetic field. (a) Calculate the normalized population difference at 4 T and liquid nitrogen temperature (77 K). (b) What magnetic flux density is equivalent to this temperature reduction?
5. In a spin-echo experiment the echo amplitude was found to have reduced by a factor of two between the 90° pulse and the first 180° pulse. If the time between the 90° pulse and the 180° pulse was 34.6 ms, calculate the spin–spin relaxation time for this system.

Further reading

A very wide range of physics is involved in the interface between physics and medicine, so it is only possible to give titles which deal with the highlights or major themes, or embrace material outside the physicist's usual reading. Most of the research mentioned in this text can be found referenced in greater detail in one or other of these books.

Webb, S. (ed.) (1988) *The Physics of Medical Imaging*, IOP Publishing Ltd, Bristol. This is a multiauthor volume, written by medical physicists for medical physicists. It considers all of the current imaging modalities, including those still in the research stage, and assumes familiarity with medical terminology. A very detailed account.

Levitan, I.B. and Kaczmarek, L.K. (1991) *The Neuron; Cell and Molecular Biology*, Oxford University Press, Oxford. This is a textbook of cellular and molecular biology which provides a very wide introduction to the neuron and its attendant structures. It is not written for physicists and it, unavoidably, contains a good deal of material exposing the chemical basis for neuronal structures. It is very accessible nevertheless.

Hille, B. (1992) *Ionic Channels of Excitable Membranes*, 2nd edn, Sinauer Associates Inc., Massachusetts. This book concentrates on the ion channels which form the elementary excitable elements in cell membranes, and give them their electrical properties. Again, this is not a text written primarily for physicists, but it is still easy to follow.

Hill, C.R .(1986) *Physical Properties of Medical Ultrasonics*, Ellis Horwood, Chichester. A wide ranging survey of ultrasound in its medical applications.

Morris, P.G. (1986) *Nuclear Magnetic Resonance Imaging in Medicine*, Oxford University Press, Oxford. A very clear account of NMR is followed by a detailed explanation of its application to magnetic resonance imaging.

Hermann, G.T. (1980) *Image Reconstruction from Projections: The Fundamentals of Computed Tomography*, Academic Press, New York. The basis for X-ray computed tomography is considered in some detail. A very thorough account.

APPENDIX A

Glossary

acute	describing disease of rapid onset
angiography	X-ray examination of blood vessels using contrast enhancing dye
antipyretic	drug that reduces fever
aorta	large blood vessel arising from the left ventricle
arteriole	small branch of an artery
ascending aorta	first part of the aorta
astrocytes	type of glial cell, *see* neuroglia
atrio-ventricular (AV) node	mass of modified heart muscle situated in the lower middle part of the right atrium
axolemma	the fine cell membrane that encloses the protoplasm of an axon
axon	a nerve fibre, single process extending from the cell body of a neuron carrying impulses
axonal dystrophy	disorder of axons due to impaired nourishment
axoplasm	cytoplasm of an axon
balloon valvuloplasty	reconstruction operation performed on the valves of the heart or a vein, using an inflatable balloon
bilateral striate cortex	primary projection area for vision
bolus	rapidly injected volume of radiographic contrast material
bundle of His	bundle of modified heart muscle fibres passing from the SV node to the septum between the ventricles

central nervous system	that part of the nervous system located in spinal column and skull
cerebral angiography	angiography (q.v.) of cerebral arteries
cerebral cortex	intricately folded outer layer of the cerebrum making up some 40% of the brain by weight
cerebrospinal fluid	clear, colourless fluid in the spaces inside and around the spinal cord and brain
cerebrum	largest and most highly developed part of the brain, composed of two cerebral hemispheres
chemical synapse	*see* synapse
chronic	describing a disease of long duration involving very slow changes
coronal	relating to the crown of the head
cutaneous	relating to the skin
deep vein thrombosis	blood clot in a deep lying blood vessel carrying blood towards the heart
dendrite	one of the shorter branching processes of the cell body of a neuron.
diabetes mellitus	disorder of carbohydrate metabolism
diastole	period between two contractions of the heart muscle when the heart fills with blood, lasts approximately 0.5 s
epidermis	outer layer of the skin
erythrocytes	red blood cells containing the pigment haemoglobin
femoral artery	main artery to the leg
glial cells	*see* neuroglia
haematocrit	the proportion (percentage) of erythrocytes in blood
hyperlipaemia	the presence in the blood of an abnormally high fat level
hypothalamus	region of the forebrain in the floor of the third ventricle linked with the thalamus above and the pituitary gland below, integrates hormonal and autonomic nervous activity
inferior	lower in the body in relation to another structure

inhibitory postsynaptic potential	a potential which hyperpolarizes a nerve membrane, i.e. moves the potential away from the threshold
intercalated disc	disc situated between other structures
intercostal space	space between the ribs
internal carotid artery	artery supplying cerebrum, forehead, nose, eye, and middle ear
ischaemia	inadequate flow of blood caused by constriction or blockage in the blood vessels
left ventricle	chamber of heart, *see* Figure 2.26
mammography	making of X-ray or IR photograph of the breast
medulla oblongata	the extension within the skull of the upper end of the spinal cord, forming the lowest part of the brain stem
miniature end plate potentials	small electrical potentials recorded at the neuromuscular junction of skeletal muscle fibres even in the absence of nerve impulses in the motor fibres
mitochondria	granules of short rods present in all cells which supply energy to the cell by stepwise hydrolysis of ATP (adenosine triphosphate)
motor axon	axon attached to a motor neuron
motor neuron	one of the neuronal units which goes to make up the nerve pathway between the brain and, say, a skeletal muscle
motor end plate	an area of specialized structure beneath the sarcolemma, where a motor nerve fibre makes functional contact with a muscle fibre
multiple sclerosis	a chronic disease of the nervous system in which the myelin sheath is damaged
myelin sheath	complex material composed of protein and phospholipids laid around axons as a sheath
myocardial infarction	death of a segment of heart muscle following interruption of blood supply
myopathy	any disease of the muscles
neoplasm	a benign or malignant expanding lesion, or tumour

neuroglia	special cells of the central nervous system; they outnumber neurons by 10 : 1
neuron	nerve cell, basic functional unit of nervous system
neurotransmitter	chemical substance released from nerve endings to transmit impulses across synapses
node of Ranvier	one of the gaps that occur at regular intervals in the myelin sheath
occipital	at the back of the head
oesophagus	the gullet, a muscular tube extending from the pharynx to the stomach
oligodendrocytes	glial cell which produces myelin sheath for neurons of central nervous system
organelle	structure within a cell which is specialized for a particular function
overperfusion	oversupply of fluid, e.g. blood
parieto-occipital	one of the major divisions of each cerebral hemisphere lying behind the frontal lobe and in front of the occipital lobe; contains the sensory cortex
pneumo-encephalography	technique used in the X-ray diagnosis of disease within the skull; air is introduced into the ventricles (cavities) of the brain
presynaptic inhibition	grading of the synaptic potential by reducing the amount of neurotransmitter
presynaptic nerve	nerve situated near or occurring before a synapse
process	a thin prominence or protuberance
protoplasm	material of which living cells are made
Purkinje cell	nerve cell found in great numbers in the cortex of the cerebellum
pyrogen	any substance or agent producing fever
rectum	terminal part of the large intestine
sagittal	describing the dorsoventral plane that extends down the long axis of the body, dividing it into right and left halves
sarcolemma	the delicate sheath enveloping a muscle fibre
Schwann cell	*see* oligodendrocyte

sino-atrial node the pacemaker of the heart, a microscopic area of specialized cardiac muscle located in the upper wall of the right atrium

soleus muscle broad flat muscle in the calf of the leg

soma the entire body excluding the germ cells

stenosis abnormal narrowing of a passage such as a blood vessel, or an opening, such as a heart valve

sternum the breastbone

synapse minute gap across which nerve impulses pass from one neuron to another

syncytium mass of protoplasm containing several nuclei

systole period of cardiac cycle when heart contracts

tibia shin bone, inner and larger bone of the lower leg

tibial nerve nerve associated with the tibia

vasoconstriction decrease in the diameter of the blood vessels, especially arteries

vasodilation increase in diameter of blood vessels, especially arteries

vasomotor centre collection of nerve cells in the medulla oblongata which receive information from sensory receptors in the circulatory system

venous thrombosis clotting of blood in a vein

ventricle a cavity in the body, as in the brain (fluid-filled), or heart

vesicle a small bladder or sac containing liquid

Wallerian degeneration pathologic process affecting peripheral nerves in which there is degeneration of the axon cylinder and myelin distal to the break in the myelin sheath

APPENDIX B

Some useful physical constants and conversion factors

Speed of light in vacuum	2.99×10^{8} m s^{-1}
Planck's constant	6.62×10^{-34} J s
Elementary charge	1.60×10^{-19} C
Magnetic flux quantum	2.06×10^{-15} Wb
Bohr magneton	9.27×10^{-24} J T^{-1}
Nuclear magneton	5.05×10^{-27} J T^{-1}
Electron mass	9.10×10^{-31} kg
Avogadro constant	6.02×10^{23} mol^{-1}
Atomic mass unit	1.66×10^{-27} kg
Boltzmann constant	1.38×10^{-23} J K^{-1}
Stephan–Boltzmann constant	5.67×10^{-8} W m^{-2} K^{-4}
Electron volt	1.60×10^{-19} J
Faraday constant	96 485 C mol^{-1}
1 Dalton	1.657×10^{-21} kg
1 Roentgen	2.58×10^{-4} C kg^{-1}
1 Gray	6.24×10^{12} MeV kg^{-1}
1 Curie	3.70×10^{10} disintegrations s^{-1}

APPENDIX C

Nobel Laureates

Nobel Laureates mentioned in this text whose work has found application, directly or indirectly, in the application of physics to medicine.

Year	*Name*	*Subject*	*Citation*
1901	W.C. Roentgen	Physics	Discovery of X-rays
1903	A.H. Becquerel, P. Curie and M. Curie	Physics	Discovery of radioactivity
1905	P.E.A. Lenard	Physics	Research on cathode rays (X-rays)
1908	G. Lippman	Physics	Colour photography (capillary electrometer)
1911	M. Curie	Chemistry	Discovery of radium
1924	W. Einthoven	Physiology or Medicine	Invention of electrocardiograph
1932	C. Sherrington	Physiology or Medicine	Discoveries with respect to the function of neurons
1939	E.O. Lawrence	Physics	Invention of cyclotron and production of radioisotopes
1952	F. Bloch and E. Purcell	Physics	Discovery of NMR
1963	A.L. Hodgkin and A. Huxley	Physiology or Medicine	Transmission of nerve impulses
1970	B. Katz	Physiology or Medicine	Chemistry of nerve transmission
1979	A.M. Cormack and G.N. Hounsfield	Physiology or Medicine	Development of X-ray CT
1991	E. Neher and B. Sakman	Physiology or Medicine	Discovery of how cells communicate

APPENDIX D

Answers to problems

1 From thermometry to thermography

1 305 K
2 0.95
3 2.76 J m^{-2}
4 72.74 W m^{-2}

2 Electrical signals from the human body

1 15 707
2 102 nmol m^{-2}
3 26.6 mW m^{-2}
4 2.92 pW

3 Neuromagnetic signals

1 3.14×10^{-14} A m
2 0.12 pT

4 X-rays and projection radiography

1 0.012 mm, 0.016 pGy
2 1.8%
3 152%
4 95.2%, 2.7%

5 Ultrasound imaging

1 30%
2 (a) 7.31×10^{6} kg m^{-2} s^{-1}, (b) 0.76 mm
3 (a) 6.66 MHz, (b) 0.35 mm
4 99.9%
5 2.56 kHz

7 Imaging with radioisotopes

1 (a) 7.86×10^{15} Bq, (b) 2.52×10^{15} Bq
2 1.39 MeV
3 (a) Secular equilibrium, (b) 2.46×10^{14} Bq, (c) 2.38×10^{13} Bq
4 46.8 mm, 16.3 mm

8 Magnetic resonance imaging: the last window

1 8.81×10^{-8} eV
2 4.2 kHz
3 1.36×10^{-5}
4 (a) 5.33×10^{-5}, (b) 15.66 T
5 99.8 ms

Index

Page numbers appearing in **bold** refer to figures and page numbers appearing in *italic* refer to tables.